TODAY'S TECHNICIAN ™

SHOP MANUAL
FOR AUTOMOTIVE HEATING &
AIR CONDITIONING

TODAY'S TECHNICIAN ™

SHOP MANUAL
FOR AUTOMOTIVE HEATING & AIR CONDITIONING

FIFTH EDITION

Mark Schnubel
Naugatuck Valley Community College
Waterbury, Connecticut

DELMAR
CENGAGE Learning·

Australia • Brazil • Japan • Korea • Mexico • Singapore • Spain • United Kingdom • United States

Today's Technician™: Shop Manual for Automotive Heating & Air Conditioning, Fifth Edition

Mark Schnubel

Vice President, Editorial: Dave Garza

Director of Learning Solutions: Sandy Clark

Executive Editor: Dave Boelio

Managing Editor: Larry Main

Senior Product Manager: Matthew Thouin

Editorial Assistant: Courtney Troeger

Vice President, Marketing: Jennifer Baker

Marketing Director: Deborah S. Yarnell

Marketing Manager: Erin Brennan

Associate Marketing Manager: Jillian Borden

Production Director: Wendy Troeger

Production Manager: Mark Bernard

Senior Content Project Manager: Cheri Plasse

For product information and technology assistance, contact us at
Cengage Learning Customer & Sales Support, 1-800-354-9706

For permission to use material from this text or product, submit all requests online at **www.cengage.com/permissions**
Further permissions questions can be emailed to **permissionrequest@cengage.com**

Library of Congress Control Number: 2011936857

ISBN-13: 978-1-133-01744-8

ISBN-10: 1-133-01744-4

Delmar
5 Maxwell Drive
Clifton Park, NY 12065-2919
USA

Cengage Learning is a leading provider of customized learning solutions with office locations around the globe, including Singapore, the United Kingdom, Australia, Mexico, Brazil, and Japan. Locate your local office at: **international.cengage.com/region**

Cengage Learning products are represented in Canada by Nelson Education, Ltd.

To learn more about Delmar, visit **www.cengage.com/delmar**

Purchase any of our products at your local college store or at our preferred online store **www.cengagebrain.com**

Notice to the Reader

Publisher does not warrant or guarantee any of the products described herein or perform any independent analysis in connection with any of the product information contained herein. Publisher does not assume, and expressly disclaims, any obligation to obtain and include information other than that provided to it by the manufacturer. The reader is expressly warned to consider and adopt all safety precautions that might be indicated by the activities described herein and to avoid all potential hazards. By following the instructions contained herein, the reader willingly assumes all risks in connection with such instructions. The publisher makes no representations or warranties of any kind, including but not limited to, the warranties of fitness for particular purpose or merchantability, nor are any such representations implied with respect to the material set forth herein, and the publisher takes no responsibility with respect to such material. The publisher shall not be liable for any special, consequential, or exemplary damages resulting, in whole or part, from the readers' use of, or reliance upon, this material.

Printed in the United States of America
1 2 3 4 5 6 7 16 15 14 13 12

CONTENTS

PHOTO SEQUENCES

Thanks to the support that the *Today's Technician*™ series has received from those who teach automotive technology, Delmar, a division of Cengage Learning and the leader in automotive-related textbooks, is able to live up to its promise to provide new editions regularly. We have listened and responded to our critics and our fans and present this new updated and revised fifth edition. By revising our series regularly, we can and will respond to changes in the industry, changes in technology, changes in the certification process, and to the ever-changing needs of those who teach automotive technology.

The *Today's Technician*™ series features textbooks that cover all mechanical and electrical systems of automobiles and light trucks (while the "Heavy-Duty Trucks" portion of the series does the same for heavy-duty vehicles). Principally, the individual titles correspond to the main areas of ASE (National Institute for Automotive Service Excellence) certification. Additional titles include remedial skills and theories common to all of the certification areas and advanced or specific subject areas that reflect the latest technological trends. Each text is divided into two volumes: a Classroom Manual and a Shop Manual.

Unlike yesterday's mechanic, the technician of today and for the future must know the underlying theory of all automotive systems and be able to service and maintain those systems. Dividing the material into two volumes provides the reader with the information needed to begin a successful career as an automotive technician without interrupting the learning process by mixing cognitive and performance learning objectives into one volume.

The design of Delmar's *Today's Technician*™ series was based on features that are known to promote improved student learning. The design was further enhanced by a careful study of survey results, in which the respondents were asked to value particular features. Some of these features can be found in other textbooks, whereas others are unique to this series.

Each Classroom Manual contains the principles of operation for each system and subsystem. The Classroom Manual also contains discussions on design variations of key components used by the different vehicle manufacturers. This volume is organized to build on basic facts and theories. The primary objective of this volume is to allow the reader to gain an understanding of how each system and subsystem operates. This understanding is necessary to diagnose the complex automobiles of today and tomorrow. Although the basics contained in the Classroom Manual provide the knowledge needed for diagnostics, diagnostic procedures appear only in the Shop Manual. An understanding of the basics is also a requirement for competence in the skill areas covered in the Shop Manual.

A coil-ring-bound Shop Manual covers the "how-to's." This volume includes step-by-step instructions for diagnostic and repair procedures. Photo Sequences are used to illustrate some of the common service procedures. Other common procedures are listed and are accompanied with fine-line drawings and photos that allow the reader to visualize and conceptualize the finest details of the procedure. This volume also contains the reasons for performing the procedures, as well as when that particular service is appropriate.

The two volumes are designed to be used together and are arranged in corresponding chapters. Not only are the chapters in the volumes linked together, the contents of the chapters are also linked. This linking of content is evidenced by marginal callouts that refer the reader to the chapter and page on which that same topic is addressed in the other volume. This feature is valuable to instructors. Without this feature, users of other two-volume textbooks must search the index or table of contents to locate supporting information in the other volume. This is not only cumbersome, but it also creates additional work for an instructor when planning the presentation of material and when making reading assignments. It is also valuable to the students; with the page references they also know exactly where to look for supportive information.

Both volumes contain clear and thoughtfully selected illustrations, many of which are original drawings or photos especially prepared for inclusion in this series. This means that the art is a vital part of each textbook and not merely inserted to increase the number of illustrations.

The page layout used in the series is designed to include information that would otherwise break up the flow of information presented to the reader. The main body of the text includes all of the "need-to-know" information and illustrations. In the wide side margins of each page are many of the special features of the series. Items that are truly "nice-to-know" information such as simple examples of concepts just introduced in the text, explanations or definitions of terms that will not be defined in the glossary, examples of common trade jargon used to describe a part or operation, and exceptions to the norm explained in the text. This type of information is placed in the margin out of the normal flow of information. Many textbooks attempt to include this type of information and insert it in the main body of text; this tends to interrupt the thought process and cannot be pedagogically justified. By placing this information off to the side of the main text, the reader can select when to refer to it.

Jack Erjavec, Series Advisor

HIGHLIGHTS OF THIS EDITION—CLASSROOM MANUAL

The Classroom Manual of this edition has been updated to include new technology used in the automotive heating and air-conditioning systems of today's vehicles while still retaining information on systems used in older vehicles that are still in use. In addition, an emphasis has been placed on updating images throughout the text with full-color photos. Charts, graphs, and line drawings are now also in full color to be more visually appealing and improve the content comprehension by the reader. The chapter sequence remains the same, but changes to content have been made to improve the flow of the material and avoid redundancy. Chapter 2 covers the basic theories required to fully understand the operation and diagnosis of the complete HVAC system. In addition, it now has a section on basic electrical theory and coverage of the application and use of digital multimeters for those readers that have a limited background in electrical applications. This is intended to improve their understanding of electrical applications material covered in latter chapters and prepare them with the electrical knowledge needed to complete future job sheets. Chapter 3 covers the automotive heating system and engine cooling system, including systems used on today's hybrid electric vehicles. The electronic thermostat used on some of today's vehicle is thoroughly explained along with the rationale behind its use.

The rest of the text is laid out in a logical order, beginning with basic air-conditioning system operating principles and progressing to diagnosis of the refrigerant system. The end of chapter questions in all chapters have been revised and updated. Updated coverage on advanced electronics has been included from the operation of electronic variable compressors and electric motor driven compressors to advanced sensors such as airborne pollutants sensor. Chapter 10 on HVAC system controls has been updated to include more information on advanced climate control systems while still including a thorough description of CAN system operation. Chapter 11, while still containing retrofit information, has been completely updated to include future trends in automotive refrigerant systems and the emergence of R-1234yf as an alternative refrigerant and other low Global Warming Potential (GWP) refrigerants that may emerge.

HIGHLIGHTS OF THIS EDITION—SHOP MANUAL

Safety information remains the first chapter of the Shop Manual and covers general safety issues as well as topics specific to automotive HVAC service. This chapter includes an in-depth discussion on High-Voltage Safety on today's hybrid and electric vehicles and the equipment necessary to service these vehicles. As with the Classroom Manual, an emphasis has been placed on updating Shop Manual images and photo sequences throughout the text with full-color photos and line art.

Chapter 3 and later chapters cover service information related to the system information covered in the corresponding Classroom Manual chapters. Many new job sheets have been added and existing job sheets have been updated with 100 percent of the NATEF tasks covered. The latest use of tools and technology has been integrated into the text, including hybrid electric compressors and the operation and use of SAE standard J2788 refrigerant recovery/recycling/recharging equipment needed to service today's small-capacity refrigerant systems. Chapter 8, Compressors and Clutches, has been revised, placing more emphasis on actual compressor system diagnosis and service. Added coverage of today's automatic climate control system service and diagnosis has been updated and includes specific examples. This edition of the Shop Manual will guide the student/technician through all the basic tasks related to automive heating and air-conditioning service and reair.

Features of this manual include:

COGNITIVE OBJECTIVES

These objectives outline the contents of the chapter and define what the student should have learned upon completion of the chapter.
Each topic is divided into small units to promote easier understanding and learning.

AUTHOR'S NOTES

This feature includes simple explanations, stories, or examples of complex topics. These are included to help students understand difficult concepts.

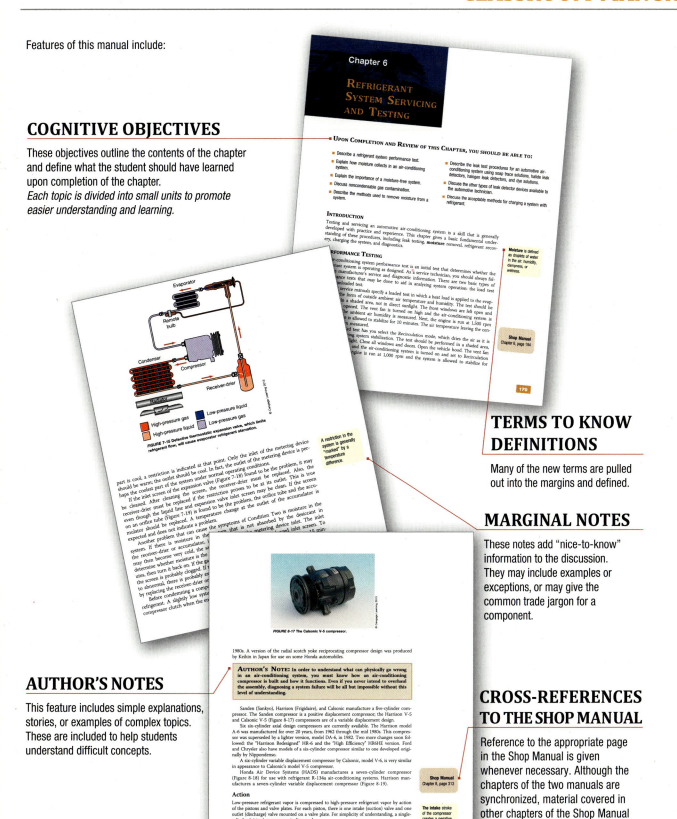

TERMS TO KNOW DEFINITIONS

Many of the new terms are pulled out into the margins and defined.

MARGINAL NOTES

These notes add "nice-to-know" information to the discussion. They may include examples or exceptions, or may give the common trade jargon for a component.

CROSS-REFERENCES TO THE SHOP MANUAL

Reference to the appropriate page in the Shop Manual is given whenever necessary. Although the chapters of the two manuals are synchronized, material covered in other chapters of the Shop Manual may be fundamental to the topic discussed in the Classroom Manual.

SUMMARIES

Each chapter concludes with a summary of key points from the chapter. These are designed to help the reader review the contents.

REVIEW QUESTIONS

Short answer essay, fill-in-the-blank, and multiple choice questions are found at the end of each chapter. These questions are designed to accurately assess the student's competence in the stated objectives at the beginning of the chapter.

A BIT OF HISTORY

This feature gives the student a sense of the evolution of the automobile. It not only contains nice-to-know information, but should also spark some interest in the subject matter.

TERMS TO KNOW LIST

A list of new terms appears after the Summary.

To stress the importance of safe work habits, the Shop Manual also dedicates one full chapter to safety. Other important features of this manual include:

PERFORMANCE OBJECTIVES

These objectives outline the contents of the chapter and identify what the student should have learned upon completion of the chapter. These objectives also correspond to the list of required tasks for NATEF certification.

Although this textbook is not designed to simply prepare someone for the certification exams, it is organized around the NATEF task list. These tasks are defined generically when the procedure is commonly followed and specifically when the procedure is unique for specific vehicle models. Imported and domestic model automobiles and light trucks are included in the procedures.

TOOLS LISTS

Each chapter begins with a list of the Basic Tools needed to perform the tasks included in the chapter. Whenever a Special Tool is required to complete a task, it is listed in the margin next to the procedure.

TERMS TO KNOW DEFINITIONS

Many of the new terms are pulled out into the margin and defined.

PHOTO SEQUENCES

Many procedures are illustrated in detailed Photo Sequences. These detailed photographs show the students what to expect when they perform particular procedures. They can also provide a student a familiarity with a system or type of equipment that the school may not have.

CUSTOMER CARE

This feature highlights those little things a technician can do or say to enhance customer relations.

CAUTIONS AND WARNINGS

Throughout the text, cautions are given to alert the reader to potentially hazardous materials or unsafe conditions. Warnings are also given to advise the student of what can go wrong if instructions are not followed or if a nonacceptable part or tool is used.

CROSS-REFERENCES TO THE CLASSROOM MANUAL

Reference to the appropriate page in the Classroom Manual is given whenever necessary. Although the chapters of the two manuals are synchronized, material covered in other chapters of the Classroom Manual may be fundamental to the topic discussed in the Shop Manual.

SERVICE TIPS

Whenever a shortcut or special procedure is appropriate, it is described in the text. These tips are generally those things commonly done by experienced technicians.

JOB SHEETS

Located at the end of each chapter, the Job Sheets provide a format for students to perform procedures covered in the chapter. A reference to the NATEF Task addressed by the procedure is referenced on the Job Sheet.

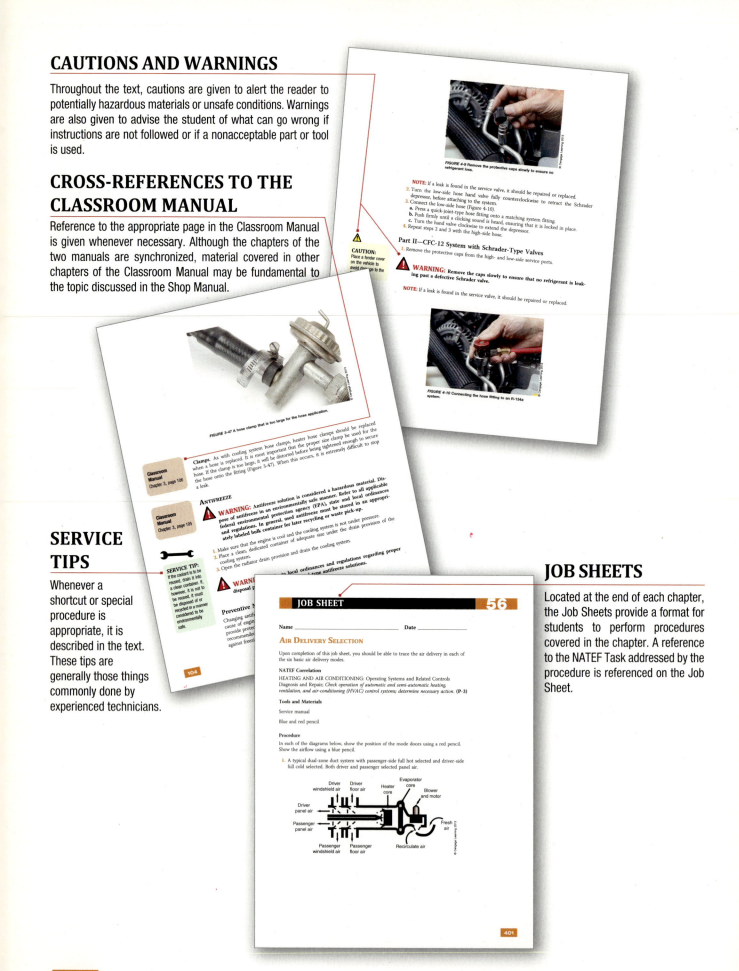

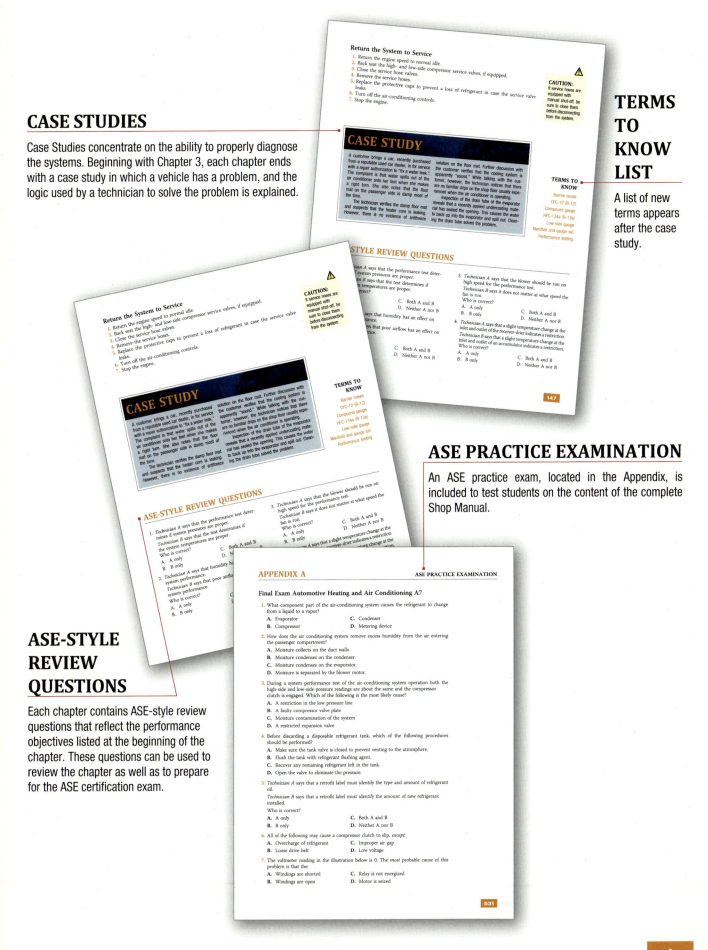

CASE STUDIES

Case Studies concentrate on the ability to properly diagnose the systems. Beginning with Chapter 3, each chapter ends with a case study in which a vehicle has a problem, and the logic used by a technician to solve the problem is explained.

TERMS TO KNOW LIST

A list of new terms appears after the case study.

ASE PRACTICE EXAMINATION

An ASE practice exam, located in the Appendix, is included to test students on the content of the complete Shop Manual.

ASE-STYLE REVIEW QUESTIONS

Each chapter contains ASE-style review questions that reflect the performance objectives listed at the beginning of the chapter. These questions can be used to review the chapter as well as to prepare for the ASE certification exam.

INSTRUCTOR COMPANION WEBSITE

The Instructor Companion Website contains all preparation tools to meet any instructor's classroom needs. It includes chapter outlines in PowerPoint with images, video clips, and animations that coincide with each chapter's content coverage, chapter presentations in PowerPoint with images, video clips, and animations, chapter tests in ExamView with hundreds of test questions, a searchable Image Library with all photos and illustrations from the text, theory-based Worksheets in Word that provide homework or in-class assignments, the Job Sheets from the Shop Manual in Word, a NATEF correlation chart, and an Instructor's Guide in electronic format.

WEBTUTOR ADVANTAGE

The WebTutor Advantage, available for Blackboard and Angel online learning management systems, includes presentations in PowerPoint with video clips and animations, end-of-chapter review questions, pre-tests and post-tests, Worksheets, Job Sheets, discussion springboard topics, and more. The WebTutor is designed to enhance the classroom and shop experience, engage students, and help them prepare for ASE certification exams.

COURSEMATE

The all new CourseMate for *Today's Technician: Automotive Heating & Air Conditioning* offers students and instructors access to important tools and resources, all in an online environment. The CourseMate includes an Interactive eBook of the core text, interactive quizzes, flashcards, as well as an Engagement Tracker tool for monitoring students' progress in the CourseMate product.

REVIEWERS

The author and publisher wish to thank the instructors who reviewed this text and offered their invaluable feedback:

Lance David
College of Lake County
Grayslake, IL

John Koehn
Pueblo Community College
Pueblo, CO

Shannon Kies
University of Northwestern Ohio
Lima, OH

Chris Marker
University of Northwestern Ohio
Lima, OH

Gary McDaniel
Metropolitan Community College - Longview
Longview, MO

William McGrath
Moraine Valley Community College
Palos Hills, IL

Mike Shoebroek
Austin Community College
Austin, TX

Ira Siegel
Moraine Valley Community College
Palos Hills, IL

Chris Wood
University of Northwestern Ohio
Lima, OH

Chapter 1

SHOP SAFETY

UPON COMPLETION AND REVIEW OF THIS CHAPTER, YOU SHOULD BE ABLE TO:

- Recognize the hazards associated with the automotive repair industry.
- Identify hazardous conditions that may be found in the automotive repair facility.
- Explain the need for a health and safety program.

- Discuss the philosophy regarding health and safety.
- Compare and identify unsafe and safe tools.
- Understand the limitations, by design, of hand tools.

GENERAL SHOP SAFETY

Many studies have been made to determine which of the school shops are the more hazardous. The automobile mechanics shop, it has been found, ranks third in frequency of accidents. It is exceeded only by the wood shop and the machine shop.

Principal hazards and injuries in the automotive shop are:

- Flammable materials
- Bruised and cut fingers
- Acid burns
- Strains and hernia
- Falls
- Eye injuries

There is little manual lifting required in the modern automotive repair shop. Most lifting, when required, is accomplished with hoists, jacks, and other lifting devices. When using such equipment, understand and follow all applicable safety procedures. When manual lifting is required, lift with the legs—not with the back (Figure 1-1). Get help for heavy or bulky objects.

Many schools have designated their shops "total eye protection areas." This means that everyone who enters the shop must wear eye protective equipment. It is essential to wear safety goggles on any job where the eyes may be endangered, such as when grinding, using compressed air, working underneath cars, or servicing an air-conditioning system.

Chemical or splash-proof goggles as well as a face shield (Figure 1-2) should be worn when servicing batteries or when boiling out or testing radiators for leaks.

Sparks near an automobile battery may create a hazard and can cause an explosion. The accumulation of hydrogen vapor at the top of the cell being charged is very explosive. Do not test a battery by "flashing" or "sparking" the terminals with a piece of wire to see if it has a charge.

> There are many airborne hazards in the automotive shop.

> Always use proper tools. Use a battery tester for testing batteries.

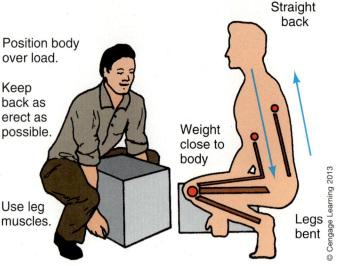

Position body over load.

Keep back as erect as possible.

Use leg muscles.

Straight back

Weight close to body

Legs bent

© Cengage Learning 2013

FIGURE 1-1 Use legs, not back, for lifting.

Always wear eye protection.

© Cengage Learning 2013

FIGURE 1-2 A face shield may be worn over goggles for more protection.

Use only extension lights, cords, and sockets that are in good condition. Portable lights should be protected by a rubber- or neoprene-covered steel guard. All portable lights should have a third-wire ground. Do not place cords or wires across the floor where they may become a tripping hazard. Do not use portable electric tools unless they are electrically grounded with a third wire or are designated "double insulated." Be sure that the extension cables used with portable electric tools are in good condition and are of the proper size.

Do not attempt to use any power tools or equipment in the automotive shop until the proper and safe use has been fully explained by the instructor.

Hammers with broken handles, defective screwdrivers, and greasy tools can all be the cause of serious accidents. Keep all tools clean and free from grease.

Never engage in horseplay of any kind in the auto shop. This includes running, scuffling, and throwing tools or materials. Never use compressed air except for the purpose for which it is intended. Horseplay with compressed air equipment or dusting off clothing

FIGURE 1-3 **Always use safety stands.**

or work benches with compressed air is extremely dangerous. Flying particles of metal or glass may be blown into the eyes or the skin. Also, compressed air blown into the skin or body openings can cause serious injury and even death.

An approved hoist should be used for work underneath a car. The proper instruction, use, and operation of a hoist should be a work assignment for each learner in any automotive technician training program.

Vehicles raised by jack, chain hoist, or end lift should always be supported with safety stands (Figure 1-3) or with other approved safety devices. Before use, these devices should be carefully inspected for damage. Never crawl or work under a vehicle that is not supported by safety stands. This precaution should even be followed for inspection purposes.

When working under a vehicle:

- Use a creeper.
- Keep legs and arms clear of passageways.
- Keep vehicle doors closed.
- Do not place tools above the technician.
- Other technicians should not work on top of the vehicle.
- Wear safety glasses, goggles, or face shield.
- Do not leave creepers, tools, or other equipment where anyone can step on or trip over them.

Burns may result from working on a car that has not cooled off, most frequently by coming in contact with the manifold, exhaust pipe, or engine coolant.

Gasoline and diesel engines should only be operated in a shop or other area where there is adequate ventilation or there are provisions to connect the exhaust to an approved system that is designed to remove harmful exhaust fumes from the work area.

Before starting the engine, make sure the car is out of gear. On cars with automatic transmissions, make sure the gear select lever is in the neutral or park position when the motor is running. Set the parking brake. Ensure that there is no one working under the hood of the vehicle.

Be especially cautious around moving parts such as the flywheel, fan blades, belt, gears, and alternator pulley. Keep long sleeves rolled up when working on any moving machinery. Do not lubricate an engine while it is running, and do not attempt to wipe moving parts of the engine. Keep hands out of the area of moving parts.

Handle fluids carefully so that they do not splash in the eyes. Use a syringe when transferring fluids. It is important that brake fluid and some synthetic lubricants not be allowed to come into contact with a painted surface. Many such fluids contain ingredients that can soften, blister, and remove paint.

FIGURE 1-4 **Always wear safety glasses.**

To avoid burns from accidental short circuits and to prevent accidental engagement of the starting motor, be sure to disconnect the battery ground cable and insulate the connection before working on the electrical system of the car.

Never consider a job complete until a check has been made to ensure that all parts that were removed have been replaced. Also, observe the following rules:

- Always refer to manufacturer's specifications.
- Keep tools clean and in good condition. Screwdriver blades should be kept sharp and square; handles should be of a nonconducting material.
- Use the proper type and size of tool. Use box wrenches in preference to open-end wrenches; use adjustable wrenches as little as possible. Do not use files as punches or chisels; they are brittle and may shatter.
- Use handles on files.
- Do not put sharp-edged tools—such as chisels, punches, and open knives—in your pockets, even temporarily; keep guards on sharp edges or points of tools in tool kits.
- Push sharp tools away from you instead of drawing them toward you.
- Whenever possible, do not hold the screw or work piece with one hand and the screwdriver or tool with the other hand.
- Keep your face away from tools.
- Wear goggles when grinding or when working on any job that may involve flying debris (Figure 1-4).
- When working around moving machinery, do not wear gloves, ties, or loose clothing that may become caught in the machine and cause you severe injury.
- Do not wear rings when working.
- Remove all loose jewelry, such as chains and watches.
- Use the proper fuel. Some vehicles use fuels other than gasoline and diesel fuel, such as liquified petroleum gas (LPG) and compressed natural gas (CNG) as well as alcohol and alcohol blends. Hydrogen gas may be used in the future.

PERSONAL SAFETY

Technicians working in the automobile repair industry may be exposed to a wide variety of **hazards** in the form of gases, dusts, vapors, mists, fumes, and noise, as well as ionizing or nonionizing radiation. In the course of their work, automotive air conditioning

Disconnect the battery ground cable before servicing the vehicle. See the manufacturer's precautions.

Refrigerant pressure can exceed 300 psig (2,068 kPa).

A **technician** is a person who has been technically trained to understand and repair the technicalities of the system on which they are working on.

Hazards are possible sources of danger that may cause damage to a structure or equipment or that may cause personal injury.

technicians may not be directly exposed to all such hazards, but they must be aware of all the potential dangers that may exist in the facility. Some of the most common hazards include:

- Asbestos
- Carbon monoxide
- Caustics
- Solvents
- Paints
- Glues
- Heat and cold
- Oxygen deficiency
- Radiation
- Refrigerants

Fibrous Material (Asbestos)

Perhaps one of the most serious of the hazards found in an automotive repair facility is the exposure to **asbestos** fibers. Asbestos has been used in brake linings and clutch friction plates for many years. Exposure to asbestos may result in asbestosis or lung cancer. The problem is so serious that work with asbestos must now be done in accordance with Standard 1910.93a of the **Occupational Safety and Health Administration (OSHA)**. Materials that are less hazardous have now replaced most asbestos applications, but asbestos must still be considered a hazard. Always follow all applicable procedures associated with good industrial hygiene any time there is a possibility of airborne fibers.

Technicians must not be exposed to unsafe levels of airborne asbestos. It is required that:

- Asbestos waste and debris must be collected in impermeable bags or containers.
- All asbestos and materials bearing asbestos must be appropriately labeled.
- Special clothing and approved **respirators** are to be worn when handling asbestos.
- Technicians handling asbestos must be given regular periodic physical checkups.
- Methods to limit technician exposure to asbestos include isolation and **ventilation** of dust-producing operations and wetting the material before handling.

 WARNING: The risk of cancer for people who smoke while working with asbestos is almost 90 times greater than for people who do not smoke.

Carbon Monoxide

Alternate fuels such as propane and CNG, diesel and gasoline-powered vehicles, and some hot work operations, such as welding, all produce **carbon monoxide (CO)**. The technician's exposure to carbon monoxide may be excessive if such operations are conducted in low-ceilinged or confined areas. Corrective action must be taken if the levels exceed safe standards. The technician must always work in a well-ventilated area to avoid exposure to excessive vapors and fumes.

Caustics, Solvents, Paints, Glues, and Adhesives

Many caustics, acids, and solvents are used in the automotive industry for cleaning operations (Figure 1-5). Epoxy paints, resins, and adhesives are used regularly in body repair and refinishing shops. Always read the cautions printed on the container label before using the contents.

Asbestos is a silicate of calcium (Ca) and magnesium (Mg) mineral that does not burn or conduct heat. It has been determined that asbestos exposure is hazardous to health and must be avoided.

Occupational Safety and Health Administration (OSHA) is a federal agency in charge of workplace safety.

Respirators are masks designed to protect the wearer from airborne contaminants and to provide clean air.

Ventilation is the act of supplying fresh air to an enclosed space, such as the inside of an automobile.

Smokers are at greater risk than nonsmokers by almost 10:1.

Carbon monoxide (CO) is a major air pollutant that is potentially lethal if inhaled, even in small amounts. It is an odorless, tasteless, colorless gas composed of carbon (C) and oxygen (O) formed by incomplete combustion of any fuel containing carbon.

FIGURE 1-5 Many hazardous solvents may be used in the automotive shop.

Some of the more common organic chemicals may cause dizziness, headaches, and sensations of drunkenness; they can also affect the eyes and respiratory tract. The use of many chemicals in this special trade industry can cause various types of skin irritations and, in extreme cases, dermatitis. Proper use and availability of appropriate protective equipment is essential. Such equipment includes:

- Gloves
- Goggles or face shields
- Aprons
- Respirators

Lack of oxygen is evidenced by dizziness and drunkenness.

Any hazardous materials that come in contact with skin should be washed off immediately. In addition, an eye wash fountain or safety shower (Figure 1-6) should be provided. For example, an exploding battery may saturate the entire body and clothing with sulfuric acid. The fastest way to reduce the effects of this type of contamination is to step into a shower while disrobing.

FIGURE 1-6 Showers should be available for personal safety.

Adequate ventilation is necessary to avoid excessive exposure to fumes and vapors during operations in confined spaces. Spray booths with a personal respiratory system are required in many areas for production spray operations involving adhesives and paints.

Heat and Cold

Consideration should be given to ensure that the temperature of the work area assigned to the technician be maintained within acceptable narrow limits. When exposed to extremes of heat and cold, it has been found that the technician's work performance can suffer because of:

- Fatigue
- Sunburn
- Discomfort
- Collapse
- Other health-related problems

Oxygen Deficiency

If proper precautions are not observed, many operations carried out in a confined space, such as the repair of an automotive air-conditioning system, can be very dangerous. Not only may the technician be exposed to various toxic gases, but the atmosphere may also be deficient in oxygen, which would immediately pose a danger to life. Other vapors, which may not be harmful in themselves, displace the oxygen essential to life. Such potential hazards should be approached only when absolutely necessary and when there are adequate procedures outlining the proper precautions and safeguards, such as:

- Air line
- Respirator
- Lifeline
- Buddy system

Radiation

Lasers, which are used for some alignment procedures, may produce intense, non-ionizing **radiation**. While it should be avoided, this minor radiation is not generally considered harmful. Welding, however, produces **ultraviolet (UV)** light that is hazardous to the eyes and skin. If proper safeguards are not observed, both ionizing and non-ionizing radiation can be very hazardous. Only qualified and trained technicians should use such equipment. It is often a requirement that such technicians be licensed by a federal, state, or local authority. Welders, for example, may be certified by the American Welding Society (AWS).

 WARNING: Refrigerants may be flammable.

Refrigerants

The primary problem that may occur during the installation, modification, and repair of an automotive air-conditioning system is the leakage of **refrigerant**. Refrigerants may be considered in the following classes:

- Nonflammable substances where the toxicity is slight, such as some hydrofluorocarbons—Refrigerant-134a (R-134a), for example. Although considered fairly safe, this refrigerant may decompose into highly toxic gases, such as hydrochloric acid or chlorine, upon exposure to hot surfaces or open flames.

Radiation is the transfer of heat without heating the medium through which it is transferred.

Ultraviolet (UV) is the part of the electromagnetic spectrum emitted by the sun (or other light source) that lies between visible violet light and x-rays.

Refrigerant is a chemical compound, such as R-134a, used in an air-conditioning system to achieve the desired chilling effect. Good refrigerants are those that boil at atmospheric pressure and temperatures and are condensable when pressurized.

FIGURE 1-7 Material safety data sheets (MSDS) are
supplied by hazardous chemical manufacturers on request.

© Cengage Learning 2013

- Toxic and corrosive refrigerants such as ammonia, often used in recreational vehicle (RV) absorption refrigerators, may be flammable in concentrations exceeding 3.5 percent by volume. Ammonia is the most common refrigerant in this category and is very irritating to the eyes, skin, and respiratory system. In large releases of ammonia, the area must be evacuated. Reentry to evaluate the situation may only be made wearing appropriate respiratory protective devices and protective clothing. As ammonia is readily soluble in water (H_2O), it may be necessary to spray water in the room via a mist-type nozzle to lower concentrations of ammonia.
- Highly flammable or explosive substances, such as propane, must be used with strict controls and safety equipment. While propane is not used as a refrigerant in mobile refrigeration, it is a fuel often found in mobile applications.

If a refrigerant escapes, action should be taken to remove the contaminant from the premises. If ventilation is used, exhaust from the floor area must be provided for gases heavier than air and, similarly, from the ceiling for gases lighter than air. For an analytical analysis of the product, consult the **material safety data sheet (MSDS)** (Figure 1-7) provided by the manufacturer.

Antifreeze

There are two basic types of antifreeze available: those with ethylene glycol (EG), and those with propylene glycol (PG).

EG-Based Antifreeze. EG-based antifreeze is a danger to animal life. Properly handled and installed, however, EG antifreeze presents little problem. If it is carelessly installed, improperly disposed of, or leaks from a vehicle's cooling system, it can be very dangerous.

EG-based antifreeze causes thousands of accidental pet deaths in the United States each year. Animals are attracted to EG antifreeze because of its sweet taste. As little as 2 ounces can kill a dog, and only 1 teaspoon is enough to poison a cat. This antifreeze can also be a hazard to small children in an undiluted quantity of as little as 2 tablespoons.

Toxicologists report that EG antifreeze inside the body is changed into a crystalline acid that attacks the kidneys. The effects are fast acting, and one must act immediately if it is suspected that an animal or child may have ingested EG antifreeze.

FIGURE 1-8 Ethylene glycol-based (A) and propylene glycol-based antifreeze (B).

The signs of EG poisoning in pets include excessive thirst and urination, lack of coordination, weakness, nausea, tremors, vomiting, rapid breathing and heart rate, convulsions, crystals in urine, diarrhea, and paralysis.

Pets do not often survive EG poisoning because owners do not usually recognize the symptoms until it is too late for treatment.

PG-Based Antifreeze. A safer alternative is an antifreeze and coolant formulated with propylene glycol. Unlike EG antifreeze, PG-based antifreeze is essentially nontoxic and hence safer for animal life, children, and the environment. PG antifreeze is classified as "Generally Recognized as Safe" (GRAS) by the United States Food and Drug Administration (U.S. FDA). Actually, propylene glycol is used in small quantities in the formula of many consumer products such as cosmetics, medications, snack food, and as a moisturizing agent in some pet food.

PG-based antifreeze coolant protects against freezing, overheating, and corrosion the same as conventional EG-based antifreeze coolants. PG-based antifreeze coolants should not be mixed with toxic EG-based antifreeze coolants (Figure 1-8) because the safety advantage will be lost.

Welding, Burning, and Soldering

Fumes from welding and other hot-work operations actually contain the metals being welded together, such as cadmium (Cd), zinc (Zn), lead (Pb), iron (Fe), or copper (Cu), as well as the filler material, flux, and the coating on the welding rods. Such operations may also generate other gases such as CO and ozone (O_3) at concentrations that may be hazardous to health. When extensive hot-work operations, such as welding, are performed in confined areas, there can be an excessive fume exposure to these materials. Ventilation or respiratory protection may be needed for certain operations. Eye protection for the welder and for other technicians working in or near the vicinity of welding operations should be provided because of the UV light produced during such operations. Engineering controls such as local exhaust ventilation are required before use of personal protective equipment as a control measure. When effective engineering controls are not feasible or while they are being instituted, personal protective equipment is required.

Flux is used to promote "wetting" during soldering.

General Hybrid Electric Vehicle Safety

Since the hybrid electric vehicle (HEV) system can use voltages in excess of 300 volts (both DC and AC), it is vital the service technician be familiar with, and follow, all safety precautions. Many high-voltage electric vehicles today use a high-voltage electric air conditioning compressor as part of the refrigerant system. Failure to perform the correct procedures can result in electrical shock, battery leakage, an explosion, or even death. The following are some general service precautions to be aware of:

- Test the integrity of the rubber insulating gloves (electrical lineman gloves) prior to use.
- Wear high-voltage (HV) insulating gloves when disconnecting the service plug and use the leather top glove during heavy or abrasive service procedures.

 WARNING: Do not attempt to test or service the system for 5 minutes after the high-voltage service plug is removed. At least 5 minutes is required to discharge the high-voltage capacitors inside the inverter module.

- Never cut the orange high-voltage power cables. The wire harness, terminals, and connectors of the high-voltage system are colored orange. In addition, high-voltage components may have a "High Voltage" caution label attached to them.
- Use insulating tools.
- Do not wear metallic objects that may cause electrical shorts.
- Wear protective safety goggles when inspecting the high-voltage battery.

 WARNING: Be sure to use the proper safety equipment when working on any high-voltage system. Failure to do so may result in a serious or fatal injury

- Follow the service information diagnostic procedures.
- Never open high-voltage components.
- Before touching any of the high-voltage system wires or components, wear insulating gloves, make sure the high-voltage service plug is removed, and disconnect the auxiliary battery.
- Remove the service disconnect prior to performing a resistance check.
- Remove the service plug prior to disconnecting or reconnecting any HV connections or components.
- Isolate any high-voltage wires that have been removed with insulation tape.
- Properly torque the high-voltage terminals.

 WARNING: When the vehicle has been left unattended, recheck that the service disconnect has not been reinstalled by a well-meaning associate.

When working on an HEV, always assume the HV system is live until you have proven otherwise; you can never be too safe. Your first mistake may be your last! If the vehicle has been driven into the service department, you know that the HV system was energized since most HEVs do not move without the HV system operating.

It is critical that the proper tools be used when working on the HV system. These include protective hand tools and a digital multimeter (DMM) with an insulation test function. The meter must be capable of checking for insulation up to 1000 volts and measuring resistance at over 1.1 mega ohms. In addition, the DMM insulation test function is used to confirm proper insulation of the HV system components after a repair is performed.

SPECIAL TOOLS

DMM capable of reading 400 Volts AC/DC

Insulating gloves

Insulating tape

FIGURE 1-9 It is recommended to follow the one-hand rule when testing any high-voltage system.

Whenever possible use the one-hand rule when servicing the HV system (Figure 1-9). The one-hand rule means working with only one hand while servicing the HV systems so that in the event of an electric shock the high-voltage will not pass through your body. It is important to follow this rule when performing the HV check out procedure since confirmation of HV system power down has not been proven yet.

Insulated Glove Integrity Test

The rubber isolating gloves (lineman gloves) that the technician must wear for protection while serving the HV system are your first line of defense when it comes to preventing contact with energized electrical components and must be tested for integrity before they are used. Also, pay attention to the date code on the gloves, gloves have an expiration date, they do not last forever. In addition for heavy services use the leather protective glove on top of the isolating gloves.

Not just any gloves will do, lineman gloves must meet current ASTM D120 specifications and NFPA 70E standards. These requirements are enforced by OSHA as part of their CFR 1910.137 regulation. These standards dictate testing, retesting, and manufacturing criteria for lineman gloves. For HV vehicles the lineman gloves must meet Class "0" requirements of a rating of 1000 volts AC (Figure 1-10). Electrical protective gloves are categorized by the amount of voltage—both AC and DC—they have been proof-tested to. In addition, the technician should wear rubber-soled shoes, cotton clothing, and safety glasses with side shields as part of their personnel protective equipment. Remove all jewelry and make sure metal zippers are not exposed. Always have a second set of isolating gloves available and let someone in the shop know their location. When preparing to work on any high-voltage vehicle, let associates know in the event they must come to your aid.

OSHA regulations require that all insulating gloves must be electrically tested before first issue and retested every 6 months thereafter by a test laboratory. For this reason most shops will discard and replace gloves after 6 months. Any unused gloves after 12 months must be retested or discarded. The manufacturers and suppliers of insulating gloves can assist in providing test laboratory locations for retest certification of the glove if you prefer over replacement. It is recommended that gloves be stored out of direct

FIGURE 1-10 Insulated rubber isolation gloves rated Class "0" must be worn when working on high-voltage system.

sunlight and away from sources of ozone (i.e., electric motors). They should be stored in a glove bag flat, never folded, and hung up rather than laid down on a flat surface.

 WARNING: Do not use insulating gloves that have been used for over 6 month or are new in the package but older than 12 months unless they have been recertified by a licensed test laboratory. Perform a daily or prior to use safety inspection of insulating gloves before working on any HV vehicle.

Daily or prior to use safety inspection of rubber insulating gloves procedure includes:

- Visually inspect rubber gloves prior to use for cracks, tears, holes, signs of ozone damage, possible chemical contact, and signs of abrasion or after any situation that may have caused damage to the gloves.
- OSHA requires a glove air inflator test.
- Blow air into glove to inflate them and seal the opening by folding the base of the glove.
- Slowly role the base of the glove toward the fingers to increase the pressure (Figure 1-11).

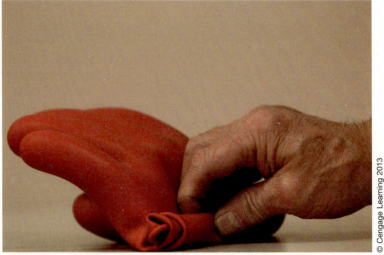

FIGURE 1-11 Insulated rubber isolation must be tested for leaks before working on high-voltage system.

FIGURE 1-12 The high-voltage service plug is generally located near the HV battery.

- Look and feel for pin holes on all surface sides.
- If a leak or damage is detected discard the glove. For extra precaution the glove should be rendered unusable (i.e., cut in half).

High-Voltage Service Plug

The HEV is equipped with a high-voltage service plug that disconnects the HV battery from the system. Usually this plug is located near the battery (Figure 1-12). Prior to disconnecting the high-voltage service plug, the vehicle must be turned off. Some manufactures also require that the negative terminal of the auxiliary battery be disconnected. Once the high-voltage service plug is removed, the high-voltage circuit is shut off at the intermediate position of the HV battery.

The high-voltage service plug assembly contains a safety interlock reed switch. The reed switch is opened when the clip on the high-voltage service plug is lifted. The open reed switch turns off power to the service main relay (SMR). The main fuse for the high-voltage circuit is inside the high-voltage service plug assembly.

However, never assume that the high-voltage circuit is off. The removal of the high-voltage service plug does not disable the individual high-voltage batteries. Use a DMM to verify that 0 volts are in the system before beginning service. When testing the circuit for voltage, set the voltmeter to the 400 VDC scale.

After the high-voltage service plug is removed, a minimum of five minutes must pass before beginning service on the system. This is required to discharge the high-voltage from the condenser in the inverter circuit.

To install the high-voltage service plug, make sure the lever is locked in the DOWN position (Figure 1-13). Slide the plug into the receptacle, and lock it in place by lifting the lever upward. Once it is locked in place, it closes the reed switch returning power to the system.

SAFETY IN THE SHOP

In general, the automotive air conditioning technician may be involved in all phases of automotive service, including electrical and mechanical repairs, relating to air conditioning malfunctions.

Some of the common occupational safety and health problems found during walk-around surveys of typical service repair facilities include:

- Poor **housekeeping**: refuse and non-salvageable materials not being removed at regular intervals; electrical cords and compressed-gas lines scattered on floors (Figure 1-14); and oily, greasy spots or water pools on floor areas.

Housekeeping is the systematic practice of maintaining an area in clean, safe working order and includes the proper storage of materials and chemicals.

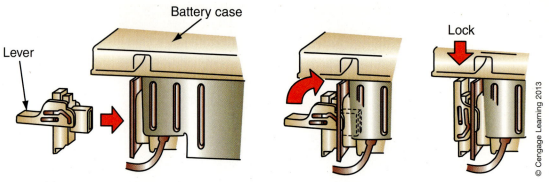

FIGURE 1-13 The high-voltage service plug lever must be placed in the down and locked position after installation.

FIGURE 1-14 Keep electrical cords, gas lines, and air hoses off the floor of the work area.

> All machine pulleys and belts should be guarded.

- Ineffective and, in many cases, nonexistent guard rails and toe-boards around open pit areas.
- Use of unsafe equipment, such as damaged creepers with missing crosspieces and broken wheels.
- The unsafe stacking of stock and other material.
- Failure to identify "safety" zones.
- Pulleys, gears, and the "point of operation" of equipment without effective barrier guards or other guarding devices or methods.
- Inadequate ventilation or unacceptable respirator programs for operations in confined spaces.
- Handling of resins, cements, oils, and solvents without protection, causing skin problems or dermatitis.
- Electrical hazards such as "U" ground prong missing from power tools (Figure 1-15); ungrounded extension cords and electrical equipment; and frayed, damaged, or misused power and extension cords.

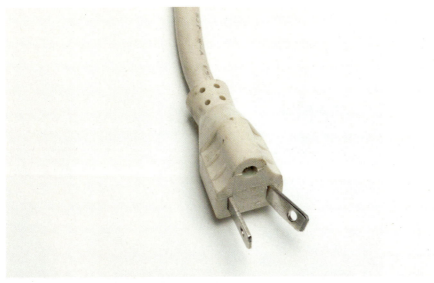

FIGURE 1-15 An unsafe male electrical plug.

■ Unsecured and improper storage of compressed-gas cylinders.
■ Fire hazards caused by improper storage and use of flammable and combustible materials and the presence of various ignition sources.
■ Improper lifting and material handling techniques.
■ Unsafe work practices that could result in burns from hot-work operations such as welding, burning, and soldering.

Although the above deficiencies cover some of the most significant problem areas, this list should not be considered exhaustive.

OSHA

OSHA was established in 1970 to ensure safe and healthful conditions for every American worker. The agency's enforcement, educational, and partnership efforts are intended to reduce the number of occupational injuries, illnesses, and deaths in America's workplaces. Since its inception, the workplace death rate has been cut in half. Still, about 17 Americans die on the job every day.

OSHA is committed to a commonsense strategy of forming partnerships with employers and their employees. They conduct firm but fair inspections, develop easy-to-understand regulations, and eliminate unnecessary rules to assist employers in developing quality health and safety programs for their employees.

State consultants, authorized and funded largely by OSHA, conduct consultation visits with employers who request assistance in establishing safety and health programs or in identifying and dealing with specific hazards at their workplaces. OSHA will also conduct unannounced inspections at work sites under its jurisdiction. An inspection is made when three or more workers are hospitalized because of injury or if a job-related death occurs. An inspection will also occur based on an employee complaint. Only half of their staff of about 2,200 are safety and health officers, so many inspections are handled by telephone and fax, often without the requirement for an on-site inspection.

In 1996 and 1997, OSHA simplified the written text outlining its regulations by eliminating almost 1,000 pages and by putting over 600 other pages into plain English. Employers must post a full-sized 10 × 16 in. (254 × 406 mm) OSHA or state-approved poster, such as that shown in Figure 1-16, where required. This is generally in a "common" area where it will be seen by all employees.

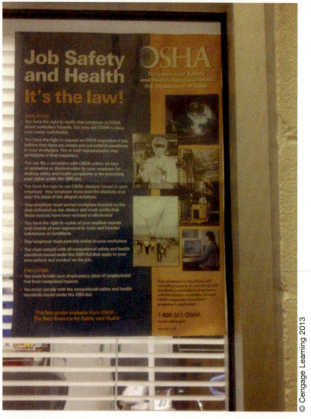

FIGURE 1-16 OSHA poster in employees' "common" area.

Health and Safety Program

Hazardous conditions or practices not covered in the standards promulgated by OSHA in 1970 are covered under the general duty clause of the act: "Each employer shall furnish to each employee a place of employment which is free from recognized hazards that are causing or are likely to cause death or serious physical harm."

A health and safety program is an effective method to help ensure a safe working environment. The purpose of such a program is to recognize, evaluate, and control hazards and potential hazards in the workplace. Hazards may be identified by:

- Performing self-inspections
- Soliciting employee input
- Interviews
- Suggestions
- Complaints
- Promptly investigating accidents
- Reviewing injury and illness records
- Other information sources

Typical examples of hazards are:

- Unsafe walking surfaces
- Unguarded machinery
- Electrical hazards
- Improper lifting
- Air contaminants

In the classroom, the instructor may assign students safety and health management responsibilities in the areas of both program development and implementation. Regular meetings and informal discussions should be held to discuss safety promotions and actual or potential hazards. To ensure program success, the participation and cooperation of all class members is essential. Leadership is also necessary. The students assigned the responsibility for carrying out the program must be delegated the proper authority and have the instructor's support. All participants must be aware of the program activities through a systematic interchange of information. Students cannot take an interest in the program if they are unaware of what is occurring. Conversely, well-informed students will likely show interest and a desire to participate.

In the learning as well as the work environment, persons may be exposed to excessive levels of a variety of harmful materials. These include:

- Gases
- Dusts
- Mists
- Vapors
- Fumes
- Certain liquids and solids
- Noise
- Heat or cold

Of the illnesses reported, respiratory problems caused by dusts, fumes, and other toxic agents are the most prevalent. Often health hazards are not recognized because materials used are identified only by trade names. A further complication arises from the fact that materials tend to contain mixtures of substances, making identification still more difficult.

The law requires suppliers to furnish MSDS on request.

To begin identifying occupational health hazards, a materials analysis or product inventory should be made of which all hazardous substances are listed and evaluated. If the composition of a material cannot be determined, the information should be requested from the manufacturer or supplier who must provide a MSDS for its product. These sheets contain safety information about materials, such as toxicity levels, physical characteristics, protective equipment requirements, emergency procedures, and incompatibilities with other substances.

A process analysis should be performed, noting all chemicals used and all products and byproducts formed. When doing such an analysis, allied activities such as maintenance and service operations should be included:

Toxic gases may be found around welding operations.

- Welding performed around chlorinated materials, such as R-12, may cause the formation of toxic gases in addition to welding fumes.
- Exhaust gases from cars and trucks with internal combustion engines contain CO.
- When certain cleaning agents are mixed, poisonous gases such as chlorine are sometimes formed.

It should be noted that skin conditions such as chemical burns, skin rashes, and dermatitis constitute over half of all occupational health problems. The use of protective creams or lotions, proper personal protective clothing (Figure 1-17) and other protective equipment, and good personal hygiene practices can often prevent these problems.

There are various control methods that can be used to prevent or reduce the technician's exposure to air contaminants. They are as follows:

- Substituting less-toxic materials.
- Isolating or placing the potentially hazardous process in a separate room or in a corner of the building to reduce the number of technicians exposed.
- Ventilating, including local exhaust ventilation, where contamination is removed at the point of generation, and general mechanical ventilation.
- Limiting the total amount of time a technician may be exposed to a health hazard via administrative controls.

FIGURE 1-17 Protective creams and lotions may be used when necessary.

- Training and educating technicians about the hazards to which they are to be exposed and how to reduce or limit that exposure (Right-to-know Laws training). See http://www.CCAR-GREENLINK.ORG for information on online training for schools.
- Practicing personal hygiene cannot be overemphasized. Technicians should wash their hands before eating. If chemicals such as caustic epoxies or resins come in contact with the skin, they should be washed off immediately.
- Avoiding eating around toxic chemicals or in contaminated areas.
- Changing clothing and washing them daily if they become contaminated with toxic chemicals, dusts, fumes, or liquids.
- Using personal protective equipment such as respirators, hearing protection devices (Figure 1-18), protective clothing, and latex gloves when appropriate.

General Philosophy. A health and safety program helps identify unsafe acts or conditions in the workplace. For many of these there may not be specific standards for rectifying the dangers they pose. Nevertheless, it is important to find a solution for these recognized problems.

FIGURE 1-18 Use personal hearing protection.

During the analysis of the workplace for health and safety problems, it may also become apparent that "the letter of the law" is not being met. If it is apparent that the "intent" of the law is being met, instead of making changes, a variance may be requested from OSHA. The application for a variance must show it is "as effective as" the OSHA standards in ensuring a safe work environment. The decision not to make changes must be made only with the concurrence of OSHA.

Even when a citation is issued, it is desirable that the employer have demonstrated the willingness to comply with the intent of the law by operating effective, ongoing safety and health programs, by correcting imminent dangers in the workplace; and by maintaining records of purchases, installations, and other compliance-promoting activities. Therefore, after an OSHA compliance visit and a citation, the employer can substantiate the intent to provide a safe and healthy workplace for employees by producing records that document that purpose and may be given the benefit of having shown "good faith," which can serve to reduce penalties.

Technician Training

The following are suggestions that can help reduce unsafe acts and practices in the shop:

- Be constantly aware of all aspects of safety, particularly good housekeeping, and the elimination of slipping, tripping, and other such hazards.
- Be knowledgeable in the maintenance and operation of any special equipment. Do not attempt to use such equipment without first having been instructed in its proper use.
- Use appropriate personal protective and safety equipment, such as safety glasses (Figure 1-19) and, whenever necessary, respiratory apparatus.
- Develop and maintain check points to be observed as part of the standard and emergency procedures.
- Be knowledgeable in the proper use of portable fire extinguishers. Know where fire extinguishers are located.
- Know who is trained and responsible for emergency first aid treatment and the procedures for reporting an emergency. At least one technician should be trained in first aid on each shift at each site.
- The Coordinating Committee for Automotive Repair (CCAR) offers safety and pollution prevention training at http://www.SP2.org.

Safety glasses must be worn at all times.

FIGURE 1-19 Use personal eye protection.

FIGURE 1-20 Remove the chuck key before turning on power to the drill press.

SAFETY RULES FOR OPERATING POWER TOOLS

The following rules apply to those who use and operate power tools. To ensure safe operation the technicians must:

- Know the application, limitations, and potential hazards of the tool used.
- Select the proper tool for the job.
- Remove chuck keys (Figure 1-20) and wrenches before turning the power on.
- Not use tools with frayed cords or loose or broken switches.
- Keep guards in place and in working order.
- Have electrical ground prongs in place (see Figure 1-10).
- Maintain working areas free of clutter.
- Keep alert to potential hazards in the working environment, such as damp locations or the presence of highly combustible materials.
- Dress properly to prevent loose clothing from becoming caught in moving parts.
- Tie back long hair or otherwise protect it from becoming caught in moving parts.
- Use safety glasses, dust or face masks, or other protective clothing and equipment when necessary.
- Do not surprise or distract anyone using a power tool.

Machine Guarding

It is generally recognized that machine guarding (Figure 1-21) is of the utmost importance in protecting the technician. In fact, it could be said that the degree to which machines are guarded in an establishment is a reflection of management's interest in providing a safe workplace.

A technician cannot always be relied on to act safely enough around machinery to avoid accidents. One's physical, mental, or emotional state can affect the attention paid to safety while working. It follows that even a well-coordinated and highly trained technician may at times perform unsafe acts that can lead to injury and death. Therefore, machine guarding is important.

Good Housekeeping Helps Prevent Fires

Maintaining a clean and orderly workplace reduces the danger of fires. Rubbish should be disposed of regularly. If it is necessary to store combustible waste materials, a covered metal receptacle is required.

Combustible material must be stored in special metal cabinets.

FIGURE 1-21 Machine guarding is essential for safety.

Cleaning materials can create hazards. Combustible sweeping compounds such as oil-treated sawdust can be a fire hazard. Floor coatings containing low-flash-point solvents can be dangerous, especially near sources of ignition. All oily mops and rags must be stored in closed metal containers (Figure 1-22). The contents should be removed and disposed of at the end of each day.

Some of the common causes of fires are:

■ Electrical malfunctions
■ Friction
■ Open flames
■ Sparks
■ Hot surfaces
■ Smoking

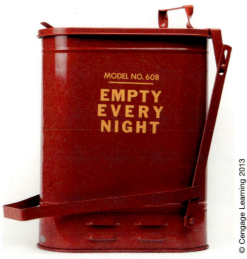

FIGURE 1-22 Oily rags are stored in metal containers approved for that purpose.

FIGURE 1-23 Aisles should be identified for safety.

The proper housekeeping and safety policies can help reduce these fire hazards. It can also help to eliminate many tripping hazards.

Walking and Working Surfaces

All areas, passageways, storerooms, and maintenance shops must be maintained in a clean and orderly fashion and be kept as dry as possible. Spills should be cleaned up promptly. Floor areas must be kept clear of parts, tools, and other debris. Areas that are constantly wet should have nonslip surfaces where personnel normally walk or work.

Every floor, working place, and passageway must be maintained free from protrusions such as nails, splinters, and loose boards.

Where mechanical handling equipment is used, such as lift trucks, sufficient safe clearances must be provided for aisles at loading docks, through doorways, and wherever turns or passage must be made. Low obstructions that could create a hazard are not permitted in the aisle.

All permanent aisles must be easily recognizable. Usually aisles are identified by painting or taping lines on the floors (Figure 1-23).

THE VALUE AND TECHNIQUES OF SAFETY SENSE

Safety sense with tools pays off. The technician should think safety whenever applying a tool to the task. Some of the tips presented in this chapter may seem to be nothing more than common safety sense; they are included because technicians who overlook them are apt to be injured.

Safety sense reminds the technician to protect against the possibility of something going wrong. Whenever tools are used, there is a risk of tools breaking or slipping. And there is also a risk that the part on which tools are used may break loose, too.

Bracing against a Backward Fall

Always pull on a wrench handle; never push on it. It is far easier to brace against a backward fall than against a sudden lunge forward should the tool slip or break. To brace against a backward fall when pulling on a wrench, place one foot well behind the other.

 WARNING: Never push a wrench.

Have you ever pulled an open-end wrench right off the nut or bolt? This danger can be minimized by using a wrench of the proper size and by making sure it is positioned so that the jaw opening faces in the direction of pull.

Haste Makes Waste

A technician attempted to make a speedy adjustment to the shift linkage while the engine was running. When the technician lost control of his wrench, he was lucky to receive nothing more serious than a few bruises.

Engines should be turned off when making adjustments whenever possible. There are, of course, adjustments that must be made with the engine running. The secret, then, is safety ... Be safe.... Take care.... Take time.... Haste makes waste.

Safety Accessories

Safety glasses are essential eye protection when metal strikes metal, such as in using punches and chisels, or when grinding metal tools or parts on a power grinder. Safety glasses are recommended for everyone working in a shop. In many situations, safety glasses and hard hats are required for anyone entering the premises.

Putting Safety Sense into Action

- Make a thorough check of the toolbox. Discard all tools that do not meet minimum safety standards. Where necessary, replace them with quality tools.
- Investigate procurement standards to make certain that only professional-quality tools are purchased by your company.
- Instruct one technician or the tool room attendant in the repair of ratchets, screwdrivers, and other tools.
- Instruct technicians in the care of hand tools at a regular departmental safety meeting. Use bad-example tools picked from toolboxes to illustrate the hazards. Advise those using ratchets to bring them into the tool room at regular periods for service.
- Set up a small stock of repair parts for ratchets, screwdrivers, pliers, and other small tools with replaceable parts.
- Make spot checks for correct tool application.
- Review shop tooling to ensure that an adequate selection of tools is readily available for all jobs. This step is important to the elimination of makeshift tool procedures.
- Investigate tool applications involving moving machinery, and correct or minimize any hazards noted.
- Incorporate safety sense tips into tool safety education programs in departmental meetings.

> Start a health and safety program in your shop.

SAFE USE OF TOOLS

The proper use of tools is an important consideration for today's technician. This notion may be divided into two major categories:

1. The use of safe tools.
2. The safe use of tools.

The two go hand in hand, for without one there cannot be the other. The formula for tool safety comprises the following three general rules:

1. Use safe tools.
2. Maintain tools in a safe condition.
3. Use the right tool for the task.

> Use the proper tool. Do not use, for example, a metric tool on an English fastener.

Rule #1: Use Safe Tools

The safe use of tools and equipment is fundamental to any automotive technician safety program. The first step in any tool safety program is to upgrade tools to maintain

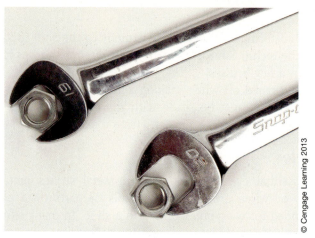

FIGURE 1-24 Tools should fit intended applications.

minimum safety standards. This requires inspecting all current tools on hand and replacing any that are defective or do not meet minimum quality standards.

Quality Standards for Tools. Tools that serve the automotive industry are put to rugged use. If they are to stand up, the tools must be designed and manufactured according to rigid quality standards.

Some of the more important points for consideration when selecting tools include:

- Tools should be made of alloy steel. Finer-grade alloys impart toughness to the metal used in the manufacture of tools. If a tool made of an alloy steel is inadvertently overloaded, it will deform before it will break, thus providing a warning to the user.
- Tools should be tempered by heat. Tool strength and lasting quality is enhanced by precision heat treatment of the metal.
- Tools should be machined accurately. If a tool is to fit the intended application accurately (Figure 1-24), without slip or binding, machining must be held to close tolerances.
- Tools should be designed for safety. Firm, safe tool control with a minimum of effort should be provided by a lightweight, balanced design. Design features should include those that prevent slipping or accidental separation of tool parts.

Unsafe Tools. Identifying and discarding unsafe tools is an important part of developing the habit of tool safety. In addition to those tools that are easily recognized as below standard, broken, or otherwise damaged, you should avoid using homemade and reworked tools. Few repair facilities are equipped to work steel into tools suitable for high-leverage automotive repair applications. Homemade tools are therefore often heavy and awkward to handle. Lighter and stronger tools, for virtually any purpose, are commercially available and should replace all homemade relics.

Grinding or otherwise reworking a tool to fit a particular application usually results in a tool that no longer measures up to safety requirements. Grinding a tool robs it of metal needed for strength. Heat created by grinding, bending, and brazing impairs the temper of the metal, which also weakens the tool. These tools, too, should be replaced.

Light-Duty Tools. There are many inexpensive tools (Figure 1-25) in the marketplace. Most are intended for amateur or do-it-yourself (DIY) mechanics or other light-duty applications. Examples include stamped or die-cast tools made of non-alloy carbon steel. These tools are not designed for professional use and are not suitable for general use in the automobile repair trade.

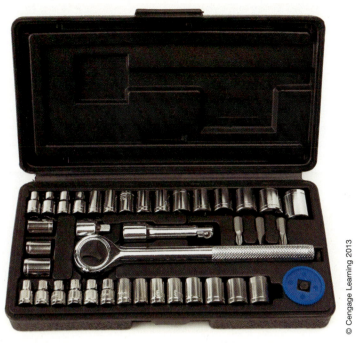

FIGURE 1-25 Light-duty tools are not intended for use in professional automotive shops.

Rule #2: Maintain Tools in a Safe Condition

When safe tools are used in the workplace, the next step is to keep them in safe condition. A routine inspection on a regular basis leads to repairing or replacing those tools that are worn or otherwise considered no longer safe. The following tips on specific tool care will help minimize the risk of personal injury due to tool failure.

Ratchets. Ratchets are mechanical devices and, as such, are subject to mechanical failure. Frequent causes of failure are worn parts and dirt. The results of failure are slippage, which in turn can lead to possible injury.

To reduce the risk of ratchet failure, a program of preventive maintenance—cleaning and lubricating the ratchet mechanism—should be performed at least once every 6 months.

Screwdrivers. Screwdrivers with worn, chipped, or broken tips (Figure 1-26) are a potential menace to the technician who works with them. Such tools have little grip on the screw head and frequently jump the slot, leaving the technician open to injury.

FIGURE 1-26 Damaged screwdrivers should be repaired or replaced.

FIGURE 1-27 Grind down the flat surfaces.

CAUTION:
If excessive grinding is required, it is best to replace the screwdriver. Only the tip of the screwdriver is tempered, and excess grinding will remove the tempered area.

Regular inspection and replacement of screwdrivers is a must because of the performance expected of these tools. The following procedure can be used to repair screwdrivers with slightly worn or nicked tips:

1. Very lightly grind down the flat surfaces using the grinder. Avoid overheating and destroying the temper of the metal (Figure 1-27). The amount of material removed should be minor. "Quench" the tip in H_2O every few seconds to keep it cool.
2. Square off the edges and the tip (Figure 1-28).
3. Test the tip for correct fit (Figure 1-29).
4. Use a file or an oil stone to remove burrs.

Wrenches and Sockets. Tools that show signs of "old age" are prime candidates for replacement. Because worn-out wrenches and sockets take only a partial "bite" on the corners of a nut, they are often likely to slip on a heavy pull.

FIGURE 1-28 Square off the edges and the tip.

FIGURE 1-29 Test the tip for correct fit.

A regular inspection of the tool box for worn tools will prevent many mishaps. Look for:

- Open-end wrenches with battered, spread-out jaw openings.
- Sockets or "box-sockets" whose walls have been battered and rounded by use.
- Tools that have been abused, such as standard thin-wall (hand-use) sockets with lapped-over metal around square drive opening (revealing their use on impact wrenches) and wrenches or handles bearing hammer marks. These tools should also be scrapped because hammer or impact shock leads to metal fatigue, which substantially weakens tools of this type.

It is important, too, that dirt and grit are not caked inside sockets. Such debris may prevent the socket from seating fully on the nut or bolt head. This would concentrate the twisting force at the very end of the socket, possibly causing the socket to break even with a moderate pull.

Keep tools clean.

Other Tools. Routine inspection of toolboxes will also uncover other unsafe tool conditions that could result in accidents:

- Hammers with cracked heads or handles
- Pliers with smoothly worn gripping sections
- Pliers with rivets or nut-and-bolt assemblies that have become sloppy

Many hand tool accidents can be traced to poor housekeeping. Safety, as well as good workmanship, dictates that tools be properly stored and cleaned.

A misplaced tool frequently is the cause of a technician's tripping or being hit by a falling object. Tools should always be kept in tote trays, boxes, or chests when not being used.

Tools with oily handles can be slippery and dangerous. Technicians should establish a habit of wiping off tools with a dry shop rag before starting each job or, better yet, before putting them away after use. With tools, as with everything else, good housekeeping makes good safety sense.

Rule #3: Use the Right Tool for the Task

Safe tools, in safe condition, are only half of the tool safety story. The other half rests with the technicians who use the tools. Every technician in the shop who works with tools should understand the type, size, and capacity of the tools they use.

Hand tools are available in an endless assortment of types, styles, shapes, and sizes—perhaps over 5,000 in all. Each tool is designed to do a certain job quickly, safely, and easily. Here are three important considerations in selecting tools for a task:

Type of Tool. The safest tool is the tool that is specifically designed for the particular task. The use of makeshift tools just to "get by" is one of the major causes of hand tool accidents.

Size and Shape of Tool. The safest tool is one that fits the job squarely and snugly. Misuse can lead to the tool's slipping or breaking. This can lead to injury.

Capacity of Tool. Every tool has a design safety limit. Exceeding the design limit can result in tool failure.

The following sections look more carefully at specific tools commonly used in the workshop.

Wrenches and Socket Wrenches

These are the safest tools for turning bolts. Some bolt-turning tools have definite safety advantages over others. Here is a list of tools suitable for bolt turning in order of preference:

1. Box-sockets and socket wrenches (Figure 1-30). These tools are preferred for bolt-turning jobs where a heavy pull is required and safety is a critical consideration. A socket or box-socket completely encircles the hex nut or bolt and grips it securely at all six corners. It cannot slip off laterally, and there is no danger of springing jaws.
2. Open-end and flare-nut wrenches (Figure 1-31). Firm, strong jaws make open-end wrenches a very satisfactory tool for medium-duty bolt-turning work. Many technicians work with combination wrenches, using the open end to speed the nut on or off and the box end for breaking loose or final tightening. Flare-nut wrenches are recommended for those jobs where sockets or box-sockets cannot be used. Flare-nut wrenches are a "must" for use in servicing automotive air conditioning hoses.
3. Adjustable wrenches. The adjustable wrench, commonly referred to by its trade name, *Crescent*, is recommended only for light-duty applications where time is an important factor and the proper tool is not readily available.

Adjustable wrenches (Figure 1-32) are prone to slip because of the difficulty encountered in setting the correct wrench size. They also have a tendency for the jaws to "work" as the wrench is being used. For these reasons, an adjustable wrench should not be considered an

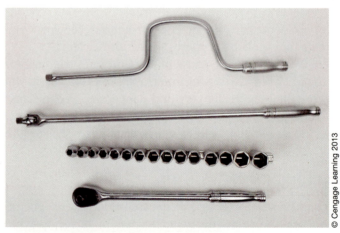

© Cengage Learning 2013

FIGURE 1-30 An assortment of sockets and drivers.

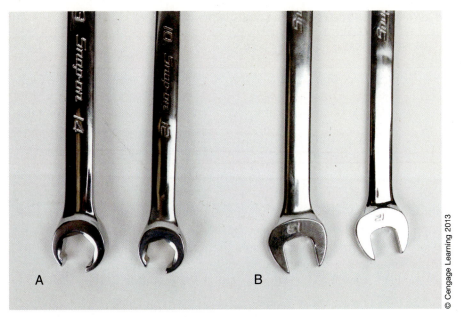

FIGURE 1-31 Flare-end (A) and open-end (B) wrenches.

© Cengage Learning 2013

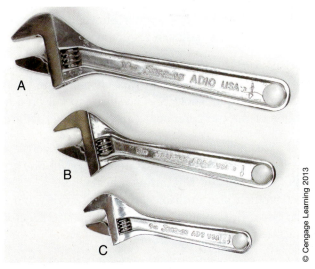

FIGURE 1-32 Adjustable wrenches: 10 in. (A), 8 in. (B), and 6 in. (C).

© Cengage Learning 2013

all-purpose tool. Though often used for that purpose, pliers are not on the list of tools recommended for bolt turning.

Overloading Wrenches and Socket Wrenches. The "safety limit" of a wrench or socket wrench is determined by the length of its handle. Use of a pipe extension or other "cheater" to move a tightly rusted nut can overload the tool past its safety limit.

When the tool being used cannot turn the nut, a heavier-duty tool is required. Both open-end and box-socket wrenches are available in a heavy-duty series that can be safely used with tubular handles from 15 in. to 36 in. in length and can be substituted for a wrench that is too light for the job.

Table 1-1 presents a breakdown of government minimum proof loads for ratchets and other socket wrench handles. These figures can be used as a guide for the safe use of socket wrenches.

TABLE 1-1 MAXIMUM TORQUE LOAD FOR SOCKET WRENCHES

Drive	Minimum Torque (in.-lb)	Proof Load (ft.-lb)	Load Approximate Equivalent
¼ in.	450	37	Pulling 100 pounds at the end of a 4.5 in. handle
⅜ in.	1,500	125	Pulling 200 pounds at the end of a 7.5 in. handle
½ in.	4,500	375	Pulling 200 pounds at the end of a 22.5 in. handle
¾ in.	9,500	792	Pulling 200 pounds at the end of a 47.5 in. handle
1 in.	17,000	1,417	Pulling 200 pounds at the end of an 85 in. handle

Use of Hammer with Wrenches. Wrenches and socket wrench handles should never be used as hammers, nor should a hammer be used on these tools to break loose a tightly seized nut or bolt. Hammer abuse weakens the metal and can cause the tool to fail under a heavy load. Sometimes, however, hammer shock is the only cure for an extra stubborn nut or bolt. In such cases, here are the suggested tools to use:

- *Sledge-type box-sockets.* They have plenty of "beef" and are especially tempered for use with a sledge hammer.
- *Cupped-anvil box-sockets.* These air-driven tools were developed to meet the requirements of heavy processing industries.
- *Impact driver.* This tool transmits a hammer blow into rotary shock. It is especially useful for smaller nuts and bolts and can also be used on screws.

Sockets for Impact Use. Sockets used with impact wrenches or impact drivers should be the thick-wall power type (Figure 1-33). Thin-wall chrome standard sockets designed for hand use will weaken under impact shock and are likely to fail on a high-leverage application. Impact abuse is one of the most frequent causes of socket failure.

> Use thick-wall sockets with an impact driver.

Selecting the Correct Size. Selection of the correct wrench size is a necessary part of safe bolt-turning work. A wrench or socket one size too large will not grip the corners of the nut securely. The result can be a bad slip during a heavy pull.

There is a correct wrench size available for virtually every nut or bolt made in the United States, Canada, Great Britain, and Europe. Wrench size is determined by measuring the nut or bolt head across the flats.

> Use a six-point hex socket on worn nuts and bolts.

Danger of "Cocking." Sockets and box-sockets should also fit squarely. When these tools are "cocked," they are likely to break even under a moderate load. This is due to "binding" that concentrates the entire strain at one point, rather than spreading it evenly over the tool. The point at which the strain is concentrated becomes vulnerable to failure.

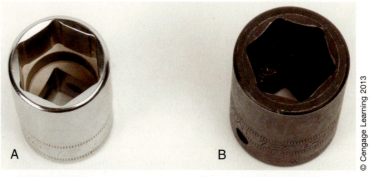

© Cengage Learning 2013

FIGURE 1-33 Thin-wall chrome standard sockets (A) are not intended for impact use. A thick-wall socket (B) is used with an impact wrench.

FIGURE 1-34 **An open-end wrench should contact the entire flat surface.**

Cocking is a frequent cause of tool failure. It can usually be avoided by using different arrangements of sockets, flex-sockets, and extensions, or by substituting box-sockets of different lengths and offsets.

Selecting Open-End Wrenches. For the most secure grip, open-end wrench jaws should contact the entire length of two flat surfaces of the nut or bolt head (Figure 1-34). When it is necessary to reach the fastening at extreme angles, there is a danger that the wrench will slide off. This can usually be avoided by the use of crowfoot, offset-head, or taper-head open-end wrenches.

Correct Style of Socket or Box-Socket. Here are some rules that govern the selection of the safest tool for the job:

- When turning a fitting where corners are rounded by wear or corrosion, single-hex sockets or box-sockets offer more protection because they grip a larger amount of the surface of the fitting.
- On square fittings, use a single-or double-square, not a double-hex.
- Where bolt clearance is a problem, avoid tool breakage by using deep, extra-length sockets.

Special-Purpose Tools. Many special-purpose tools are designed for jobs where a critical clearance problem exists and tool jaws or walls are extra thin. Examples are wrenches and sockets for removing and replacing air conditioning compressor clutch retaining nuts and bolts.

These tools are plenty safe for the job intended, but should not be used for general bolt-turning work. The usual result is overload and failure.

Pliers

Pliers (Figure 1-35) are often misused as general-purpose tools. Their use should be limited to gripping and cutting operations for which they were designed. Pliers are not recommended for bolt-turning work for two reasons: (1) because their jaws are flexible, they slip frequently when used for this purpose; and (2) they leave tool marks on the nut or bolt head, often rounding the corners so badly that it becomes extremely difficult to service the fittings in the future.

Keep Pliers Jaws Parallel. For a firm, safe grip with a minimum of effort, pliers jaws should be as nearly parallel as possible. Use of the right size pliers and proper positioning make this possible.

FIGURE 1-35 Typical pliers.

Avoid Overloading Cutting Pliers. To avoid overloading the tool, the user should select the pliers that will cut a wire using the strength of only one hand. *Another tip:* The inside of the cutting jaws should point away from the user's face to prevent injury from flying cuttings.

Screwdrivers

Torx bit drivers are also referred to as star bits.

The screwdriver is not an all-purpose tool, although some attempt to use them as such in place of lining-up punches, chisels, and prybars. The usual result is a damaged tool and a possible injury. The use of screwdrivers should be limited to screw turning only.

Correct Tool for Phillips Screws. It is very common for those not familiar with tools to try to turn a Phillips screw with a standard tip screwdriver designed for use on slotted screw heads. They usually end up with the tool slipping off the job, a nicked screwdriver tip, and a hopelessly chewed-up fastener. Only a Phillips tip screwdriver (Figure 1-36) should be used on a Phillips screw.

Phillips versus Reed and Prince. The tools to turn these "look-alike" screws are not interchangeable (Figure 1-37). A screwdriver of one type will not seat properly in the other screw head.

FIGURE 1-36 A Phillips screwdriver is used with a Phillips screw.

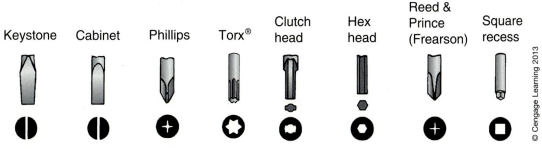

FIGURE 1-37 Typical screw head types.

TABLE 1-2 PHILLIPS SCREWDRIVER SELECTION GUIDE

Phillips	Machine Screw Diameter	Sheet-Metal Screw Diameter
1	#4 and smaller	#4
2	#5 to #10	#5 to #10
3	#12 to 5/16 in.	#12 to #14
4	3/8 in. and larger	

Selecting the Right Size. Selection of screwdriver tip size is an important factor in the safe use of these tools. An oversized tip will tend to jump from the slot. A screwdriver with an undersized tip is also likely to twist out. In either case, a slip of the tip can result in a trip to the first aid kit.

Here is an easy rule to remember: Use the largest screwdriver that will fit snugly in the slot. The length of the tip should be the same as that of the slot. Table 1-2 may be used as a guide in determining which size Phillips should be used for a particular fastener.

Lining Up with the Screw. Screwdrivers should line up with the screw on which they are being used to provide sufficient contact between the tip and the screw head. To avoid an out-of-line application, substitute a different length or an offset-type screwdriver.

Punches

There are several types of punches (Figure 1-38), each designed to do certain jobs properly and safely. Misuse often ends up in tool breakage. The jobs for which each punch tool was designed are:

- *Starter* or *drift punch:* For starting tightly jammed pins and bolts and for driving pins clear through a hole after they have been started.
- *Pin punch:* A speedy combination punch for starting and driving pins through holes, to be used only on light-duty jobs.

Correct Punch or Chisel Size. The greatest tool life and safety will result from the selection of the chisel whose cutting edge is the same width or wider than the area to be cut. This avoids unnecessary strain on a small chisel trying to do a big job.

When punches are used, the largest punch that will fit the job without binding should be used. Use of an undersized punch is apt to result in wedging the part being driven as well as tool failure.

Pullers

The puller is the only quick, easy, and safe tool for forcing a gear, wheel, pulley, or bearing off a shaft (Figure 1-39). Use of prybars or chisels often causes the part to cock on the shaft, making it even more difficult to remove. Also, with the wrong tools for the job, the

> Use the correct size punch and chisel.

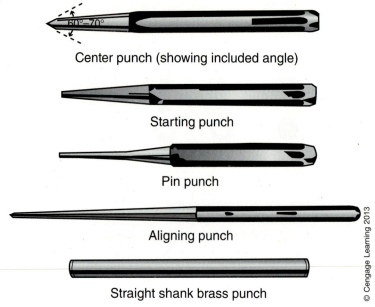

Center punch (showing included angle)

Starting punch

Pin punch

Aligning punch

Straight shank brass punch

© Cengage Learning 2013

FIGURE 1-38 Typical punches.

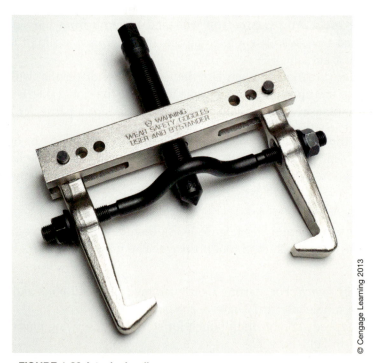

© Cengage Learning 2013

FIGURE 1-39 A typical puller.

operator must exert a great deal of force that is difficult to control, thereby creating an unnecessary hazard.

When a puller is used, the technician enjoys a mechanical advantage that reduces the amount of force required. Furthermore, the puller is so designed that the force that is used is always under control.

Selecting the Correct Size. Selection of the correct size puller can prevent serious accidents. Some important considerations are:

■ The jaw capacity of the puller should be such that when the tool is applied to the job, the jaws press tightly against the part being pulled.

- In pulling gears, the jaws should be wide enough to cover as many gear teeth as possible to minimize the danger of breakage.
- Use a puller with as large a pressure screw as possible, but avoid using one that is larger than the hole in the part that is to be pulled.
- Power capacity of pullers is stated in tons. To avoid the danger of overloading, it is best to use the largest capacity puller that will fit the job.

SUMMARY

- The many hazards associated with the automotive repair industry include exposure to asbestos, carbon monoxide, radiation, caustics, solvents, glues, and paints.
- Hybrid system can use voltage in excess of 300 volts—both AC and DC—it is critical that you as a service technician be familiar with and follow all safety and service precautions.
- Always remove the high-voltage service plug prior to servicing any HV system and place the plug where it cannot be accidently reinstalled by someone else, such as on your toolbox.
- Do not attempt to test or service the high-voltage system of an HEV for five minutes after the high-voltage service plug is removed.
- Confirm that the high-voltage circuits have 0 V using a DMM before performing any service procedure on the high-voltage system of an HEV. Never assume that the HV circuit is off.
- Test the integrity of your lineman gloves (isolation gloves) prior to servicing any high-voltage system. Do not use linemen gloves that fail a leak test, visual inspection, or are over 6 months old.
- The Occupational Safety and Health Administration (OSHA) was established to ensure safe and healthful conditions for every American worker.
- Hazardous conditions—such as cluttered floors, missing guardrails, lack of safety zones, inadequate ventilation, and improper storage of combustible materials—are often found in the repair facility.
- A health and safety program is an effective method to assist in providing for safe working conditions.
- Through the use of a health and safety program, unsafe acts and conditions become apparent and may be corrected.
- Substandard tools, those not made of alloy steel, should be replaced with tools made of industry-standard, high-alloy steel.
- Tools must never be "worked" beyond their design capabilities.

TERMS TO KNOW

Asbestos

Carbon monoxide (CO)

Hazards

Housekeeping

Material safety data sheet (MSDS)

Occupational Safety and Health Administration (OSHA)

Radiation

Refrigerant

Respirators

Technician

Ultraviolet (UV)

Ventilation

ASE-STYLE REVIEW QUESTIONS

1. *Technician A* says to lift heavy objects using the legs, not the back.
   Technician B says that one should get help when lifting heavy objects.
   Who is correct?
   A. A only
   B. B only
   C. Both A and B
   D. Neither A nor B

2. *Technician A* says that ethylene glycol is essentially a nontoxic anti-freeze and hence is safer for children and animals.
   Technician B states that toxicologists report that propylene glycol inside the body is changed into a crystalline acid that attacks the kidneys.
   Who is correct?
   A. A only
   B. B only
   C. Both A and B
   D. Neither A nor B

3. All of the following statements are true concerning high-voltage system safety, EXCEPT:
   A. Turn the power switch to the OFF position prior to performing a resistance check.
   B. Do not attempt to test or service the system for 5 minutes after the high-voltage service plug is removed.
   C. Test lineman gloves for damage and leaks prior to use.
   D. Disconnect the motor generators prior to turning the ignition off.

4. *Technician A* says that exhaust gases from internal combustion engines contain carbon monoxide gas.
   Technician B says that exhaust gases from internal combustion engines contain phosgene gas.
   Who is correct?
   A. A only   C. Both A and B
   B. B only   D. Neither A nor B

5. The shop's safety program is being discussed.
   Technician A says that the program will be ineffective if the technician does not work with care.
   Technician B says that the program will be ineffective if the rules are not observed.
   Who is correct?
   A. A only   C. Both A and B
   B. B only   D. Neither A nor B

6. Tool quality is being discussed.
   Technician A says that mechanics' tools should be made of alloy steel.
   Technician B says that mechanics' tools should be tempered.
   Who is correct?
   A. A only   C. Both A and B
   B. B only   D. Neither A nor B

7. *Technician A* says that an OSHA is a state agency responsible for Occupational Standards for Heating and Air Conditioning.
   Technician B says that an OSHA poster must be displayed in the employees' common area, such as the break room.
   Who is correct?
   A. A only   C. Both A and B
   B. B only   D. Neither A nor B

8. *Technician A* says the wire harness, terminals, and connectors of the high-voltage system are identified by red.
   Technician B says to remove the service plug prior to disconnecting or reconnecting any HV connections or components.
   Who is correct?
   A. A only   C. Both A and B
   B. B only   D. Neither A nor B

9. *Technician A* says that a flare nut wrench may be used on a flare nut.
   Technician B says that an open-end wrench may be used on a flare nut.
   Who is correct?
   A. A only   C. Both A and B
   B. B only   D. Neither A nor B

10. *Technician A* says that placing one foot behind the other braces oneself against a fall if the wrench slips while pulling.
    Technician B says that placing one foot in front of the other braces oneself if the wrench slips while pushing.
    Who is correct?
    A. A only   C. Both A and B
    B. B only   D. Neither A nor B

Name _____ **Date** _____

PERSONNEL AND SHOP SAFETY ASSESSMENT

As a service professional, one of your first concerns should be safety. Upon completion of this job sheet you should have an increased awareness of personnel and shop safety items. As you take this personnel assessment and survey your shop answering the following questions, you will learn to evaluate your workplace and personnel safety.

Procedure

Give a brief description following each step. Always wear eye protection when working or walking around a service facility.

1. Before you evaluate your work place you must first evaluate yourself. Are you dressed for work? ☐ YES ☐ NO
   a. If yes, why do you believe your attire is appropriate? _____

   _____

   b. If no, what must you correct to be properly attired? _____

   _____

2. Are your safety glasses OSHA approved? ☐ YES ☐ NO
   a. Do you have side shields? ☐ YES ☐ NO
3. Are you wearing leather boots or shoes with oil resistant soles? ☐ YES ☐ NO
   a. Does your footwear have steel toes? ☐ YES ☐ NO
4. Is your shirt tucked into your pants? ☐ YES ☐ NO
5. If you have long hair is it tied back or under a hat? ☐ YES ☐ NO

Next carefully inspect your shop, noting any potential hazards.
Note: A hazard is not necessarily a safety violation but in an area of which you must be aware (i.e., pothole in parking lot).

6. Are there safety areas marked around grinders and other machinery?
   ☐ YES ☐ NO
7. What is the air pressure in the shop set at? _____
8. Where are the tools stored in the shop? _____

   _____

9. What recommendations would you make to improve the tool storage? _____

   _____

   _____

10. Have you been instructed on proper use of the shop vehicle hoist (lift)?
☐ YES   ☐ NO
   a. If not ask your instructor to demonstrate vehicle lift use.

11. Where is the first aid kit located? _____

12. Where is the eye wash station located? _____

13. List the location of the exits. _____

14. What is the emergency evacuation plan and where are you to assemble once you evacuate the building? _____

_____

_____

_____

15. Where are the MSDSs located? _____

16. Does the shop have a hazardous spill response kit?   ☐ YES   ☐ NO

Name _____ Date _____

COMPARE AND IDENTIFY SAFE AND UNSAFE TOOLS

Upon completion of this job sheet, you should be able to identify unsafe tools.

Tools and Materials

Miscellaneous and assorted hand tools

Procedure

Lay out the tools and separate them into two groups—safe and unsafe. Briefly describe and inventory the tools as follows:

TOOL	SAFE	UNSAFE	WHY (UNSAFE)
1. Screwdrivers	_____	_____	_____
2. Pliers	_____	_____	_____
3. Open-end wrench	_____	_____	_____
4. Box-end wrench	_____	_____	_____
5. Punch/chisel	_____	_____	_____
6. Hammer	_____	_____	_____
7. Socket wrench	_____	_____	_____
8. Snapring tools	_____	_____	_____
9. *_____	_____	_____	_____
10. *_____	_____	_____	_____

* Other (describe) _____

_____

What can the results be of using an unsafe tool such as a:

11. Hammer? _____

12. Screwdriver? _____

13. Pliers? _____

14. Wrench: _____

   Open-end? _____

   Box-end? _____

   Socket? _____

15. Punch or chisel? _____

Instructor's Response _____

_____

_____

Name _____ **Date** _____

THE NEED FOR HEALTH AND SAFETY

Upon completion of this job sheet, you should be capable of participating in a health and safety program.

Tools and Materials

None required

Procedure

Briefly describe your plan to eliminate or avoid the following health and safety hazards:

1. Engine exhaust fumes. _____
   _____

2. Caustic chemicals. _____
   _____

3. Liquid refrigerant. _____
   _____

4. Hot engine parts. _____
   _____

5. Cooling fan start without notice. _____
   _____

6. Coolant boilover. _____
   _____

7. Oil spill on floor. _____
   _____

8. A discharged fire extinguisher. _____
   _____

9. Spontaneous combustion. _____
   _____

10. Electrical shock. _____
    _____

Instructor's Response _____

_____

_____

Name _____ **Date** _____

IDENTIFY AND CORRECT HAZARDOUS CONDITIONS

Upon completion of this job sheet, you should be able to identify hazardous conditions and to make recommendations for correction.

Tools and Materials

None required

Procedure

Inspect your work area, identify five hazardous or potentially hazardous conditions, and briefly describe your plan to prevent or eliminate them.

1. _____

   _____

2. _____

   _____

3. _____

   _____

4. _____

   _____

5. _____

   _____

Inspect adjoining areas, identify five hazardous or potentially hazardous conditions, and briefly describe your plan to prevent or eliminate them.

6. _____

   _____

7. _____

   _____

8. _____

   _____

9. _____

   _____

10. _____

    _____

Instructor's Response _____

_____

_____

Chapter 2

Typical Shop Procedures and Tools

Upon Completion and Review of this Chapter, you should be able to:

- Identify the responsibilities of the employer and employee.

- Identify required and alternative services and the special tools required.

- Discuss how to use and interpret service information procedures and specifications.

- Compare the English and metric system of measurement as related to automotive technologies.

Shop Rules and Regulations

There are many rules and regulations that are imposed in an automotive repair facility. Some have to do with the Occupational and Safety Health Administration (OSHA), safety of the customer or technician, and fire and local ordinances; others are at the discretion of management. In any event, it is expected that everyone associated with the facility will help ensure that all rules and regulations are followed. For example, it is generally posted that customers are not permitted in the service area (Figure 2-1). Reasons given include insurance requirements, fire codes, or local ordinances. The real reason, some feel, is that customers simply get in the way, ask foolish questions, and slow down production. Understandably, the average customer does not have a

CAUTION

AUTHORIZED PERSONNEL ONLY

**Service Bays Are A Safety Area.
Eye Protection Is Required At All Times.**

**Insurance regulations prohibit customers
in the service bay area during work hours.
We suggest you check out our everyday low prices until
your technician has completed the work on your car.
Thank you for your cooperation.**

© Cengage Learning 2013

FIGURE 2-1 Rules and regulations are posted to provide for safety.

knowledge of the operation of an automobile and, much less, the routine procedures associated with an automotive repair facility. The actual reason, then, is for customer safety.

Employee-Employer Relationship

It is very important that a good rapport exists between the technician and those for whom he or she works. This is known as employer-employee (facility owner or representatives-technician) relationship. Note that the word *representatives* is plural. That means that the beginning technician may have to be accountable to and answer to several supervisors. Many look at this as a great advantage, for after finishing formal training, the real learning experiences begin—on the job.

Employer-employee relationships are a two-way street. There are certain assumed obligations of both parties. Having a good understanding of these obligations eliminates any problems that may arise relating to what you may expect or what your employer may expect of you.

Work Area

First, and perhaps foremost, you are entitled to a clean, safe place to work (Figure 2-2). There should be **facilities** for your personal **hygiene** and accommodations for any handicap that you may have.

Opportunity

The work environment should provide you an opportunity to successfully advance. This could be in the form of in-house training or an incentive to further your studies in vocational education programs. Today, it is desirable for a technician to obtain a 2-year Associate of Science degree in automotive technology. This can be a general (generic) program or it could be manufacturer-specific such as the General Motors Automotive Service Education Program (ASEP). Others, such as Chrysler, Honda, Ford, and Toyota, have similar programs. Opportunity includes fair treatment. This means that all technicians must be considered and treated equally without *prejudice* or *favoritism*.

Supervision

There should be a competent and qualified supervising technician who can lead you in the right direction if you have problems, suggest alternative methods, and tell you when you are correct.

Facilities are environments created and equipped to service a particular function, such as a specialty garage used to service motor vehicles.

Hygiene refers to a system of rules and principles intended to promote and preserve health.

OSHA-required "Right-To-Know" Law should be posted.

Available positions should be posted.

FIGURE 2-2 The service bay area should be well organized.

© Cengage Learning 2013

Wages and Benefits

There are other aspects of your employment, other than how much you will be paid, that should be known. For example, how often and on what day are you paid? Is it company policy to hold back pay? If you are being paid a commission, do you have a guarantee? What are the **fringe benefits**? If there is a health plan, what is the employee's contribution, if any? Is there a waiting period before being eligible? Does the company have a tool purchase plan? If there is a paid vacation plan, find out the details; generally, 1 week is given after 1 year and 2 weeks after 3 years or more. And, finally, while it does not seem important now, is there a retirement plan? If so, get the details; if not, it is time to consider a personal plan for the future.

Employee Obligations

There are also employee obligations. They are really more simple than employer obligations, for all you have to do is be a caring, loyal employee. This probably begins with your ability to follow directions (Figure 2-3). Remember, you are being paid to follow instructions. Doing it your way may not be in the best interest of the employer. If you have any questions or any doubts, *ask.* "I didn't know what you meant" is not a good response after something is botched.

Attitude. Proceed with a positive attitude. *Your* attitude can have an effect on your fellow technicians as well as your supervisors. Saying, "That's not the way we did it in school" does not portray a positive attitude.

Responsibility. Be responsible and take pride in your work. Regardless of the task assignment, remember that someone has to sweep the floor—it may be you. Never forget that your primary responsibility is to make your employer a profit. Always be busy and productive. Be willing to learn and take advice from the senior technicians. You may be surprised how little time it takes to become one of their peers.

Dependability. Be dependable. Repair orders for the following day are often scheduled in advance. Habitual lateness or absenteeism cannot be tolerated.

© Cengage Learning 2013

FIGURE 2-3 Listen to and follow the directions of your supervising master technician.

Take pride in everything you do and be dependable.

Pride. Remember, the most important technician in that facility is you, and you work for the most important and best company in town—perhaps anywhere.

Failure to adhere to these four basic ideals will affect your long-term success and could result in termination of employment.

SERVICE TOOLS

Special tools are needed to perform service, testing, and many repair procedures on most automotive air-conditioning systems. In addition to common mechanics' hand tools, such as pliers, screwdrivers, wrenches, and socket sets, you will need a **manifold and gauge set** with hoses, a refrigerant **can tap**, a thermometer, and safety glasses or other suitable eye protection.

In addition to the technicians' tools, the shop must have the required refrigerant recovery, recycle, and recharge systems for all refrigerant types serviced at their service facility in order to perform refrigerant service. The Society of Automotive Engineers (SAE) has set standards for recovering and recycling refrigerant, and the Environmental Protection Agency (EPA) has acknowledged these standards in Section 609 of the Clean Air Act. Other equipment the shop may supply includes antifreeze recovery, recycle, and recharge systems; electronic scales; electronic leak detectors; and electronic thermometer. With the proliferation of alternative blends of refrigerants, a refrigerant gas electronic purity identifier should also be considered a necessary piece of equipment to avoid contamination of service equipment.

Manifold and Gauge Set

The manifold and gauge set (Figure 2-4) generally consists of a manifold with two hand valves and two gauges, a compound gauge and a pressure gauge, and three hoses. There are several types of manifold and gauge sets. Regardless of the type, they all serve the same purpose.

FIGURE 2-4 The manifold and gauge set.

© Cengage Learning 2013

Note that the manifold has fittings for the connection of three hoses. The hose on the left, below the compound gauge, is the low-side hose. On the right, below the pressure gauge, is the high-side hose. The center hose is used for system service, such as for evacuating and charging (see Chapter 5). Some manifolds may be equipped with two center hoses: a small hose, generally ¼ in. or 6 mm, for refrigerant recovery and charging; and a large hose, generally ⁵/₁₆ in. or 8 mm, for evacuation.

Regardless of the type selected, a separate and complete manifold and gauge set with appropriate service hoses having unique fittings are required for each type of refrigerant that is to be handled in the service facility. This means that a minimum of three sets are generally required: one set for R-12(CFC-12) refrigerant, one set for R-134a(HFC-134a) refrigerant, and a set for contaminated refrigerant.

Low-Side Gauge. The **low-side gauge**, also referred to as the compound gauge, will indicate either a vacuum or pressure (Figure 2-5). Generally, this gauge will be calibrated from 30 in. Hg (0 kPa absolute) vacuum to 250 psig (1,724 kPa) pressure. Actually, the pressure calibration is to 150 psig (1,035 kPa) with respect to 250 psig (1,724 kPa) maximum. That means that pressures to 150 psig (1,035 kPa) may be read with reasonable accuracy, while pressures to 250 psig (1,724 kPa) may be applied without damage to the gauge movement. The low-side gauge is found at the left of the manifold.

Pressure Gauge. The **high-side gauge** (Figure 2-6) is usually calibrated from 0 psig (0 kPa) to 500 psig (3,448 kPa). Insomuch as the high side of the system will never go into a vacuum, pressures below 0 psig (0 kPa) are not indicated on the high-side gauge. The high-side gauge is also often referred to as the pressure gauge. The high-side gauge is found at the right of the manifold and is identifiable by a red housing.

The Manifold. Note the "circuits" in the manifold. When both hand valves are closed (Figure 2-7), both the low- and high-side hose ports are connected only to the low- and high-side gauges. If, in this case, the hoses are connected from the manifold to an air-conditioning system, the low-side gauge will indicate the pressure on the low side of the system. The high-side gauge will indicate the pressure on the high side of the system. The gauges will always show respective system pressures when the manifold hand valves are closed.

The **low-side gauge** is the left-side gauge on the manifold used to read refrigerant pressure in the low side of the system and is identifiable by a blue housing.

The **high-side gauge** is the right-side gauge on the manifold used to read refrigerant pressure in the high side of the system.

FIGURE 2-5 The low-side (compound) gauge.

FIGURE 2-6 The high-side (pressure) gauge.

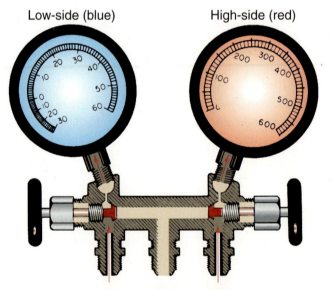

FIGURE 2-7 Manifold circuit with both hand valves closed.

SERVICE TIP: The hand valves should be closed finger tight; excessive force will damage the valve seat. The hand valves should only be open to either add or remove something from the system. Never open the high-side valve while the vehicle is running.

If either the low- or high-side manifold hand valve is cracked open and the center hose is not connected to anything, refrigerant will escape from the system. This is a procedure known as purging.

If the center utility hose is connected to a recovery station or other closed system and the low-side or high-side manifold hand valve is opened, both gauges will still indicate system pressure. If, however, both manifold hand valves are cracked, the pressures will equalize in the manifold and neither of the gauges will accurately indicate system pressure.

The center hose is used to evacuate, recover, or charge the air-conditioning system. Procedures for this service and system problems relating to gauge pressure indications are given in Chapter 6 of this manual as well as the classroom text.

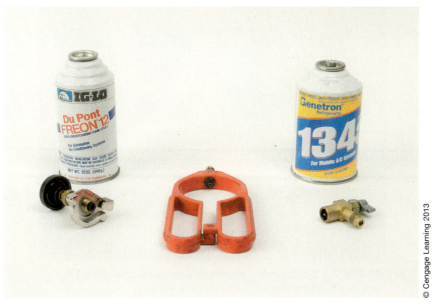

FIGURE 2-8 Typical can taps with "pound" cans of refrigerant.

Can Tap

The can tap is used to dispense refrigerant from a "pound" can. The true 1-pound can of refrigerant has not been produced since the late 1960s. "Pound" disposable cans actually contain 12 oz. (340 g) or 14 oz. (397 g) of refrigerant today. Disposable can taps can be flat top or the more popular HFC-134a screw top (Figure 2-8). Small refrigerant cans are also available with a dye charge to aid leak detection. The screw-top tap may also be used for single-charge cans of refrigerant oil. Some can taps are designed to fit either type. To install the can tap on either flattop or screw-top cans using a universal-type can tap, proceed as follows:

1. Wear suitable eye protection.
2. Hold the can at arm's length, in an upright position.
3. Affix the clamp-type fixture on the can top.
4. Turn the can tap handle fully counterclockwise (ccw).
5. Screw the handle assembly into the clamp fixture.

When ready to dispense refrigerant as outlined in Chapter 6, turn the can tap handle fully clockwise (cw). This pierces the can. The can tap should not be removed until the contents of the can have been dispensed.

It should be noted that small "pound" cans of refrigerant are not legally sold in some states. Other state requirements limit sales of small containers of refrigerant to certified and properly licensed shops and technicians only.

Safety Glasses

There are several types of safety glasses available (Figure 2-9). A safety shield-type goggle may be used with or without eyeglasses. It is important to note that the glasses or goggles selected be a type that is approved for working with liquids or gases, meeting the ANSI Z87.1-1989 standard.

While removing an air-conditioning system compressor not long ago, a technician accidentally allowed an open-end wrench to come into contact with the battery terminals. The resulting short caused a spark, which, in turn, caused the battery to explode. The technician was wearing safety goggles to keep his prescription glasses in place; he never thought they might serve an even more important purpose. Although he suffered facial burns, his

Manufacturers' specifications must be used. A service procedure for a 2006 Silverado, for example, may be different from a 2012 Silverado.

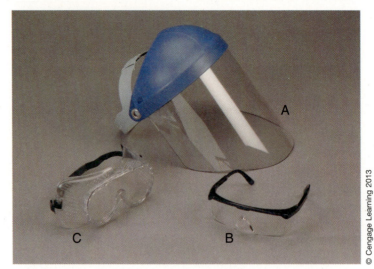

FIGURE 2-9 Typical safety glasses and goggles: face shield (A), goggle (B), safety glasses with side shield (C).

eyes were protected by the safety goggles. Wearing prescription glasses alone would have offered very little protection since they generally have no peripheral shielding.

HAND TOOLS

Common hand tools, such as wrenches, pliers, screwdrivers, punches, and hammers, are necessary. Other tools, such as a ⅜-inch drive socket set, are helpful but not a necessity. Many of the hand tools referred to throughout this text may be found around the house. If not, it is suggested that they only be purchased as needed. It is not necessary, for example, to purchase a complete set if only a few sizes are needed.

SPECIAL TOOLS

Basically, three special tools will expand your service and repair capabilities considerably. It is helpful to have a thermometer, a leak detector, and a **vacuum pump**.

> A **vacuum pump** is a mechanical device used to evacuate the refrigeration system to rid it of excess moisture and air.

Thermometer

A glass-type or a dial-type thermometer may be used. The glass type is usually less expensive, but it is more easily broken. Regardless of the type, it is suggested that the temperature range be from 0°F to 220°F (−17.8°C to 104.4°C) (Figure 2-10). Inexpensive thermometers purchased in housewares or automotive departments in large department stores may not be as accurate as a refrigeration thermometer. For this reason, they are not recommended. For more accurate and reliable service, an electronic digital thermometer is recommended.

Leak Detector

In most cases, leaks can be detected by the use of a soap solution. A good dishwashing liquid mixed with an equal amount of clean water and applied with a small brush will indicate a leak by bubbling. A commercially available product, such as "Leak Finder," can also be used.

Electronic. Though considerably more expensive, electronic leak detectors, called halogen leak detectors, are desirable because they offer great sensitivity and can pinpoint a leak as slight as 0.15 (4 g) per year. It should be noted that halogen leak detectors are available that can be used to test either CFC-12 or HFC-134a refrigerants (Figure 2-11). When refrigerant vapor enters a halogen leak detector's search probe, the device emits an audible or visual signal.

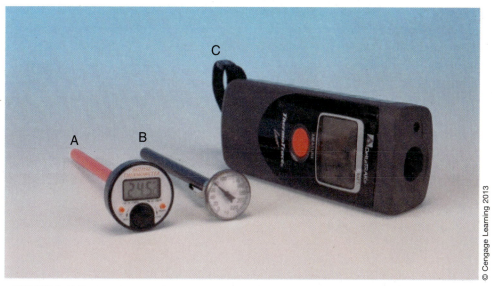

FIGURE 2-10 Typical thermometers: digital pocket thermometer (A), dial pocket thermometer (B), and infrared electronic thermometer (C).

FIGURE 2-11 A typical electronic leak detector.

Fluorescent. Fluorescent leak detectors (Figure 2-12) are becoming increasingly more popular. A fluorescent dye is injected into the system where it remains without affecting cooling performance. When a leak is suspected, an ultraviolet lamp will quickly and efficiently pinpoint the problem area. Many manufacturers now use refrigerant containing a fluorescent dye for the initial charge of the air-conditioning system.

Nitrogen. If time permits, many technicians prefer to hold a standing pressure test using nitrogen to determine the integrity of an air-conditioning system after extensive leak repairs have been made. This test is especially helpful if the leak was difficult to locate. It provides added assurance that the leak was located and repaired.

 WARNING: Nitrogen is under very high pressure. Make no attempt to disperse nitrogen without having proper pressure regulators in place.

FIGURE 2-12 A typical fluorescent leak detector.

> Take care of the special tools and equipment provided by the service facility. Treat them like they were your own.

To perform this test, the air-conditioning system is evacuated and then pressurized to 100 psig (689.5 kPa) and allowed to "rest" overnight. Since the nitrogen is dry and stable, the pressure should be within a few pounds (kiloPascals) of 100 psig (689.5 kPa) the following morning. Nitrogen poses no threat to the environment and may be purged from the air-conditioning system to the atmosphere. The air-conditioning system is then evacuated to remove any residual nitrogen and air from the system.

 WARNING: A halogen leak detector must not be used in a space where explosives, such as gases, dust, or vapor, are present. Use halogen leak detectors in a well-ventilated area only. By-products of decomposing CFC and HCFC refrigerants, hydrochloric and hydrofluoric acid, are a health hazard.

 WARNING: Take care not to inhale these fumes. To minimize the danger, work in a well-ventilated area when leak checking an air-conditioning system.

Vacuum Pump

It is necessary to remove as much moisture and air from the system as possible before charging it with refrigerant. This is best accomplished with the use of a vacuum pump (Figure 2-13). The vacuum pump is one of the most expensive pieces of service equipment required. It is usually provided by the service facility.

Some refrigerant recovery equipment has a vacuum pump incorporated into the equipment. High-volume service facilities, however, cannot generally tie up an expensive piece of equipment for the time required to adequately evacuate an automotive air-conditioning system. So, many service technicians will purchase a stand-alone electric or compressed-air-operated vacuum pump. The shop air-powered vacuum pump is a low-cost option that many technicians choose to increase their productivity and free the recovery/recycle machine for another vehicle in the shop.

FIGURE 2-13 A typical high-vacuum pump.

REFRIGERANT IDENTIFIER

To determine what type refrigerant is in a system, a refrigerant identifier (Figure 2-14), should be used prior to servicing the refrigeration system of any vehicle. The refrigerant identifier is used to identify the purity and quality of a gas sample taken directly from a refrigeration system or a refrigerant storage container. The identifier, such as Rotunda's Refrigerant Analyzer, will display:

- R-12: If the refrigerant is CFC-12 and its purity is better than 98 percent by weight.
- R-134a: If the refrigerant is HFC-134a and its purity is 98 percent or better by weight.
- FAIL: If neither CFC-12 nor HFC-134a have been identified or if it is not at least 98 percent pure.
- HC: If the gas sample contains hydrocarbon, a flammable material. A horn will also sound.

After the analysis is completed, the identifier will automatically purge the sampled gas and be ready for the next sample for analysis. If neither CFC-12 nor HFC-134a has been

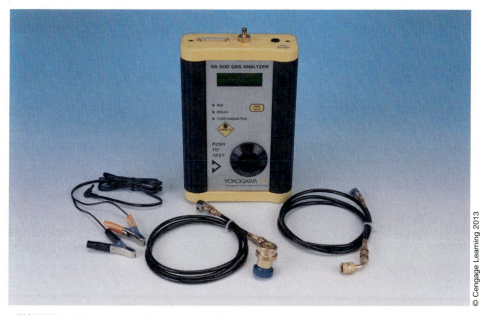

FIGURE 2-14 A typical refrigerant identifier.

identified or either is not at least 98 percent pure, consider the refrigeration system to be contaminated and perform all service accordingly.

It is important to always refer to the equipment manufacturers' instructions for specific and proper tool usage and to specific local regulations relating to refrigerant handling.

 WARNING: **If the sample reveals that the refrigeration system contains a flammable hydrocarbon, do not service the system unless extreme care is taken to avoid personal injury.**

The high cost of the equipment required for removal and storage, as well as the proper disposal of contaminated refrigerant, often discourages the customer from having repairs made. There are not many sites in the United States that dispose of contaminated refrigerants. Disposal is usually accomplished by burning at a very high temperature and it requires expensive equipment.

For more information about contaminated refrigerant disposal on a local level, check with the local automotive refrigerant supplier. If they can offer no assistance, and often they cannot, consult a major commercial refrigeration supply house.

OTHER SPECIAL TOOLS

Other special tools available to the service technician are generally supplied by the service facility. These tools include a refrigerant recovery and recycling system, an antifreeze recovery and recycling system, an electronic thermometer, and an electronic scale. Also, special testers are available for automatic temperature control (ATC) testing, and special tools are available for servicing and repairing compressors.

The following is a brief description of these tools. They are covered in more detail in the appropriate chapters of both the classroom text and shop manual.

Refrigerant Recovery and Recycle System

The service center must have a recovery, recycle, and recharge machine for each type refrigerant to be serviced. To address the issue of small capacity refrigerant systems the EPA adopted SAE standard J2788, effective December 2007, covering the accuracy level of refrigerant recovery, recycling, and recharging equipment produced after that date. Standard J2788 supersedes the previous standard J2210. Refrigerant recovery, recycling, and recharging equipment must now be certified J2788 compliant. This equipment must be capable of measuring and displaying the amount of refrigerant recovered to an accuracy level of +/– 1 oz. When recharging a refrigerant system with an SAE J2788 compliant machine it must charge a system with +/– 0.5 oz of accuracy. It should be noted that refrigerant service equipment built prior to the SAE J2788 standard taking effect may have displays down to a tenth of a pound but this does not mean they are accurate to a tenth of a pound. Always consult the specific specifications information for the model and manufacturer of the equipment you are using, especially if it was manufactured prior to 2008.

Even a minor error in refrigerant charge level (weight) can affect air conditioning system performance. This is particularly true of small capacity systems, those under 1.5 lbs of refrigerant. Air conditioning systems today are smaller than ever before, more efficient and still offer outstanding performance. In addition, systems manufactured today are much less susceptible to leaks and can go five years or longer on the original factory charge. Some refrigerant recovery, recycle, and recharge machines, however, may be used for both CFC-12 and HFC-134a. A system, similar to the one shown in Figure 2-15, is a single-pass system with an onboard microprocessor that controls the evacuation time as well as the amount of refrigerant charged into the system. The mixing of CFC-12 and HFC-134a

FIGURE 2-15 A typical refrigerant recovery/recycle machine.

refrigerants is prevented by the use of a sliding lock-out panel allowing only one set of manifold hoses to be connected at any time. Also, the fittings on the hoses prevent them from being connected to the wrong port. Each type of refrigerant has a separate dedicated set of hoses and recovery tank. A self-clearing loop removes residual refrigerant from the machine before connecting the other set of hoses and recovery tank.

Having made an initial pass through the filter-drier on its way to the recovery tank, the recovered refrigerant in the tank is always clean and ready to reuse. The refrigerant is then recirculated through the filter-drier during evacuation to provide the cleanest possible refrigerant with no extra time or procedures involved. Other desirable features of a recovery, recycle, and recharge system include an automatic air purge, a high-performance vacuum pump, and an automatic shut-off when the tank is full.

Antifreeze Recovery and Recycle System

An antifreeze recovery and recycle machine, such as Prestone's ProClean Plus™ Recycler (Figure 2-16), is a self-contained system that drains, fills, flushes, and pressure tests the cooling system. It can also be used to recycle coolant. A typical cooling system drain, recycle, and refill takes about 20 minutes.

This particular system adds additives during the recycle phase to bond heavy metals, such as lead (Pb) and other contaminants. This renders them into nonleachable solids that are not hazardous as defined by the EPA. Additives separate the contaminants so they can easily be removed. Also, inhibitors are added to protect against corrosion and acid formation. The recycled coolant exceeds ASTM and SAE performance standards for new antifreeze. This eliminates the problems of waste disposal such as costs and the necessity of storing and hauling used coolant.

Other onboard functions of the illustrated machine include standard coolant exchange, flushing procedures, pressure testing for leaks, and vacuum fill for adding coolant to an empty system. Its tank-within-a-tank design holds 40 gallons of used coolant for recycling in the inner section, and up to 60 gallons of recycled coolant in the outer section.

Electronic Thermometer

A single-or two-probe, handheld electronic thermometer, such as shown in Figure 2-17, is commonly used in automotive air-conditioning system diagnosis and service. The

FIGURE 2-16 A typical antifreeze recovery/recycle machine.

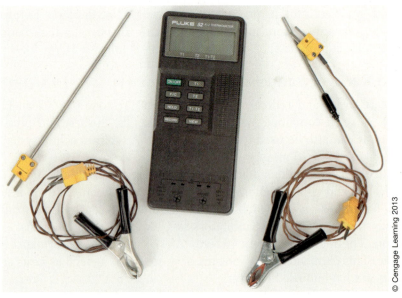

FIGURE 2-17 A typical electronic thermometer.

two-probe model is used to quickly and accurately measure superheat as required for critically charging some HFC-134a air-conditioning systems.

The range for most battery-powered digital electronic thermometers is generally on the order of −50°F to 2,000°F (−46°C to 1,093°C), and they have an accuracy greater than ±0.3 percent with a switchable resolution of 1.0 to 0.1 degree in either the °C or °F scale.

FIGURE 2-18 **A typical electronic scale.**

The desirable effective operating range for a thermometer for automotive air-conditioning system use is 32°F (0°C) to 120°F (48.9°C). The digital readout for a handheld electronic thermometer should be no less than ½ in. (12.7 mm) for easy reading.

Electronic Scale

An electronic scale (Figure 2-18) may be used for CFC-12 or HFC-134a refrigerants to deliver an accurate charge by weight, manually or automatically. Automatic charging is generally accomplished by programming the amount of refrigerant to be charged into the onboard solid-state microprocessor. The charge is stopped, and an audible tone signals that the programmed weight has been dispersed. A liquid crystal display is used to keep track of refrigerant dispersed. Some models have a switchable pounds/kilograms readout with a resolution of 0.05 lb. (0.02 kg). The 10 in. (2.5 cm) scale platform will handle up to a 50 lb. (23 kg) bulk tank of refrigerant and is equipped with control panel fittings and two hoses to accommodate both CFC-12 and HFC-134a refrigerants.

Automatic Temperature Control (ATC) Testers (Scan Tool)

There are many different types of testers available. The scan tool (Figure 2-19) is very popular and is used to enhance troubleshooting efforts to quickly locate the root of a problem. They are available in a wide variety of brands, prices, and capabilities. One good feature is that not only can a scan tool be used to retrieve trouble codes, some allow the technician

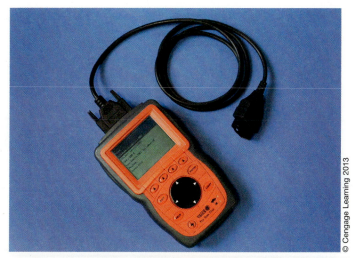

FIGURE 2-19 **A typical scanner for automatic temperature control (ATC) testing.**

to monitor and view sensor and computer information. This feature, known as *serial data* or *the data stream*, helps to pinpoint a heating, ventilation, or air conditioning (HVAC) problem. A scan tool can sometimes even take the role of a manifold and gauge set by obtaining system pressure readings through transducers in refrigerant lines.

Depending on the vehicle, the scan tool, and the software, the serial data that is obtained from an ATC system can include information such as blend door position and blower motor voltage. Some scan tools have a feature known as *bidirectional function* that enables the technician to activate various air-conditioning system components, such as the cooling fan and compressor clutch relay. Some scan tools have a *recorder* mode that is very useful in diagnosing intermittent problems. To use the recorder mode, the technician hooks up the scan tool and drives the vehicle. When the intermittent problem is experienced, the technician pushes the appropriate button and the malfunction will generally be captured and stored in short-term memory. Back at the shop, the stored information can be retrieved for evaluation.

Compressor Tools

There are several special tools required for compressor clutch and shaft seal service (Figure 2-20). Clutch plate tools are used to remove the clutch plate to gain access to the shaft seal. They are also used for reinstalling the clutch plate after service. These tools should be compact in design for working in close quarters so that it may often be possible to service the compressor clutch and shaft seal without having to remove the compressor from the vehicle.

Basically, a shaft seal service kit includes an adjustable spanner wrench, clutch plate remover/installer, snapring pliers, ceramic seal remover/installer, seal seat remover/installer, shaft seal protector, seal assembly remover/installer, thin wall socket, O-ring remover, and O-ring installer. For clutch pulley and bearing service, service tools will include a pulley puller, pulley installer, bearing remover/installer, and rotor and bearing installer.

Special compressor service tools are designed to fit a particular application. Though some are interchangeable, most are not. For example, the seal seat remover/installer used on GM's models R4 and A6 compressors may also be used on Diesel Kiki models DKS-12 and DKS-15 compressors, as well as on Sanden/Sankyo model 507, 508, and 510 compressors. If there is any doubt about the application of any particular tool, do not use force. If it does not fit freely, it may not be the correct tool for the task.

FIGURE 2-20 Special tools required for compressor service.

© Cengage Learning 2013

Often, good service tools are found at a very low cost at flea markets and garage sales. Nevertheless, the cost of having the required tool is often more than offset by the savings in the time required to accomplish a task.

SOURCES OF SERVICE INFORMATION

There are many sources for information available to the automotive technician today. These include service manuals produced by, but not limited to, manufacturers, Mitchell, Haynes, and Chilton. Service information comes in model-specific print form, CD, or DVD format. Today, the print form of service information is costly, is not as readily available, and is space prohibitive. These reasons and more have given way to computer-based systems. Mitchell-on-Demand and ALLDATA are two of the largest suppliers of CD/DVD-based and online information systems used by service facilities, offering systems and subscriptions for coverage of all the major automotive manufacturers, both domestic and import. These systems may also include estimating and shop management software. With the complexity of today's vehicles, these full-coverage information systems have become a required tool, and many large service centers subscribe to both.

The Internet has also become a major source of service information with both Mitchell-on-Demand and ALLDATA offering access to the most current updated information via an Internet connection. There are also Internet organizations for technicians such as International Automotive Technician's Network (iATN), Identifix (www.identifix.com) as well as numerous web pages. One note of *caution* regarding the wealth of information available on the Internet is to rely on original equipment and aftermarket manufacturers' "in-house" publications and other reliable sources of information. The Internet contains both accurate and inaccurate information, so find sources that other technicians use and recommend.

Also, the local library generally has an automotive book section. One should not forget the valuable information that appears in the monthly publications by the Mobile Air Conditioning Society (MACS) as well as other automotive trade magazines made available to their members.

Finally, if you are taking any secondary or postsecondary automotive classes, be sure to take notes and to save all of the handouts provided by your instructor, which often contain valuable information not available from other sources. Be sure to index your class notes and the handouts in a notebook for future reference. Although it may not seem so at first, this information can become more relevant and important for your ongoing study and practice and can serve as valuable reference material.

CONTENT OF SERVICE INFORMATION

It is not possible to provide information on all the various specifications or **service procedures** that may be performed on the many different makes and models of automobiles in service today. Both experienced and inexperienced technicians rely on service information to outline procedures and specifications for repairs and diagnostics. Service information is generally written in a straightforward, easy-to-follow format. They generally provide information in a step-by-step sequence based on manufacturer recommendations.

> **Service procedures** are suggested routines for the step-by-step act of troubleshooting, diagnosis, and repairs.

SERVICE MANUAL PROCEDURES AND SPECIFICATIONS

Service Procedures

To find a particular service procedure in any service source, it is first necessary to know the exact vehicle year, make, and model. The following sequence is based on the use of a computer-based information system; the steps, though not exactly the same, are similar in

FIGURE 2-21 Vehicle identification number (VIN) is observed through the windshield.

a print-based service manual. It may be necessary to refer to the **VIN (vehicle identification number)** (Figure 2-21). The VIN identifies specific information about the vehicle. Part of the VIN is the WMI (world manufacturer identifier) code. The first three positions in the VIN uniquely identify the maker of the vehicle. The vehicle descriptor section is positions 4 through 8 of the VIN and identifies specific characteristics of the vehicle. The eighth position identifies the engine size, while the tenth position identifies the year the vehicle was built. The vehicle identifier section is the last eight positions of the VIN and is used for the identification of a specific vehicle. In addition, the last five characters are always numeric and contain no letters. It is a good practice to fill in the vehicle year, make, and model as well as the VIN on the back of the repair order and to confirm that the information on the front of the repair ticket is correct.

First select the vehicle year, make, and model (Figure 2-22). Next, locate the general group from the systems list (Figure 2-23) for the procedure you are looking for. Since we are concerned with automotive air conditioning, select heating and air conditioning and locate the procedure for replacing the heater blower motor. Most service application software also includes a "Help" section if you are unsure how to proceed, or you may choose to go to the Table of Contents section.

The proper removal and installation procedures are outlined. If these step-by-step procedures are followed, there should be no problem with the removal and installation of the blower motor assembly; the key to any service is to follow the steps as listed.

SERVICE INFORMATION

To view service information,
First select the Year, Make, and Model of the vehicle and click on "NEXT"

Or click here to enter a Vehicle Identification Number (VIN)

YEAR: MAKE: MODEL:

Select year ▼   Select make ▼   Select model ▼   Next >

Reset Vehicle Selection

FIGURE 2-22 Typical vehicle selection screen on computer-based information systems.

FLUID CAPACITIES

APPLICATION	SPECIFICATION	
	METRIC	ENGLISH
Automatic transmission		
Pan removal	7.0 liters	7.4 quarts
Complete overhaul	9.5 liters	10.0 quarts
Dry	12.7 liters	13.4 quarts
Engine cooling system		
3.4 L engine	10.7 liters	11.3 quarts
3.8 L engine	11.0 liters	11.7 quarts
Engine oil		
3.4 L engine		
With filter change	4.3 liters	4.5 quarts
Without filter change	3.75 liters	4.0 quarts
3.8 L engine		
With filter change	4.3 liters	4.5 quarts
Without filter change	3.75 liters	4.0 quarts
Fuel tank	64.0 liters	17.0 gallons
Power steering system	0.70 liters	1.5 pints

© Cengage Learning 2013

[< Back] [Forward >]   [Print]

FIGURE 2-23 Typical example of a fluid capacities chart found in service information.

Specifications

Specifications are found in the same manner as service procedures. Suppose, for example, that the cooling system capacity must be known in order to properly add 50 percent antifreeze solution for maximum winter protection. This information is generally located in the section entitled "Maintenance and Lubrication" and the "Powertrain" section. Locate the engine size for the vehicle you are servicing, and reference the cooling system capacity listed in the specification chart. After the system has been thoroughly flushed and drained, the cooling system capacity must be divided by two in order to determine the correct amount of coolant to add to obtain a 50/50 mixture.

> **Specifications** provide information on system capacities. This information is also generally given in the owner's manual.

 WARNING: Neither the manufacturer's service manuals nor any school text, such as this one, can anticipate all conceivable ways or conditions under which a particular service procedure may be performed. It is therefore impossible to provide precautions for every possible hazard that may exist. The technician must always exercise extreme caution and pay heed to every established safety practice when performing automotive air conditioning service procedures.

REPAIR ORDER

The vehicle repair order (Figure 2-24) is a legal document that must filled out for every vehicle worked on, no matter how insignificant the repair may seem. This applies to

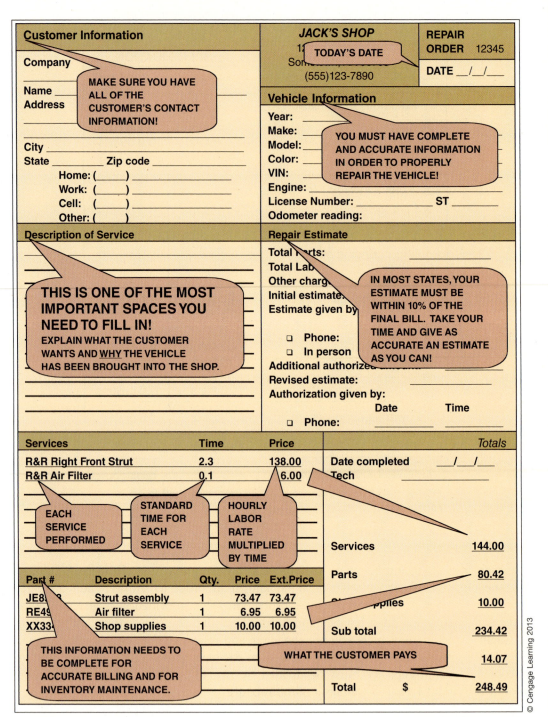

FIGURE 2-24 This repair shows a recent history (top) and the customer's complaint (center).

Trade Jargon:
When a vehicle must return to the repair facility for a repeat repair due to a part failure or improperly performed repair, this repeat repair is referred to as a "come-back." If the come-back was due to technician error or negligence, the technician is generally not compensated for the repeat repair.

vehicles being worked on at the premises or on the road as part of roadside assistance offered by some repair facilities under the auspices of the business license. In addition, a repair order must be completed whether the repair is at no charge (free) or is a repeat repair (come-back) in the same manner as a routine repair. It must be remembered that even when a repair is performed for free, the repair facility is still legally obligated, just as it would be for any fee-based repair. In addition, a repair order should also be completed if an employee vehicle is going to be serviced, even if the employee is going to perform the service on his or her own vehicle. And last but not least, the repair order should be signed, authorizing the work and agreeing to the original cost estimate.

Being a legal document, every notation a technician makes on the repair order must be accurate and complete. The technician should also verify that the VIN, make, model, and mileage, as well as other specific information, are correct. Any existing physical vehicle damage (e.g., dent, cracked glass) should be noted at the time the repair order is filled out so as to avoid controversy when the customer returns to pick up the vehicle.

While a vehicle is in for service, additional maintenance or repairs may be suggested to the owner to address concerns noted by the technician or as part of routine maintenance suggested by a manufacturer. Sometimes customers decide not to approve this additional work. The recommendations should be noted on the repair order and the customer should sign, acknowledging his or her refusal to address the shop recommendations. This is especially important if the repair is required for vehicle safety or system failure. Some service facilities will even tow the vehicle back to the owner's home at no cost to the owner to limit the shop's liability.

THE METRIC SYSTEM

The United States is slowly but surely joining the rest of the world in a uniform system of physical measurement known as the metric system.

In the metric system, speed is measured in kilometers per hour (km/h), pressure is measured in **kiloPascals** absolute (kPa absolute) or kiloPascals (kPa), liquid is measured in liters (L), temperature in degrees Celsius (°C), length in millimeters (mm) or meters (m), and weight in grams (g) or kilograms (kg). Whenever practical, measures in this text are given in both English and metric units. The metric equivalent is given in parentheses following the English measure. For example: The freezing point of water at atmospheric pressure is 32°F (0°C) and its boiling point is 212°F (100°C).

There are several terms commonly used in the English system of measure that are the same as those used in the metric system of measure. Those most familiar to the automotive technician include ohm, volt, and ampere.

Some English standard measures cannot be converted to the standard metric measure. For example, there is no standard metric measure for ⅜ in. An English ⅜ in. measure is actually equal to 9.53 mm in the metric measure. Therefore, if you need to remove ⅜ in. cap-screws from a plate, a 9 mm wrench would be too small, while a 10 mm wrench would be too large.

For the purpose of conversion, the metric equivalents in this text are held to one or two decimal places. For example, a temperature of 69°F converts to 20.5555555555°C. Little is gained by carrying the conversion to three or even two decimal places. The difference between one and ten decimal places in this example amounts to only 8/100°F (0.08°F) or 44/1,000°C (0.044°C), hardly worth consideration for the purpose of practical application.

As in the example given, the conversion of 69°F to °C will then be given as 20.6°C, rounded off to the nearest decimal place.

Pressure in the metric system is measured in terms of the Pascal (Pa). One pound per square inch (1 psi) is equal to 6.895×10^3 Pascals. For a more practical application, the kiloPascal (kPa) is used. One pound per square inch (1 psi) is equal to 6.895 kiloPascals (6.895 kPa). This equation applies to both the absolute (psia) and atmospheric (psig) pressure conversions.

A conversion chart for English-to-metric and metric-to-English values is given in Figure 2-25.

KiloPascals is a unit of measure in the metric system. One kiloPascal (kPa) is equal to 0.145 pound per square inch (psi) in the English system.

The metric system is known as "Systéme International d'Unites," a French term that literally translates to "International System of Units."

METRIC TO ENGLISH		
Multiply	**By**	**To Get**
Celsius (°C)	1.8 (+32)	Fahrenheit (°F)
gram (g)	0.035 3	ounce (oz.)
kilogram (kg)	2.205	pound (lb.)
kilometer (km)	0.621 4	mile (mi.)
kilopascal (kPa)	0.145	lb/in.^{2} (psi)
liter (L)	0.264 2	gallon (gal.)
meter (m)	3.281	foot (ft.)
milliliter (mL)	0.033 8	ounce (oz.)
millimeter (mm)	0.039 4	inch (in.)

ENGLISH TO METRIC		
Fahrenheit (°F)	(−32) 0.556	Celsius (°C)
foot (ft.)	0.304 8	meter (m)
fluidounce (fl. oz.)	29.57	milliliter (mL)
gallon (gal.)	3.785	liter (L)
inch (in.)	25.4	millimeter (mm)
mile (mi.)	1.609	kilometer (km)
ounce (oz.)	28.349 5	gram (g)
pound (lb.)	0.453 6	kilogram (kg)
lb/in.^{2} (psi)	6.895	kilopascal (kPa)

© Cengage Learning 2013

FIGURE 2-25 English/metric conversion chart.

SUMMARY

TERMS TO KNOW

Can tap

Facilities

Fringe benefits

The high-side gauge

Hygiene

KiloPascals

Manifold and gauge set

Service procedures

Specifications

The low-side gauge

Vacuum pump

VIN

■ Shop rules and regulations are put in place by management to create and maintain a safe and pleasant environment for both employees and customers.

■ A manifold and gauge set is a manifold block complete with gauges and charging hoses used to check and service refrigerant gas in an air-conditioning system.

■ There are many special tools required for servicing heating and air-conditioning systems. The EPA, under the Clean Air Act, requires some of these tools if you plan on servicing mobile refrigerant systems. You must have a dedicated refrigerant recovery system for each type of refrigerant serviced at your facility. For all practical purposes, this means you need to be able to capture and recover both R-12 and R-134a refrigerants.

■ Many of the specialty tools that are required to service today's air-conditioning systems are supplied by your employer, but some technicians choose to purchase some of these tools themselves, such as manifold gauge sets and vacuum pumps.

■ Computer literacy has also become a required skill for today's service technician due to the proliferation of service information in digital format. It is not uncommon to find a shop that has few (if any) paper service manuals and relies solely on CD/DVD and Internet-based service information.

ASE-STYLE REVIEW QUESTIONS

1. *Technician A* says that rules are made to protect the customer.
   Technician B says that rules are made to protect the technician.
   Who is correct?
   A. A only
   B. B only
   C. Both A and B
   D. Neither A nor B

2. *Technician A* says that the repair order is a legal document that must be filled out for every vehicle worked on.
   Technician B says that a repair order does not need to be filled out for a no charge (free) repair.
   Who is correct?
   A. A only
   B. B only
   C. Both A and B
   D. Neither A nor B

3. *Technician A* says the low-side gauge on the manifold gauge set is on the left and has a green housing.
   Technician B says the high-side gauge on the manifold gauge set is on the left and has a red housing.
   Who is correct?
   A. A only
   B. B only
   C. Both A and B
   D. Neither A nor B

4. *Technician A* says that you only need one manifold gauge set and it may be used for any type of refrigerant you need to service.
   Technician B says a vacuum pump is an essential piece of equipment for servicing a refrigerant system.
   Who is correct?
   A. A only
   B. B only
   C. Both A and B
   D. Neither A nor B

5. *Technician A* says a service information system may include estimating and shop management software.
   Technician B says service information you find on the Internet contains both accurate and inaccurate information.
   Who is correct?
   A. A only
   B. B only
   C. Both A and B
   D. Neither A nor B

6. *Technician A* says that safety glasses should be approved for gases.
   Technician B says that safety glasses should be approved for liquids. Who is correct?
   A. A only
   B. B only
   C. Both A and B
   D. Neither A nor B

7. *Technician A* says to determine the type of refrigerant used in a system, a refrigerant identifier should be used prior to servicing the vehicle's refrigerant system.
   Technician B says that when working on a refrigerant system you should work in a well ventilated area.
   Who is correct?
   A. A only
   B. B only
   C. Both A and B
   D. Neither A nor B

8. *Technician A* says that the English/metric measurement unit for pressure is psig/kPa.
   Technician B says that the English/metric measurement unit for temperature is Fahrenheit/Centigrade.
   Who is correct?
   A. A only
   B. B only
   C. Both A and B
   D. Neither A nor B

9. *Technician A* says that a vacuum pump is used to remove moisture from a system.
   Technician B says that an air pump is used to remove air from a system. Who is correct?
   A. A only
   B. B only
   C. Both A and B
   D. Neither A nor B

10. *Technician A* says that you are entitled to a clean, safe place to work.
    Technician B says as an employee your obligation is to be a loyal, caring individual, and have the ability to follow directions.
    Who is correct?
    A. A only
    B. B only
    C. Both A and B
    D. Neither A nor B

Name _____ Date _____

IDENTIFY THE RESPONSIBILITIES OF THE EMPLOYEE

Upon completion of this job sheet, you should understand the obligations of an employee.

Tools and Materials

Pad and pencil

Procedure

1. There are four primary employee obligations: attitude, responsibility, dependability, and pride. Can you think of any other obligations that may be an asset to the employee? If so, list them in the space provided.

   _____

2. Give several examples of employee attitude in the workplace.

   _____   _____

   _____   _____

3. Give several examples of employee responsibility in the workplace.

   _____   _____

   _____   _____

4. Give several examples of employee dependability in the workplace.

   _____   _____

   _____   _____

5. Give several examples of employee pride in the workplace.

   _____   _____

   _____   _____

6. Give examples of the other obligations you thought of in step 1.

   _____   _____

   _____   _____

Instructor's Response _____

_____

_____

Name _____   Date _____

USE A MANUFACTURER'S SERVICE MANUAL

Upon completion of this job sheet, you should understand how to use a service manual.

Tools and Materials

A late-model vehicle
A service manual for the vehicle or computer-based information system

NATEF Correlation

HEATING AND AIR CONDITIONING: A/C System Diagnosis and Repair; *Research applicable vehicle and service information, such as heating and air-conditioning system operation, vehicle service history, service precautions, and technical service bulletins.* **(P-1)**

HEATING AND AIR CONDITIONING: A/C System Diagnosis and Repair; *Locate and interpret vehicle and major component identification numbers (VIN, vehicle certification labels, and calibration decals).* **(P-1)**

Procedure

Assume that you are to replace a circuit breaker in the vehicle you selected.

1. What vehicle did you select?
   Make _____ Model _____ Year _____ VIN _____

2. Which service manual do you have?
   Title _____ Year _____

3. Does the service manual cover the vehicle that you selected? _____
   If not, explain: _____

4. Locate the Group Index and determine which group includes the circuit breaker.
   What group did you select? _____

5. Is the circuit breaker in the group that you selected? _____
   If not, what group did you find it in? _____
   _____

6. Using information found in the manual, were you able to find the circuit breaker? ____
   If not, what problems were encountered? _____
   _____

7. Were you able to determine the difference between the circuit breaker and the hazard signal flasher unit? _____ Were they similar? _____

8. How were they different? _____
   _____

Instructor's Response _____

_____

_____

Name _____ **Date** _____

COMPARE THE ENGLISH AND METRIC SYSTEMS OF MEASURE

Upon completion of this job sheet, you should understand the English and metric systems of measure.

Tools and Materials

Miscellaneous and assorted nuts and bolts
Set of English (fractional inch) open-end wrenches
Set of metric (millimeter) open-end wrenches

Procedure

Ask your instructor to identify for you a ¼-28 bolt and a ⁵/₁₆-24 bolt.

1. What size wrench fits the head of the ¼-28 bolt? _____

2. What size wrench fits the head of the ⁵/₁₆-24 bolt? _____

3. Using the formula given in the shop manual, convert ¼ in. to metric millimeters.
   a. First, convert ¼ in. to a decimal value. What is the decimal value? _____

   b. Next, multiply the decimal value by the formula. _____

4. Is there a metric wrench the size determined in step 3b? _____ Explain: _____
   _____

5. Is there a metric fastener the size as determined in step 3b? _____ Explain: _____
   _____

6. Is there a metric fastener close to the size determined in step 3b? _____ Explain: _____
   _____

7. Will the English wrench fit the fastener selected in step 6? _____ Explain: _____
   _____

8. Are the two fasteners, determined in steps 1 and 6, interchangeable? Explain:
   _____

9. Is the metric fastener identified in step 6 interchangeable with the ⁵/₁₆-24 English
   fastener? _____ Explain: _____

10. Is the metric fastener identified in step 6 closer in size to the ¼-28 fastener or the
    ⁵/₁₆-24 fastener? _____

Instructor's Response _____

_____

_____

Chapter 3

DIAGNOSIS AND SERVICE OF ENGINE COOLING AND COMFORT HEATING SYSTEMS

BASIC TOOLS
Basic mechanic's tool set

UPON COMPLETION AND REVIEW OF THIS CHAPTER, YOU SHOULD BE ABLE TO:

- Identify the major components of the automotive engine cooling and comfort heating system.
- Compare the different types of radiators.
- Discuss the function of the coolant pump.
- Explain the need for a pressurized cooling system.
- Describe the advantage of a thermostat in the cooling system.
- Understand the procedures used for testing the various cooling system components.
- Recognize the hazards associated with cooling system service.
- Understand troubleshooting procedures for determining the malfunction of cooling system components.

A typical gasoline engine is only about 15 percent efficient; only about 15 percent of the energy is used to move the vehicle. That means that 85 percent of all energy developed by the engine is wasted in friction and heat—heat that must be removed.

While the heat of combustion may reach as high as 4,000°F (2,200°C), most of it is expelled when the exhaust valve opens. This results in an actual net engine temperature range from about 750°F (410°C) to about 1,500°F (815°C). This is still a great deal of heat, and it must be removed. The coolant, a mixture of water (H_2O) and ethylene glycol, is the liquid used to transfer this heat from the engine to the radiator.

THE COOLING SYSTEM

The cooling system (Figure 3-1) is made up of several components, all of which are essential to its proper operation. They are the radiator, pump, pressure cap, thermostat, cooling fan, heater core, hoses and clamps, and coolant.

The most common cooling system problems are a result of a leaking system. A sound system seldom presents a problem. Leaks are generally easy to find using a pressure tester. Several types are available. The following is a typical procedure for pressure testing a cooling system:

1. Allow the engine and coolant to cool to **ambient temperature**.
2. Remove the pressure cap. Note the pressure range indicated on the cap (Figure 3-2).
3. Adjust the coolant level to a point just below the bottom of the **fill neck** of the radiator.
4. Attach the pressure tester (Figure 3-3).
5. While observing the gauge, pump the tester until a pressure equal to the cap rating is achieved. If the pressure can be achieved, proceed with step 6. If the pressure cannot be achieved, make a visual inspection for leaks.

> **Classroom Manual**
> Chapter 3, page 65

A common leak point is due to loose hose clamps.

Ambient temperature is the temperature of the surrounding air.

The **fill neck** is the part of the radiator on which the pressure cap is attached.

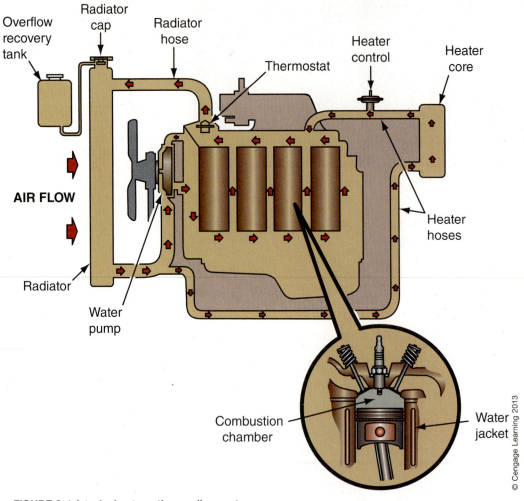

Overflow recovery tank

Radiator cap

Radiator hose

Thermostat

Heater control

Heater core

AIR FLOW

Radiator

Water pump

Heater hoses

Combustion chamber

Water jacket

© Cengage Learning 2013

FIGURE 3-1 A typical automotive cooling system.

© Cengage Learning 2013

FIGURE 3-2 A pressure cap showing the pressure rating.

FIGURE 3-3 Attach a pressure tester to the radiator neck.

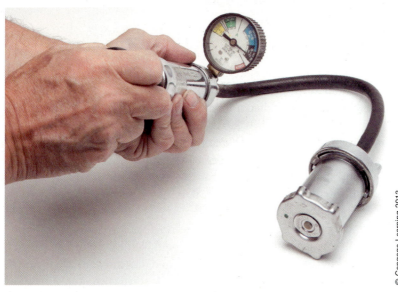

FIGURE 3-4 Pump the tester until a pressure equal to the cap rating is noted on the gauge.

6. Let the system stand for 5 minutes. Recheck the gauge. If pressure drops rapidly, proceed to step 7. If the pressure is the same as in step 5, the system is all right, and it is safe to proceed to step 8. If the pressure has dropped slightly, repressurize the system to the top end of the pressure rating and proceed to step 7. If no leak is detected, proceed to step 8.

   Prior to removing the pressure tester form the system; de-pressuring the tester according to tool manufacturers recommendation.

7. Locate the source of the leak by visibly inspecting all connections. Also inspect the passenger compartment for signs of coolant on the floor. Repair the source of the leak, then repeat steps 4 through 6 to verify the repair.

8. Check the radiator pressure cap. The cap should be able to hold the pressure noted on the cap (Figure 3-4). If the cap fails the test, replace the cap.

RADIATORS

The purpose of the radiator is to **dissipate** heat that is picked up by the coolant in the engine into the air passing through its fins and tubes. This is accomplished by natural or forced means. Natural means are created by ram air as the vehicle is in motion. Forced means are created by an engine- or electric motor-driven fan.

Classroom Manual
Chapter 3, page 66

A heat exchanger is a device that causes heat to move from one medium to another: e.g., fluid to air, air to air, air to fluid, or fluid to fluid.

Dissipate is to disperse, scatter, reduce, weaken, or use up.

FIGURE 3-5 A typical automotive radiator.

Although radiators (Figure 3-5) are often neglected by the vehicle owner, they generally have a long service life. The service life is greatly increased if periodic and scheduled routine maintenance, outlined in the owner's manual, is performed. Eventually, however, most radiators will develop a leak or become clogged with rust or corrosion due to lack of attention.

For most radiator repairs, the radiator must be removed from the vehicle and "shopped out" to a specialty shop that is equipped with the proper specialized tools, equipment, and knowledgeable service technicians to perform this type of repair. If a radiator is badly damaged, such as would result from a collision, it is often less expensive to replace it with a new unit.

The following is a typical procedure for removing a radiator from the vehicle. The actual procedure for any particular make or model vehicle is provided in the manufacturer's specifications. If there is an engine-mounted cooling fan (Figure 3-6), start with step 1. If there is an electric cooling fan (Figure 3-7), start with step 2.

The shroud is a design consideration of the engine cooling system and, as such, should not be removed.

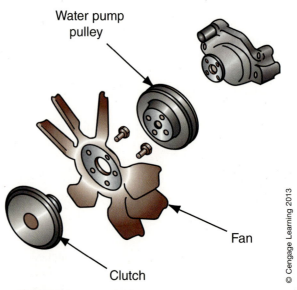

Water pump pulley

Fan

Clutch

FIGURE 3-6 A typical engine-mounted belt-driven cooling fan.

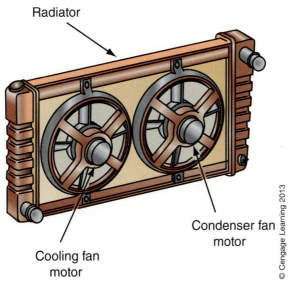

Radiator

Condenser fan motor

Cooling fan motor

© Cengage Learning 2013

FIGURE 3-7 A typical electric motor-driven cooling fan.

1. If the fan is equipped with a shroud, remove the attachments and slide the shroud toward the engine. Proceed to step 3.
2. Remove the electrical connector from the fan motor. Remove the fasteners from the fan assembly brackets and lift the fan assembly out. Care must be taken not to damage the radiator cooling fins.
3. Carefully remove the upper and lower coolant hoses from the radiator.
4. Remove the transmission cooler lines (if the vehicle has an automatic transmission) from the radiator. Plug the lines to prevent transmission fluid loss.
5. Remove the radiator attaching bolts and brackets.
6. Carefully lift out the radiator.

ELECTROCHEMICAL ACTIVITY

As noted in the Classroom Manual, electrochemical activity and electrolysis can take place in the cooling system with devastating results. Many components can be damaged, resulting in their eventual failure, such as radiator and heater cores, the water pump, hoses, and even head gaskets. The following test will check for stay voltage in the cooling system.

1. First set the digital volt-ohmmeter (DVOM) to the direct current (DC) millivolt scale.
2. Connect the negative probe to the negative post of the battery and submerge the positive lead into the coolant at the radiator filler neck, making sure the probe does not touch any metal.
3. Note the meter reading. The voltage reading should be below 0.10 volt. If higher voltage readings are obtained, proceed to step 4. If the voltage is at or below 0.10 volt, the system is okay. Electrolysis can also be caused by stray voltage from electrical components.
4. Start the vehicle, turn on the accessories, and retest. If the voltage is above 0.10 volt, methodically shut off all systems to isolate the circuit-causing voltage.

COOLANT PUMP

The coolant pump (Figure 3-8) may be thought of as the heart of the cooling system. Its purpose is to move the coolant through the system as long as the engine is running.

CAUTION:
Take care not to damage the delicate fins of the radiator when removing it.

A centrifugal pump is a variable-displacement pump. Restricting the flow of coolant does not harm the pump.

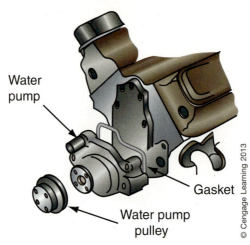

Water pump

Gasket

Water pump pulley

© Cengage Learning 2013

FIGURE 3-8 A typical coolant pump.

To replace a coolant pump, it is often necessary to remove accessories, such as the power steering pump, air conditioning compressor, alternator, or the air pump, to gain access. It is advisable to refer to the specific manufacturer's service manuals when replacing the coolant pump. The following, however, is a typical procedural outline.

1. Remove the radiator as previously outlined, if necessary, to gain access to the coolant pump.
2. Loosen and remove all belts. If there is an engine-mounted fan, proceed with step 3. If there is an electric fan, proceed with step 4.
3. Remove the fan and fan/clutch assembly.
4. Remove the coolant pump pulley.
5. Remove accessories as necessary to gain access to the water pump bolts.
6. Remove the lower radiator hose from the water pump.
7. Remove the bypass hose, if equipped.
8. Remove the bolts securing the water pump to the engine.
9. Tap the water pump lightly, if necessary, to remove it from the engine.
10. Clean the old gasket material from all surfaces. Take care not to scratch the mating surfaces.
11. Install new gaskets and seals and coat bolts with thread sealant (liquid Teflon), if specified by the manufacturer. Reverse the removal steps to install a new water pump assembly and torque the fasteners to the manufacturer's specifications.

⚠️

CAUTION:
For reassembly, note the size and length of the bolts removed. The bolts may be both English and metric and all may not be the same length.

Classroom Manual
Chapter 3, page 72

If a pressure cap fails to hold pressure or if it fails to release high pressure, it must be replaced.

PRESSURE CAP

A radiator pressure cap (Figure 3-9) is necessary to maintain the desired engine temperature without coolant loss. A pressure cap is usually designed to operate in the 14–17 psi (97–117 kPa) pressure range.

Pressure caps may be tested using a cooling system pressure tester and an adapter. The requirement is that a pressure cap not leak at a pressure below what it is rated and that it must open at a pressure above what it is rated.

To pressure test a radiator cap, proceed as follows:

1. Attach the adapter to the pressure tester (Figure 3-10).
2. Install the pressure cap to be tested (Figure 3-11).

FIGURE 3-9 Typical radiator pressure caps.

FIGURE 3-10 Attach the adapter to the pressure tester.

FIGURE 3-11 Install the pressure cap.

3. Pump the pressure tester to the value marked on the pressure cap (Figure 3-12).
4. Did the cap hold pressure? If yes, proceed with step 5. If no, replace the cap.
5. Pump to exceed the pressure rating of the cap.
6. Did the cap release pressure? If yes, the cap is good. If no, replace the cap.

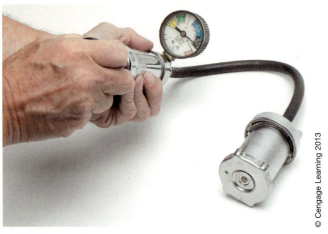

FIGURE 3-12 Pump the tester to a pressure equal to the cap rating.

FIGURE 3-13 A typical cooling system thermostat.

Classroom Manual

Chapter 3, page 84

The thermostat operating temperature is a part of the design consideration of the engine and should not be altered.

THERMOSTATS

The purpose of the thermostat (Figure 3-13) is to trap the coolant in the engine until it reaches its operating temperature. It then restricts the flow of coolant leaving the engine until overall coolant temperature is at or near operating temperature. Even when fully open, the thermostat provides a restriction to coolant flow in order to create a pressure difference. This pressure difference prevents water pump cavitations and forces coolant circulation through the cylinder block and head(s).

A typical procedure for removing a thermostat follows.

Thermostat Removal

1. Reduce the engine coolant to a level below the thermostat.
2. Remove the bolts holding the thermostat housing onto the engine (Figure 3-14). It is not necessary to remove the radiator hose from the housing.
3. Lift off the thermostat housing (Figure 3-15). Observe the pellet-side down position of the thermostat to ensure proper replacement. Do not install the thermostat backward.
4. Lift out the thermostat (Figure 3-16).
5. Clean all the old gasket material from the thermostat housing and engine-mating surface. Many thermostats today use O-ring seals, while other thermostats have centers

FIGURE 3-14 Remove the bolts holding the thermostat housing.

FIGURE 3-15 Lift off the thermostat housing.

FIGURE 3-16 Lift out the thermostat.

that are offset. Be sure the replacement thermostat physically matches the one being removed. Install the new thermostat in the same position as the one that was removed, and torque it to the manufacturer's specifications.

There are several ways to test a thermostat. Many, however, believe that if a thermostat is suspect, it should be replaced. The labor cost for the time required to test a thermostat often outweighs the cost of a thermostat.

Following is a procedure for testing a thermostat that has been removed from the engine.

Thermostat Testing

 WARNING: The water (H_2O) temperature may reach 212°F (100°C). Wear suitable protection.

1. Note the condition of the thermostat.
2. Is the thermostat corroded or open? If no, proceed with step 3. If yes, replace the thermostat.
3. Note the temperature range of the thermostat.
4. Suspend the thermostat in a heatproof glass container filled with water.
5. Suspend a thermometer in the container. Neither the thermostat nor the thermometer should touch the container or touch each other.
6. Place the container and contents on a stove burner and turn on the burner.
7. Observe the thermometer. The thermostat valve should begin to open at its rated temperature value (Figure 3-17). If it begins to open more than ±3°F its rated value, replace the thermostat.
8. Is the thermostat fully opened at approximately 25°F (11°C) above its rated value? If yes, the thermostat is all right. If not, replace the thermostat.

> **CUSTOMER CARE:** There is generally no service interval stated for the replacement of the cooling system thermostat. But, as a preventive maintenance item, thermostat replacement should be suggested when the engine coolant is changed.

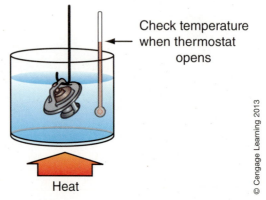

Check temperature when thermostat opens

Heat

© Cengage Learning 2013

FIGURE 3-17 Checking thermostat operation.

PULLEYS

Pulleys require periodic inspection. Pulley problems that may occur are damage due to collision or defective bearings in an **idler** or **drive pulley**. In all cases, repair is straightforward; replace the faulty bearing or pulley. Plastic pulleys used on serpentine belt systems will wear and develop grooves and cracks over time; inspect and replace them when necessary.

BELTS AND TENSIONER

There are two types of belts used in the automotive engine cooling and air-conditioning system: the **serpentine belt** (Figure 3-18) and the **V-belt** (Figure 3-19). Photo Sequence 1 illustrates a typical procedure for servicing the serpentine drive belt.

Prior to removal of the serpentine belt locate the routing diagram similar to the one shown in Figure 3-18. This diagram is often located under the hood in the engine bay, often on the radiator support cover. The diagram may also be located in the vehicle service information. Some technicians choose to draw a sketch of the belt routing prior to removal.

A belt tension gauge may be used to ensure proper belt tensioning for those systems with manual adjustment. Many late-model vehicles have an automatic belt tensioner, a spring-loaded idler pulley. If the belt is manually adjusted, it is suggested that a new belt again be tensioned after about 15 minutes of operation to allow time for initial seating and stretching. The belt should then be checked every 5,000 miles (8,000 kilometers) or so.

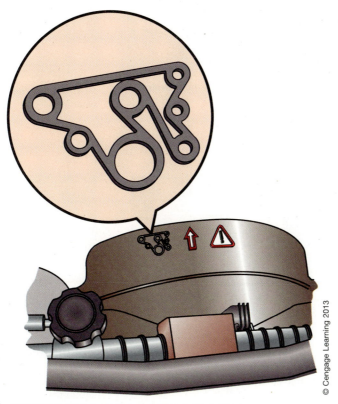

FIGURE 3-18 A typical serpentine belt routing.

© Cengage Learning 2013

An **idler** is a pulley used to tension or reroute the direction of a belt.

A **drive pulley** transmits or inputs power into a component.

Replace any belt that appears to be worn, frayed, or damaged.

A **serpentine belt** is a flat or multi V-grooved belt that winds through all of the engine accessories to drive them off the crankshaft pulley with both sides of the belt being drive surfaces.

A **V-belt** is a belt designed to run in a single V-shaped groove of a drive or idler pulley with only the tapered surface being the drive surface. Most systems require several V belts to drive all of the engine accessories.

TYPICAL PROCEDURE FOR SERVICING THE SERPENTINE DRIVE BELT

P1-1 Release the belt tension, following specific instructions provided in the manufacturer's shop manual.

P1-2 Remove the belt.

P1-3 Inspect the pulleys for nicks, cracks, bent sidewalls, corrosion, or other damage.

P1-4 Place a straightedge across pulleys to check for alignment.

P1-5 Turn each pulley one-half revolution and repeat step 4.

P1-6 Inspect the drive belt. Check for wear.

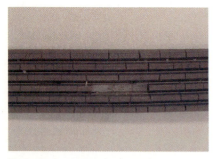

P1-7 If rib sections are missing, the belt should be replaced, and pulley's should be inspected.

P1-8 Replace the belt by reversing steps 1 and 2.

P1-9 Do not use belt dressing on a serpentine belt. Belt dressing may soften the belt and cause deterioration.

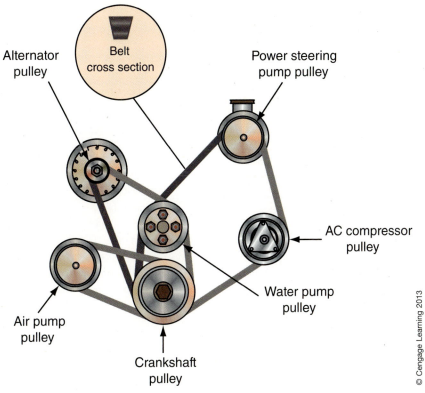

FIGURE 3-19 A typical V-belt routing.

© Cengage Learning 2013

Automatic Belt Tensioner

The drive belts on most late-model engines are equipped with a spring-loaded automatic tensioner. An automatic belt tensioner may be used with all belt configurations, such as with or without power steering and air conditioning.

Belt-driven engine accessories are often replaced due to noise or other problems only to learn that the automatic tensioner was at fault. Table 3.1 (on the next page) is an aid in diagnosing drive belt problems. If an automatic spring tensioner shows any sign of binding during belt replacement, the tensioner needs to be replaced.

Replacing A Belt Tensioner

The following procedure is typical for replacing an automatic belt tensioner. Always follow the particular manufacturer's recommended procedures for each vehicle.

1. Attach a socket wrench to the mounting bolt of the automatic tensioner pulley bolt (Figure 3-20).
2. Rotate the tensioner assembly clockwise (cw) until the belt tension has been relieved.
3. Remove the belt from the idler pulley first, then remove the belt from the other pulleys.
4. Disconnect and remove and set aside any components hindering tensioner removal.
5. Remove the tensioner assembly from the mounting bracket.

 WARNING: Because of high spring pressure, do not disassemble the automatic tensioner.

6. Remove the pulley bolt and remove the pulley from the tensioner.
7. Install the pulley and pulley bolt in the tensioner. Tighten the bolt to 45 ft.-lb (61 N·m).

TABLE 3-1 TROUBLESHOOTING DRIVE AND ACCESSORY BELT PROBLEMS

Problem	Possible Cause	Possible Remedy
Belt slipping	Belt too loose	Replace or tighten belt
	Coolant or oil on belt or pulley	Clean pulleys and replace belt
	Accessory bearing failure (seized)	Replace faulty bearing
	Belt hardened and glazed	Replace belt
Belt squeal when accelerating	Belt glazed or worn	Replace and tension belt
Belt squeak at idle	Belt too loose	Replace or tighten belt
	Dirt or paint embedded in belt	Replace and tension belt
	Misaligned accessory pulleys	Align pulleys
	Improper pulley	Replace pulley
Noise (rumble heard or felt)	Belt slipping	Tighten belt
	Defective bearing	Replace bearing
	Belt misalignment	Align belts
	Improper belt	Install proper belt
	Accessory-induced vibration	Locate cause and correct
	Resonant frequency vibration	Vary belt tension or replace belt
Belt rolls over (V-belt)	Broken cord in belt	Replace and tension belt
	Belt too loose	Replace or tighten belt
	Belt too tight	Replace or loosen belt
Belt jumps off	Broken cord in belt	Replace and tension belt
	Belt too loose	Replace or tighten belt
	Belt too tight	Replace or loosen belt
	Pulleys misaligned	Align pulleys
	Improper pulley	Replace pulley
Belt jumps grooves	Belt too loose	Replace or tighten belt
	Belt too tight	Replace or loosen belt
	Improper pulley	Replace pulley
	Foreign objects in pulley grooves	Clean grooves or replace pulley
	Pulley misaligned	Align pulley
	Broken belt cordline	Replace belt
Broken belt	Excessive tension	Replace belt and adjust tension
	Belt damaged during installation	Replace belt
	Severe misalignment	Replace and align belt
	Bent or damaged bracket or brace	Repair as required and replace belt
	Pulley or bearing failure	Repair as required and replace belt
Rib chunking (rib separation)	Foreign objects embedded in pulley	Remove objects and replace belt
	Installation damage	Replace belt
Rib or belt wear	Pulley misalignment	Align pulleys
	Abrasive environment	Clean and replace belt
	Rusted pulleys	Clean rust or replace pulleys
	Sharp or jagged pulley groove tips	Replace pulley
	Rubber deteriorated	Replace belt
Belt cracking between ribs	Belt mistracked from pulley groove	Replace belt
	Pulley groove tip has worn away rubber	Replace belt
Backside of belt separated	Contacting stationary object	Correct problem and replace belt
	Excessive heat	Replace belt
	Fractured splice	Replace belt
Cord edge failure	Excessive tension	Replace belt and adjust tension
	Contacting stationary object	Correct problem and replace belt
	Incorrect pulley	Replace pulley and belt
	Incorrect belt	Replace belt

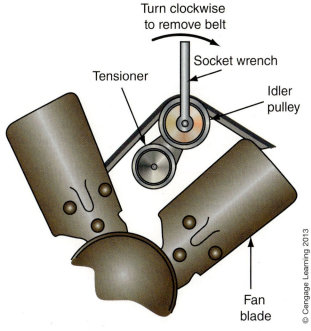

Turn clockwise
to remove belt

Socket wrench

Tensioner

Idler
pulley

Fan
blade

© Cengage Learning 2013

FIGURE 3-20 Rotate the tensioner clockwise (cw) to loosen the belt.

8. Install the tensioner assembly to the mounting bracket. An **indexing tab** (Figure 3-21) is generally located on the back of the tensioner to align with the slot in the mounting bracket. Tighten the nut to 50 ft.-lb (67 N·m).
9. Replace any components removed in step 4.
10. Position the drive belt over all pulleys, except the idler pulley.
11. Using a socket wrench on the pulley mounting bolt of the automatic tensioner, rotate the tensioner cw.
12. Place the belt over the idler pulley and allow the tensioner to rotate back into position. It should spring back smoothly and with adequate tension pressure on the belt.

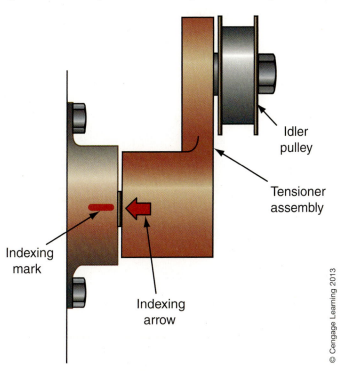

Idler
pulley

Tensioner
assembly

Indexing
mark

Indexing
arrow

© Cengage Learning 2013

FIGURE 3-21 Tensioner indexing tab.

An **indexing tab** is a mark or protrusion on mating components to ensure that they will be assembled in their proper position.

CAUTION:
When installing the serpentine accessory drive belt, the belt must be routed correctly. If not, the water pump may rotate in the wrong direction (Figure 3-22), causing the engine to overheat.

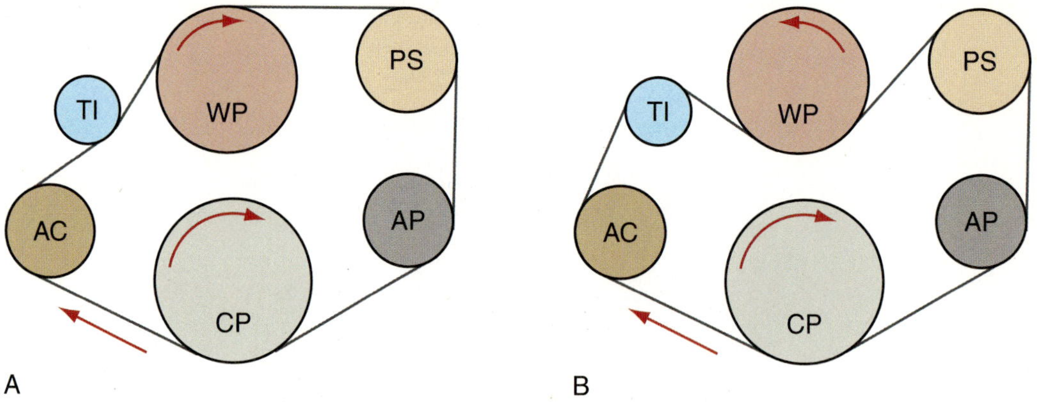

LEGEND:

AC - A/C Compressor Pulley   PS - Power Steering Pulley
AP - Alternator Pulley   TI - Tensioner/Idler Pulley
CP - Crankshaft Pulley   WP - Water Pump Pulley

FIGURE 3-22 If the belt is not installed correctly, the water pump will be turned in the opposite directions: clockwise (A) and counterclockwise (B).

Belt Failure Troubleshooting

A variety of critical engine components stop working when a serpentine belt fails. These components may include the water pump, alternator, air conditioning compressor, and power steering pump to name some of the more common belt-driven accessories. It is important to be able to identify the reason for a belt failure to ensure that the replacement belt will last and avoid premature failure. Often by observing the belts appearance we can determine the potential causes for the failure or damage. Serpentine belts may be constructed of either neoprene or ethylene propylene diene M-class rubber, which is more commonly called EPDM. The majority of belts manufactured today are made of EPDM. Below is a list and photos of some common belt failures, which outline their cause and the solutions recommended by belt manufacturers.

Cracking. Cracks that are small and perpendicular to the ribs of the belt that occur frequently around its circumference (Figure 3-23) is probably one of the most common

FIGURE 3-23 Cracks are caused by continuous flexing and bending in high heat conditions and is the most common reason for belt replacement.

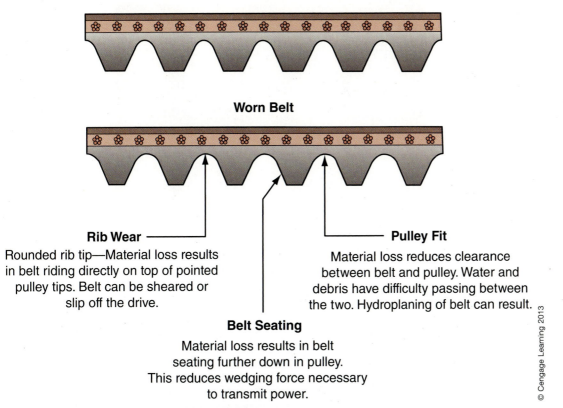

New Belt

Worn Belt

Rib Wear
Rounded rib tip—Material loss results in belt riding directly on top of pointed pulley tips. Belt can be sheared or slip off the drive.

Pulley Fit
Material loss reduces clearance between belt and pulley. Water and debris have difficulty passing between the two. Hydroplaning of belt can result.

Belt Seating
Material loss results in belt seating further down in pulley. This reduces wedging force necessary to transmit power.

FIGURE 3-24 EPDM belts are designed to last 100K miles, but over time the material will wear on the sides and valleys. A 10 percent loss of material is enough to cause performance issues.

reasons for belt replacement. These cracks are caused by continuous flexing and bending in high heat conditions found under the hood as the belt bends around pulleys during normal operation. On neoprene belts eighty percent of the belt life is gone if three or more cracks are found in a three-inch segment of the belt and the belt should be replaced. As a common practice with all belt failures check belt tensioner and idlers pulleys for bearing wear and operation. Belts made of EPDM should not show cracking for 100,000 miles. But EPDM belts may show rib height material wear.

EPDM Belt Wear. Belts made of EPDM wear with age and often require replacement between 80,000–100,000 miles under normal operating conditions. The EPDM belt is more elastic than a standard neoprene belt and resists cracking even at higher mileage, so looking for visible cracks is not an effective method for deterring belt wear and aging. A better indicator of when to replace EPDM belts is rib wear (Figure 3-24). To aid in determining belt wear a belt wear depth gauge is available (Figure 3-25). One free source for this belt wear depth gauge is Gates Belts (http://www.gatesprograms.com/beltwear). As little as 5–10 percent wear can effect operation; if belt wear is indicated, replacement is required.

Abrasion. Abrasion may cause a shiny or glazed appearance on one or both sides of the belt. In severe cases cord fibers may be visible (Figure 3-26). It is caused by a belt making contact with an object such as a flange bolt or foreign object. It may also be caused by improper belt tension, pulley surface wear, or a binding pulley bearing. Inspect belt tensioner for proper operation, replace if necessary.

Improper Installation. A belt rib on either edge begins separating from main body plies (Figure 3-27). If left unnoticed the outer covering may separate causing the belt to

FIGURE 3-25 A simple belt wear gauge can be used to detect EPDM belts for groove wear.

© Cengage Learning 2013

FIGURE 3-26 Abrasion is due to belt making contact with an object such as a flange bolt or foreign object.

© Cengage Learning 2013

FIGURE 3-27 A belt rib on either edge begins separating from main body plies caused by improper replacement procedure.

© Cengage Learning 2013

FIGURE 3-28 Pilling occurs as belt material is sheared off from the ribs building up in the grooves.

unravel. It is caused by improper replacement procedure when one of the belt ribs is placed outside of one of the pulley grooves during installation. Verify that the belt has the same number of ribs as its companion pulley grooves. Once damage has occurred the belt must be replaced. Ensure that all belt ribs fit properly in all pulley groves when replacing.

Pilling. Pilling occurs as belt material is sheared off from the ribs building up in the grooves (Figure 3-28). It is generally caused by insufficient belt tension, pulley misalignment or wear. It is more common on diesel engines, but can happen on any power train. Pilling can lead to belt noise and vibration. The pulley grooves should be inspected for buildup, the alignment should be checked, and the belt will need to be replaced.

It is wise to replace the drive belt as well as the automatic belt tensioner when any pulley-driven component is replaced. By only replacing one system component vibration is increased dramatically (Figure 3-29). This vibration may be both felt and heard leading to customer complaints and the performance and longevity of the new component may be compromised. The green line on the graph indicates vibration minimization with a new belt, tensioner, and component. The red graph line indicates vibration caused by an ageing drive belt, tensioner, and belt-driven components at 150,000 miles. The blue line indicates the vibration results of just replacing one belt-driven component and reusing the old belt and tensioner. The old belt and tensioner do not dampen new component vibration effectively. Excessive vibration leads to excessive noise and component bearing wear. Today all belt drive components are closely integrated and each depends on the belt and tensioner to keep the entire system at optimum operating performance. It is critical that you as a technician check every component in the system; this includes idler pulleys if used.

FANS

Coolant pump-mounted fans (Figure 3-30) occasionally require service. They are damaged due to metal fatigue, collision, road hazards, and abuse. Any condition that causes an out-of-balance pump-mounted fan will result in early pump failure. A check for fan problems is a rather simple task.

Classroom Manual Chapter 3, page 95

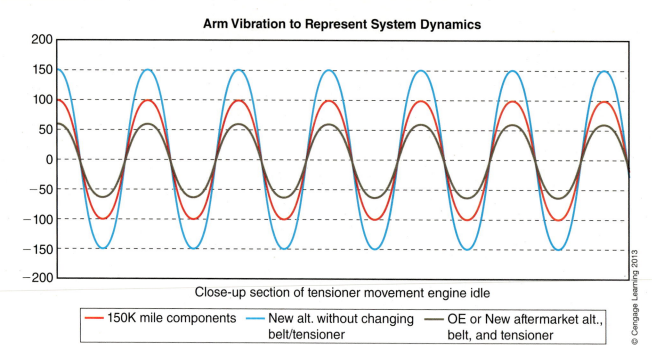

Arm Vibration to Represent System Dynamics

Close-up section of tensioner movement engine idle

— 150K mile components — New alt. without changing belt/tensioner — OE or New aftermarket alt., belt, and tensioner

© Cengage Learning 2013

FIGURE 3-29 When a belt driven component is replaced it is advisable to also replace the belt drive and automatic tensioner to limit system vibration.

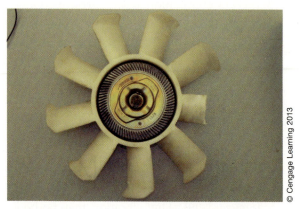

© Cengage Learning 2013

FIGURE 3-30 A damaged coolant pump-mounted fan.

If there is any doubt as to the physical condition of the fan, replace it.

1. Remove the belt(s).
2. Visually inspect the fan for cracks, breaks, loose blades, or other damage. Is the fan sound? If yes, proceed with step 3. If no, replace the fan.
3. Hold a straightedge across the front of the fan. Are all blades in equal alignment? If yes, proceed with step 4. If no, replace the fan.
4. Slowly turn the fan while looking for any out-of-true conditions or any other damage.
5. Turn the fan fast and look for out-of-true conditions.
6. If the fan fails either test (step 4 or 5), it must be replaced. Due to high operating speeds, it is not recommended that repairs be attempted to an engine cooling fan blade.

Classroom Manual
Chapter 3, page 96

Fan Clutch

Most air conditioned vehicles with a coolant pump-mounted fan have a fan clutch (Figure 3-31). This device adds about 6 lb (2.7 kg) to the coolant pump shaft, further increasing the need for a balanced fan.

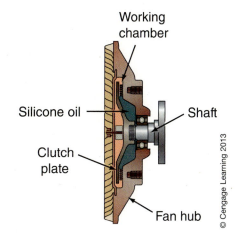

FIGURE 3-31 A typical fan clutch.

There are several basic tests for troubleshooting a fan clutch. First, make certain that the engine is cold and that it cannot be accidentally started while inspecting the fan clutch.

1. Visually inspect the clutch for signs of fluid loss.
2. Check the condition of the fan blades.
3. Check for a slight resistance when turning the fan blades. Spin the fan; if it rotates more than twice on its own power, the clutch is bad.
4. Check for looseness in the shaft bearing.
5. Install a timing light and tachometer to the engine and place the thermometer between the radiator fins and fan clutch. Start the engine and turn on the air conditioner. Bring the engine speed up to 2,000 rpm and observe the fan. It should be rotating slowly. Place a piece of cardboard in front of the radiator to speed up the warm-up time. Note the temperature at which the fan begins to rotate faster, generally 150–195°F (65–90°C). Compare this temperature to the manufacturer's specifications for engagement temperature.

If the fan clutch fails any of these tests, it should be replaced. There are no repairs for a faulty fan clutch.

Flexible Fans

Flexible fans are covered in detail in the Classroom Manual. They are subject to the same problems and are tested in basically the same manner as rigid fans.

Electric Fans

Electric cooling fans (Figure 3-32) are used because there is more precise control over their operation. They may be turned on and off by temperature- and pressure-actuated switches, thereby regulating engine coolant and air conditioning refrigerant temperatures at a more precise level.

 WARNING: Electric engine cooling fans may start and operate at any time and without warning. This may occur with the ignition switch off or on.

Follow the schematic in Figure 3-33 for testing and troubleshooting a typical engine cooling fan system.

If in doubt, replace the fan clutch.

Classroom Manual
Chapter 3, page 99

Classroom Manual
Chapter 3, page 99

The ground connection must be established to a metal part of the body that is not isolated from electrical ground.

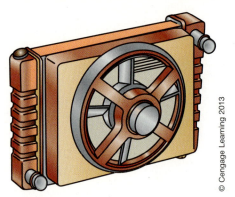

FIGURE 3-32 A typical electric cooling fan.

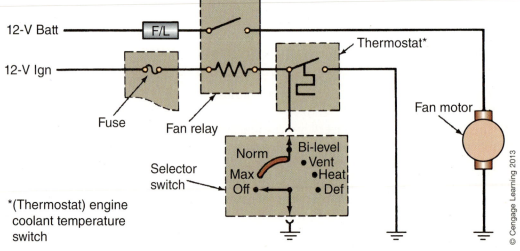

12-V Batt ——— F/L

12-V Ign ———

Thermostat*

Fuse

Fan relay

Selector switch

Norm
Max
Off

Bi-level
• Vent
• Heat
• Def

Fan motor

*(Thermostat) engine coolant temperature switch

FIGURE 3-33 An electrical schematic of an engine cooling fan system.

The **fan relay** is an electromagnetic switch that controls the cooling and auxiliary fan motors.

Make certain that the engine is up to normal operating temperature.

The **fan relay** is an electromagnetic switch that controls the cooling and/or auxiliary fan motors.

1. Start the engine and bring the coolant up to operating temperature.
2. Turn on the air conditioner.
3. Disconnect the cooling fan motor electrical lead connector (Figure 3-34). Verify fan spins freely.
4. Make sure that the ground wire is not disturbed. If the ground wire is a part of the electrical connector, establish a ground connection with a jumper wire (Figure 3-35).
5. Connect a test lamp from ground to the hot wire of the connector (Figure 3-36). Make sure the lamp is good.
6. Did the lamp light? If yes, proceed with step 7. If no, check for a defective **fan relay** or temperature switch. It is also possible that the engine is not up to sufficient temperature to initiate cooling fan action.
7. Connect a fused jumper wire from the battery positive (+) terminal to the cooling fan connector (Figure 3-37). Make sure the fuse is good.
8. Did the fan start and run? If yes, the fan is all right. Check for poor connections at the fan motor and repair them if necessary. If no, proceed with step 9.
9. Again check the fuse in the jumper wire. Is it blown? If yes, the motor is shorted and must be replaced. If no, the motor is open and must be replaced.

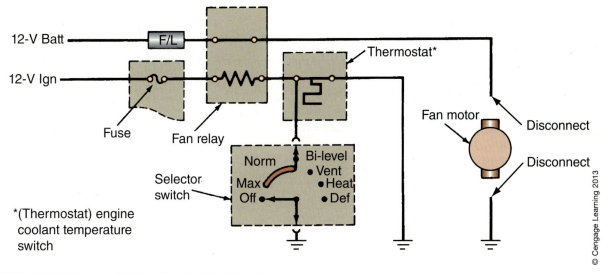

FIGURE 3-34 Disconnect the cooling fan electric motor.

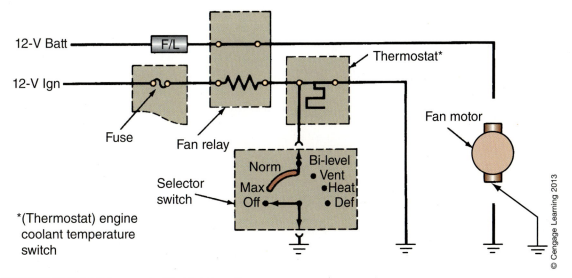

FIGURE 3-35 Establish ground with a jumper wire.

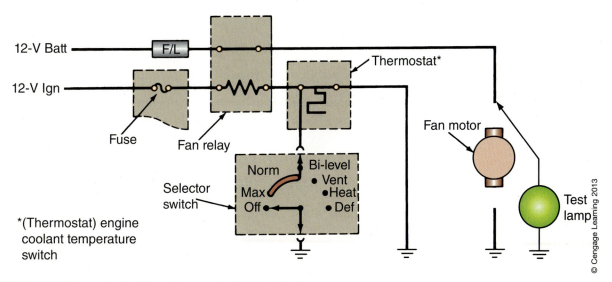

FIGURE 3-36 Connect a test lamp from ground to the hot wire.

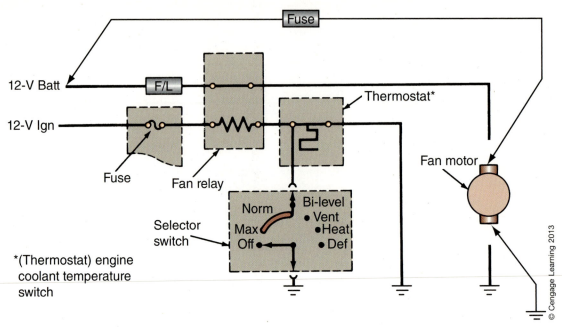

FIGURE 3-37 Connect a fused jumper wire from the battery positive (+) terminal to the cooling fan connector.

Classroom Manual

Chapter 3, page 103

HOSES AND CLAMPS

Engine cooling system hoses and clamps should be replaced every few years. This should become part of a good preventive maintenance program. If done on a periodic schedule, more expensive repairs, such as those caused by an overheating engine, are not as likely to occur.

Replace all hoses if any are found to be defective.

SERVICE TIP: Some engines have additional hoses such as a small bypass hose between the coolant pump and the engine block, hoses used to carry coolant to heat the throttle body on fuel injected engines, and short hoses used to interconnect coolant carrying components on certain engines (Figure 3-38). Do not overlook these hoses when checking the cooling system.

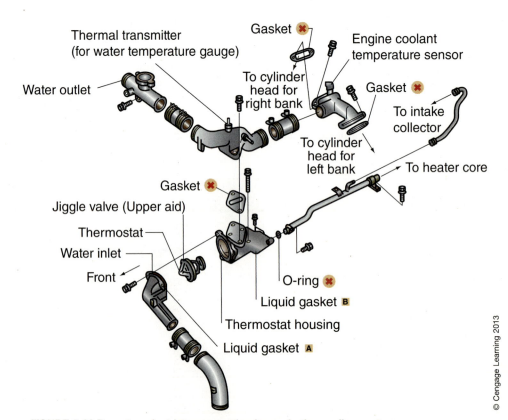

FIGURE 3-38 Do not neglect interconnecting hoses in the cooling system.

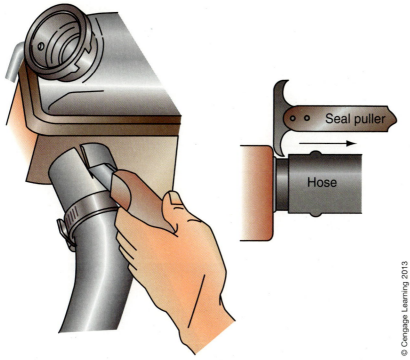

FIGURE 3-39 To avoid damage to fittings, hose should be slit and loosened with a seal remover or hose tool.

CAUTION:
Do not use unnecessary force when removing the hose end from the radiator or heater core. Use a razor knife to make a lengthwise incision (Figure 3-39), and peel the hose off the connection.

Hoses

Carefully check all cooling system hoses when a vehicle is being serviced. The following is a simple checklist for this service:

1. Check for leaks, usually noted by a white, green, or rust color at the point of the leak.
2. Check for swelling, usually obvious when the engine is at operating pressure.
3. Check for chafing, usually caused by a belt or other nearby component.
4. Check for a soft or spongy hose that would indicate chemical deterioration.
5. Check for a brittle hose indicating repeated heating, usually near the coolant pump.
6. Squeeze the hose. If its outer layer splits or flakes away, replace the hose.
7. Squeeze the lower radiator hose. If the reinforcing wire is missing (due to rust or corrosion), replace the hose.

Replacing a Hose. It is a good practice to replace all of the radiator and heater hoses if any of them are found to be defective. It is not always possible to convince the customer that this should be done.

1. Slide the hose clamp back at both ends of the hose (Figure 3-40).
2. Firmly but carefully twist and turn the hose to break it loose from the coolant pump and radiator. Using a box cutter to slice through the hose will help facilitate its removal.
3. Remove the hose (Figure 3-41).

Hose Clamps

Many technicians feel that original equipment hose clamps are only good for one-time use. In almost all cases, it is an accepted practice to replace the clamps when replacing a hose. They are inexpensive and are good insurance against an early failure.

SERVICE TIP:
Some lower radiator hoses contain a spring. The lower hose is the intake hose to the engine cooling system and under suction the spring prevents the lower hose from collapsing.

FIGURE 3-40 Slide the radiator hose clamp back.

FIGURE 3-41 Remove the hose.

A **constant tension hose clamp** is a spring tension clamp designed to maintain a consistent clamping tension.

At least one manufacturer, however, recommends that a **constant tension hose clamp**, used on many cooling systems (Figure 3-42), be replaced with an original equipment clamp. A number or letter is stamped into the tongue of constant tension clamps for identification. If replacement is necessary, use only an original equipment clamp with matching number or letter.

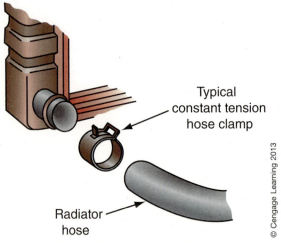

Typical constant tension hose clamp

Radiator hose

FIGURE 3-42 A typical constant tension hose clamp.

FIGURE 3-43 A typical hose clamp tool.

© Cengage Learning 2013

Classroom Manual
Chapter 3, page 103

The recovery tank is often referred to as an overflow tank or a surge tank.

A special clamp tool (Figure 3-43) is available for use in removing and replacing constant tension hose clamps; however, slip joint pliers may also be used. In either case, eye protection should be worn when servicing constant tension clamps.

RECOVERY TANK

The only problem that one may experience with a recovery tank is an occasional leak. Since the recovery tank is not a part of the pressurized cooling system, it can often be successfully repaired with hot glue. If it is found to be leaking, proceed as follows:

1. Remove the tank from the vehicle.
2. Thoroughly clean the tank inside and outside.
3. Use a piece of sandpaper to roughen up the surface at the area of the leak.
4. Use a hot glue gun and make several small beads of hot glue at the point of the leak. Cover the area thoroughly, overlapping each successive bead.
5. If it is accessible, repeat steps 3 and 4 inside the tank.

A heat exchanger is an apparatus in which heat is transferred from one medium to another on the principle that heat moves from an object with more heat to an object with less heat.

HEATER SYSTEM

The comfort heater system is actually a part of the engine cooling system. The heater core, a small radiator-type **heat exchanger**, is located in the case/duct system of the heater/air conditioner unit.

Heater Core

Most failures of the heater core (Figure 3-44) are due to a leak. This is easily detected by noting a wet floor carpet just below the case on the passenger side of the vehicle or if fogging of the windshield is occurring (moisture coming from ducts). Replacement of the heater core, unfortunately, is not so simple. Because of the many different variations of installation, it is necessary to follow the manufacturer's shop manual instructions for replacing the heater core.

The following is a typical procedure only and is not intended for any particular make or model vehicle:

1. Remove the coolant.
2. Remove the access panel(s) or the split heater/air conditioning case to gain access to the heater core.
3. Loosen the hose clamps and remove the heater coolant hoses.
4. Remove the cable and vacuum control lines (if equipped).
5. Remove the heater core, securing brackets and clamps.
6. Lift the core from the case.

Classroom Manual
Chapter 3, page 105

A wet floor carpet can also be caused by a leaking sea around the windshield.

CAUTION: Do not use force. Take care not to damage the fins of the heater core when removing and replacing it.

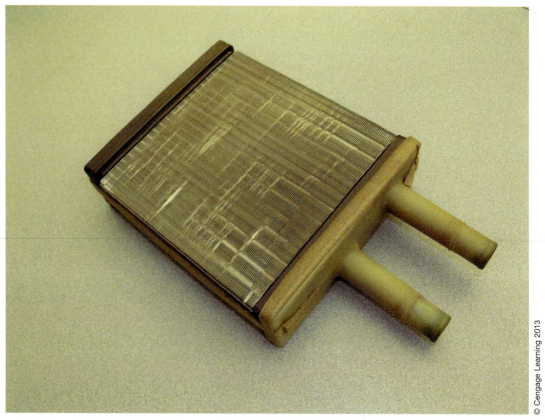

FIGURE 3-44 **A typical heater core.**

Control Valve

The control valve is a cable-, vacuum-, or electrically-operated shut-off valve used at the inlet of the heater core to regulate coolant flow through the core. Other than a leak, which is usually obvious, the valve fails due to rust or corrosion. To replace the valve:

1. Remove the coolant to a level below the control valve.
2. Remove the cable linkage, vacuum hose(s), or electrical connector from the control valve (Figure 3-45).
3. Loosen the hose clamps and remove the inlet hose from the control valve.
4. Remove the heater control valve, as applicable. Remove the outlet hose from the heater core. Remove the attaching brackets or fasteners from the control.
5. Inspect the hose ends removed. If they are hard or split, cut 0.5–1.0 in. (12.7–25.4 mm) from the damaged ends. Better yet, replace the hoses.

Hoses and Clamps

Heater hoses and clamps are basically about the same as radiator hoses and clamps except they are generally smaller in diameter. It is a practice of some technicians to use a hose that is too large for the application and overtighten the hose clamp to stop the leak. A hose clamp that is too large for the hose is often distorted when tightened sufficiently to secure the hose.

Hoses. Heater hoses are replaced in the same manner as radiator hoses. It is much easier to use the wrong size hose, however. For example, a ¾ in. hose fits very easily onto a ⅝ in.

FIGURE 3-45 Typical heater flow control valves.

FIGURE 3-46 A hose clamp overtightened to compensate for a hose that is too large for the application.

fitting. The hose clamp then must be overtightened (Figure 3-46) to squeeze the hose onto the fitting sufficiently to prevent a leak. It is not so easy to slide a ⅝ in. hose onto a ⅝ in. fitting. The intent, however, is to use the proper size hose for the application.

It is good practice to replace all heater hoses if any are found to be defective. The following is a typical procedure:

1. Remove the coolant to a level below that of the hoses to be replaced.
2. Loosen the hose clamp at both ends of the hose.
3. Turn and twist the hoses to break them loose.
4. Remove the hose. Do not use unnecessary force when removing the hose end from the heater core.

FIGURE 3-47 A hose clamp that is too large for the hose application.

Classroom Manual
Chapter 3, page 106

Clamps. As with cooling system hose clamps, heater hose clamps should be replaced when a hose is replaced. It is most important that the proper size clamp be used for the hose. If the clamp is too large, it will be distorted before being tightened enough to secure the hose onto the fitting (Figure 3-47). When this occurs, it is extremely difficult to stop a leak.

ANTIFREEZE

Classroom Manual
Chapter 3, page 105

 WARNING: Antifreeze solution is considered a hazardous material. Dispose of antifreeze in an environmentally safe manner. Refer to all applicable federal environmental protection agency (EPA), state and local ordinances and regulations. In general, used antifreeze must be stored in an appropriately labeled bulk container for later recycling or waste pick-up.

1. Make sure that the engine is cool and the cooling system is not under pressure.
2. Place a clean, dedicated container of adequate size under the drain provision of the cooling system.
3. Open the radiator drain provision and drain the cooling system.

 WARNING: Refer to local ordinances and regulations regarding proper disposal procedures for ethyl glycol-type antifreeze solutions.

Preventive Maintenance

Changing antifreeze/coolant annually helps to prevent cooling system failure, the primary cause of engine-related breakdowns. The antifreeze/coolant, depending on mix ratio, can provide protection for the cooling systems from −84−276°F (−64−135.6°C). The generally recommended ratio of 1:1 (50 percent antifreeze and 50 percent water) provides protection against freezing with an ambient temperature as low as −34°F (−36.7°C) and provides pro-

SERVICE TIP:
If the coolant is to be reused, drain it into a clean container. If, however, it is not to be reused, it must be disposed of or recycled in a manner considered to be environmentally safe.

tection from rust and corrosion for the cooling system metals. This includes protection for the thin, lightweight aluminum radiators found in many late-model vehicles.

 WARNING: Store used coolant in a properly labeled container. Do not pour used coolant down a drain. Ethylene glycol antifreeze is a very toxic chemical. Do not dispose of coolant into the sewer system or ground water. This is illegal and ecologically unsound.

Extended Life

Extended-life antifreeze generally provides freeze-up and boil-over protection for up to 5 years or 150,000 miles (241,350 kilometers). High-mileage drivers, as well as those who do not have or take the time for regular vehicle preventive maintenance (PM), should rely on the long-lasting protection of an extended-life antifreeze/coolant. Such mixtures are silicate and phosphate free, providing extended protection against rust and corrosion to all metals of the cooling system. Extended-life antifreeze, such as Havolin's Extended Life DEX-COOL™, manufactured from ethylene glycol (EG), meets all of the compatibility requirements for the extended-life antifreeze used in most General Motors (GM) vehicles since 1996.

Low Tox

Low-tox antifreeze contains propylene glycol (PG) and is less toxic than EG types. A low-tox antifreeze/coolant mixture offers similar freeze-up and boil-over protection while providing rust and corrosion protection for all cooling system parts, including the aluminum used in radiators. Also, low-tox antifreeze/coolant provides an added margin of safety if accidentally ingested by pets or wildlife.

Drain and Flush. If replacing existing antifreeze/coolant with any other type of antifreeze/coolant, first completely drain and flush the cooling system. Be aware that if antifreeze solutions are mixed, the intended protection of either solution may be lost, particularly the added margin of safety afforded by the PG formulas.

Both types of antifreeze, EG and PG, are biodegradable. It is the rate at which they degrade, however, that is important. EG, though considered more toxic than PG, degrades the fastest. For more specific information, request a material safety data sheet (MSDS) from the manufacturer of the particular product that is being used. All used EG and PG coolants must be recycled for reuse or disposed of in a manner consistent with applicable local, state, and national regulations. Both used EG and PG coolants, though considered toxic, are not considered as a hazardous waste by the Environmental Protection Agency (EPA) unless they contain more than five parts per million (5 ppm) lead (Pb). Lead is a hazardous by-product that leaches out of the solder used to secure the tubes and header tanks.

Coolant Recovery/Recycle. Several companies manufacture antifreeze recovery/recycle/recharge machines. Some systems connect to the vehicle's cooling system and inject new or recycled coolant that forces the old coolant out of the system. This type of machine is ideal for field use where electricity is not available. The pump is powered by the vehicle's battery while the engine is running.

FLUSH THE COOLING SYSTEM

When replacing an existing antifreeze/coolant with an extended-life antifreeze/coolant, the cooling system must first be completely drained and flushed (see Photo Sequence 2). As mentioned earlier, this is necessary to gain the full benefits of the longer-lasting formula.

DRAINING AND REFILLING THE COOLING SYSTEM

P2-1 Ensure that the engine is cold, and slowly remove the radiator cap. **CAUTION:** If the radiator cap is removed from a hot cooling system, serious personal injury may result.

P2-2 Place a drain pan of adequate size under the radiator drain cock.

P2-3 Install one end of a tube or hose on the draincock and position the other end in the drain pan.

P2-4 Open the radiator draincock and allow the radiator to drain until the flow stops.

P2-5 Place a drain pan of adequate size under the engine.

P2-6 Remove the drain plug from the engine block and allow the engine block to drain until the flow stops. **NOTE:** There may be more drainage from the radiator at this time.

P2-7 Close the radiator draincock and replace the engine block drain plug.

P2-8 Remove the pans and dispose of the coolant in a manner consistent with local regulations.

P2-9 Add sealant pellets, if required. Some car lines require that sealant pellets be added to the radiator whenever the cooling system is drained and refilled with fresh coolant. Failure to use the correct sealant pellets may result in premature water pump leakage.

P2-10 Premix the antifreeze with clear water to a 50:50 ratio. **NOTE:** Distilled water is required by some manufacturers.

P2-11 With a large funnel in the radiator fill hole, slowly pour in the coolant mixture. **NOTE:** Refer to the manufacturer's specifications for the cooling system capacity.

P2-12 Fill the cooling system to within about 1 in. (25.4 mm) below the fill hole.

P2-13 Start the engine and let the cooling system warm up. **NOTE:** When the thermostat opens, the coolant level may drop. After the thermostat opens, add coolant until the level is up to the fill hole.

P2-14 Replace the radiator cap.

P2-15 Check the coolant level in the recovery reservoir and add coolant if needed.

For example, if an extended-life antifreeze/coolant is currently being used in a vehicle and a regular-type antifreeze is added to the cooling system, the extended-life protection will be lost. When adding antifreeze, always add the same type as that previously used in the cooling system. If the type is unknown, it is generally recommended to drain, flush, and refill the cooling system. The procedure that follows is typical:

NOTE: Often instructions are given in the owner's manual, and one may be told to open an air bleed valve on the engine or to remove a heater hose to purge air that may have entered the engine during draining.

1. Drain the cooling system following steps 1 through 5 of Job Sheet 7.
2. Close the drain valve.
3. Recycle or dispose of the used coolant according to local laws and regulations.

NOTE: Label the container clearly as used antifreeze. Do not use beverage containers to store antifreeze, new or used. If stored, keep the containers away from children and animals. If the antifreeze is to be disposed, do so promptly and properly.

4. Flush the cooling system to clean the engine block of any scale, rust, or other debris before refilling with new antifreeze/coolant.

NOTE: A cooling system flush may be used to remove stubborn rust, grease, and sediment, which may not be removed by plain water alone.

5. Remove the radiator cap and fill the radiator with cleaner (if used) and water.
6. Run the engine with the heater on HI and the temperature gauge reading normal operating temperature for the time recommended on the flush product label.

NOTE: An infrared (IR) thermometer is a handy tool to use for determining when "normal" engine-operating temperature has been reached.

7. Stop the engine and allow it to cool.
8. Again, open the drain valve and drain the cooling system.
9. Close the drain valve and refill the radiator with plain water.
10. Run the engine for about 15 minutes at normal engine-operating temperature.
11. Stop the engine and allow it to cool, open the drain valve, and redrain the cooling system.
12. Close the drain valve and refill the cooling system, following steps 6 through 10 of Job Sheet 7.

NOTE: Check the owner's manual for the cooling system capacity, as well as the service manual for any special service instructions. In some cooling systems, the location of the heater core is higher than the cooling system filler neck. These systems are prone to air locks, especially if you are not using an automated cooling system drain and fill machine. One method to lessen the chance of air pockets is to jack up the front of the vehicle until the filler neck is the high point in the cooling system, and then fill the system. Another option is to use a cooling system evacuator and fill adapter. This is an inexpensive, handheld device (Uview Airlift II is one such system) used to place the cooling system under a vacuum, which is then used to draw fresh coolant into the system, thereby eliminating air locks.

13. Once the radiator is filled, run the engine at normal operating temperature with the heater on HI for 15 minutes to mix and disperse the coolant fully throughout the cooling system.
14. Shut off the engine and allow the cooling system to cool. Thermocycling is the process of allowing the cooling system to reach operating temperature and then be allowed to cool down again. Recheck hose clamp tension.

15. Check the coolant level and concentration. Adjust, if necessary.

16. After a few days of driving, recheck the cooling system. A hydrometer may be used for EG testing, and test strips may be used for PG testing.

NOTE: If additional coolant is required, use a premixed 50/50 percent concentrate. If additional coolant is needed, it is suggested that the cooling system be leak tested.

Hybrid Electric Cooling System Service

Many normal cooling system maintenance procedures on the hybrid electric engine are similar to the conventional service procedures on a non-hybrid platform, but there are exceptions that can become both hazardous and frustrating. One item that is found on a hybrid platform is the coolant heat storage tank that is designed to store hot coolant for up to 3 days (Figure 3-48). In addition to being hot, the coolant is stored under pressure and serious burns may result if manufacturer's service procedures are not followed during a cooling system service or repair procedure. The cooling system on some platforms is also tied to the inverter assembly, which increases the potential of air being trapped in the cooling system during service. In order to purge air from the cooling system following service, there is a bleed screw and the use of a scan tool is required to cycle the auxiliary electric coolant pump. The service procedure for draining and filling the cooling system of a hybrid electric vehicle is outlined in the manufacturer's service manual. The following is an outline of the basic steps required but is not intended to substitute for the manufacturer's recommended procedure:

Classroom Manual
Chapter 3, page 80

1. When servicing the cooling system of a hybrid electric vehicle equipped with a coolant heat storage tank and an auxiliary water pump, first disconnect the coolant heat storage auxiliary water pump connector.

2. Remove the radiator cap and drain the engine coolant from both the coolant heat storage tank and the radiator assembly.

3. Perform the required service.

4. Connect the coolant heat storage auxiliary water pump connector.

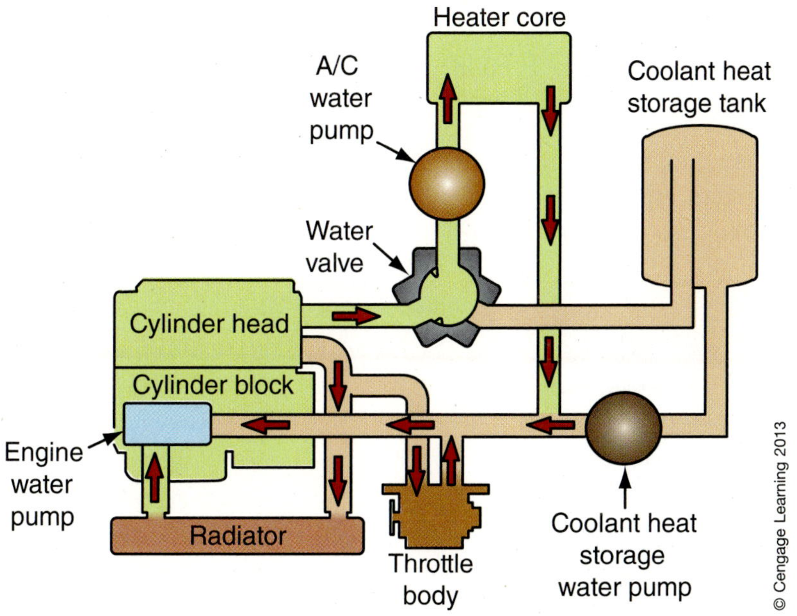

FIGURE 3-48 Hybrid electric vehicle hot coolant storage tank and water pump system.

5. While refilling the cooling system with the specified coolant, use a scan tool to operate the coolant heat storage auxiliary water pump at 30-second intervals to help the inflow of coolant into the tank to avoid air pockets from being trapped in the cooling system. It will also be necessary to open the cooling system bleed plug during this procedure. Repeat this procedure until no air escapes from the bleed plug.

6. Start and run the engine for 1 to 2 minutes.

7. Turn off the engine and top off the coolant level if necessary.

DTC P1151 Coolant Heat Storage Tank

On a Toyota Hybrid platform, a diagnostic trouble code (DTC) P1151 recommends that the coolant heat storage tank be replaced. But be aware that the service manual also indicates that this code may be set if there is an air bubble in the cooling system. To avoid replacing the coolant storage tank unnecessarily, check for the sound of air bubbles flowing through the heater core from the passenger compartment. If there is no air present in the cooling system, you will not be able to hear the sound of water flowing when the coolant pump is running. If air is present you will hear the sound of rushing water.

If air is trapped in the cooling system it will be necessary to bleed the air out of the system as outlined in the manufacturer's service information. After the air is removed, the DTC must be cleared and the vehicle must be driven for two trips. If the code returns after the air has been bled and the vehicle has been driven for two trips, the coolant heat storage tank must be replaced.

 WARNING: The coolant heat storage tank is designed to store hot coolant at 176°F (80°C) for up to 3 days. Severe burns may result if proper manufacturer service procedures are not followed during service.

TROUBLESHOOTING THE HEATER AND COOLING SYSTEM

The following procedure is given as a quick reference to enable the service technician to isolate many of the conditions that can cause improper engine cooling system or heater operation. This procedure is given in three parts: engine overcooling, engine overheating, and loss of coolant.

The customer's complaint would usually be for an overheating condition. If the problem is due to a loss of coolant, the customer may complain that coolant or water must be added frequently.

It should be noted that cooling system problems are often caused by, or may cause, air-conditioning system problems. Conversely, air conditioning problems may be caused by, or may cause, cooling system problems.

ENGINE OVERCOOLING	
Possible Cause	**Possible Remedy**
1. Thermostat missing	Replace the thermostat
2. Thermostat defective	Replace the thermostat
3. *Defective temperature sending unit	Replace the sending unit
4. *Defective dash gauge	Replace the dash gauge
5. *Broken or disconnected wire	Repair or replace the wire
6. *Grounded or shorted cold indicator wire (if equipped with a cold lamp)	Repair or replace the wire

*These symptoms indicate overcooling, though the engine temperature may be within safe limits.

ENGINE OVERHEATING

Possible Cause	Possible Remedy
1. Collapsed radiator hose	Replace the radiator hose
2. Coolant leak	Locate and repair the leak
3. Defective water pump	Replace the water pump
4. Loose fan belt(s)	Torque fan belts to specs
5. Defective fan belt(s)	Replace the fan belts
6. Broken belt(s)	Replace the belt(s)
7. Fan bent or damaged	Replace the fan
8. Fan broken	Replace the fan
9. Defective fan clutch	Replace the fan clutch
10. Exterior of radiator dirty	Clean the radiator
11. Dirty bug screen	Clean or remove the screen
12. Damaged radiator	Repair or replace the radiator
13. Engine improperly timed	Service the engine
14. Engine out of tune	Service the engine
15. *Temperature sending unit defective	Replace the sending unit
16. *Dash gauge defective	Replace the dash gauge
17. *Grounded or shorted indicator wire	Repair or replace the wire

*These symptoms indicate overheating, though the engine temperature may be within safe limits.

LOSS OF COOLANT

Possible Cause	Possible Remedy
1. Leaking radiator hose	Replace the radiator hose
2. Leaking heater hose	Replace the heater hose
3. Loose hose clamp	Tighten the clamp
4. Leaking radiator (external)	Repair or replace the radiator
5. Leaking transmission cooler (internal)	Repair or replace the radiator
6. Leaking coolant pump shaft seal	Repair or replace the coolant pump
7. Leaking gasket(s)	Replace the gaskets
8. Leaking core plug(s)	Replace the core plug(s)
9. Loose engine head(s)	Re-torque the head(s)
10. Warped head(s)	Replace the head(s)
11. Excessive coolant	Adjust the coolant level
12. Defective radiator pressure cap	Replace the cap
13. Incorrect pressure cap	Replace the cap
14. Defective thermostat	Replace the thermostat
15. Incorrect thermostat	Replace the thermostat
16. Rust in system	Flush the system and add rust protection
17. Radiator internally clogged	Clean or replace the radiator
18. Heater core leaking	Repair or replace the core
19. Heater control valve leaking	Replace the valve

An overheated engine may result in poor to no cooling from the air conditioner.

An air-conditioning system malfunction may result in an overheating engine condition.

TERMS TO KNOW

Ambient temperature

Constant tension hose clamp

Dissipate

Drive pulley

Fan relay

Fill neck

Heat exchanger

Idler

Indexing tab

Serpentine belt

V-belt

CASE STUDY

A customer brings his car into the shop because the temperature gauge does not operate. It remains on cold all of the time, regardless of engine heat conditions.

The lead wire to the sending unit is disconnected, and a testlight is used to probe for voltage. The testlight comes on when the ignition switch is placed in the ON position. When the lead is connected to ground (−) through a 10 Ω resistor, the dash unit needle moves to the full hot position. This is a normal operation according to the service manual.

The diagnosis is that the sending unit is defective. It is replaced after approval by the customer and the temperature gauge system is returned to normal operation.

ASE-STYLE REVIEW QUESTIONS

1. Engine overcooling is being discussed:

   Technician A says that a thermostat stuck open could be the cause of this condition.

   Technician B says that a missing thermostat could be the cause of this condition.

   Who is correct?

   A. A only
   B. B only
   C. Both A and B
   D. Neither A nor B

2. *Technician A* says that when pressure testing a cooling system, it should hold pressure for 5 minutes.

   Technician B says that a wet carpet may indicate a heater core leak.

   Who is correct?

   A. A only
   B. B only
   C. Both A and B
   D. Neither A nor B

3. All of the following may cause the back side of a serpentine belt to separate *except:*

   A. Contacting stationary object
   B. Excessive heat
   C. Fractured splice
   D. Pulley misalignment

4. Coolant loss is being discussed:

   Technician A says that a missing thermostat could be the problem.

   Technician B says a heater control valve stuck open may be the problem.

   Who is correct?

   A. A only
   B. B only
   C. Both A and B
   D. Neither A nor B

5. *Technician A* says that as a neoprene serpentine belt ages and wears cracks will form on the belt ribs and that if there are more than 3 cracks in a 3-inch span, the belt should be replaced.

   Technician B says that serpentine belts made of EPDM resist cracking and instead exhibit wear to the belt ribs similar to tire wear, and that a depth gauge should be used to assess belt wear.

   Who is correct?

   A. A only
   B. B only
   C. Both A and B
   D. Neither A nor B

6. *Technician A* says that antifreeze should be changed every 2 years.

   Technician B says that extended-life coolant may last up to 5 years.

   Who is correct?

   A. A only
   B. B only
   C. Both A and B
   D. Neither A nor B

7. An overheating condition is being discussed:

   Technician A says that replacing the thermostat with one of a lower temperature rating will reduce the coolant temperature.

   Technician B says that replacing the pressure cap with one of a lower rating will reduce the coolant temperature.

   Who is correct?

   A. A only
   B. B only
   C. Both A and B
   D. Neither A nor B

8. All of the following conditions could cause poor passenger compartment heating system performance except:
   A. A restricted heater core.
   B. A heater control valve stuck in the closed position.
   C. An air pocket in the heater core.
   D. A thermostat stuck in the closed position.

9. Electric cooling fans are being discussed:
   Technician A says that they are independent of the ignition switch and may start and run without notice at any time.
   Technician B says that they are generally protected by a shroud and pose no safety problem.
   Who is correct?
   A. A only
   B. B only
   C. Both A and B
   D. Neither A nor B

10. Cooling system service is being discussed:
    Technician A says that ethylene glycol antifreeze must be disposed of in a manner prescribed by the Environmental Protection Agency (EPA) and local ordinances.
    Technician B says that after hoses have been replaced, clamp tension should be rechecked after the cooling system has been allowed to thermocycle.
    Who is correct?
    A. A only
    B. B only
    C. Both A and B
    D. Neither A nor B

ASE CHALLENGE QUESTIONS

1. The tank on a radiator has ruptured:
   Technician A says that a thermostat that is stuck in the open position could be the cause.
   Technician B says that a faulty radiator cap could be the cause.
   Who is correct?
   A. A only
   B. B only
   C. Both A and B
   D. Neither A nor B

2. A vehicle is in for repair with a complaint of poor heat output. During testing and diagnosis air is found to be trapped in the heater core. Which of the following is the most likely cause?
   A. A faulty head gasket
   B. A stuck open thermostat
   C. A radiator cap with a failed pressure relief valve
   D. A faulty water pump

3. The least likely problem associated with a cooling system that had the thermostat removed is:
   A. Poor heater performance
   B. Erratic computer engine control
   C. Lower-than-normal operating temperature
   D. Loss of coolant

4. The most likely use for an engine coolant temperature switch is to electrically energize the:
   A. Compressor clutch
   B. Cooling system fan motor
   C. Blower motor
   D. Coolant "hot" warning light

5. If the crankshaft pulley turns clockwise (cw), all of the following statements about the illustration below are true, *except:*
   A. The compressor turns clockwise (cw)
   B. The idler pulley turns counterclockwise (ccw)
   C. The water pump turns clockwise (cw)
   D. The alternator pulley turns clockwise (cw)

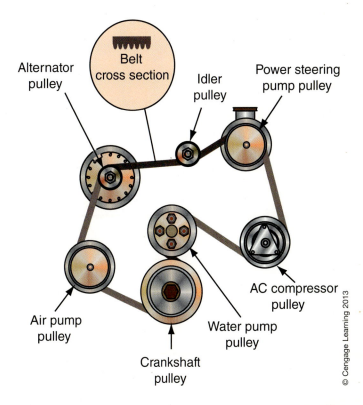

Alternator pulley · Belt cross section · Idler pulley · Power steering pump pulley · AC compressor pulley · Water pump pulley · Crankshaft pulley · Air pump pulley

© Cengage Learning 2013

Name _____ Date _____

Drain and Fill Coolant

Upon completion of this job sheet, you should be able to remove and replace cooling system coolant.

NATEF Correlation

Determine coolant condition; drain and recover coolant. **(P-1)**

Tools and Materials

Late-model vehicle

Shop manual

Two pans

Safety glasses or goggles

Hazardous waste container

Funnel

Rubber hose

Hand tools, as required

Describe the vehicle being worked on.

Year _____ Make _____ Model _____

VIN _____ Engine type and size _____

Procedure

Follow the procedure outlined in the service manual. Photo Sequence 2 may also be used as a guide wherever applicable. Write a brief description of each step.

1. Ensure that the engine is cold, and slowly remove the radiator cap.

 WARNING: **If the radiator cap is removed from a hot cooling system, serious personal injury may result.**

2. Place a drain pan of adequate size under the radiator draincock and install one end of a tube or hose on the draincock. Position the other end in the drain pan. Open the radiator draincock and allow the radiator to drain until the flow stops.

3. Place a drain pan of adequate size under the engine. Remove the drain plug from the engine block and allow the engine block to drain until the flow stops.

4. Close the radiator draincock and replace the engine block drain plug.

_____

_____

5. Remove the pans and dispose of the coolant in a manner consistent with local regulations.

6. Using your shops service information system such as ALLDATA or similar look up and record the vehicle manufacturer's cooling system capacity.

    Record cooling system capacity: _____

7. Premix antifreeze with clear water to a 50:50 ratio.

8. With a large funnel in the radiator fill hole, slowly pour in the coolant mixture. (Refer to the manufacturer's specifications for the cooling system capacity.) Fill to about 1.0 in. (25.4 mm) below the fill hole.

9. Start the engine and let the cooling system warm up. When the thermostat opens, the coolant level may drop. If it drops, add coolant until the level is up to the fill hole. At what Temperature did the thermostat open?_____

_____

10. Follow the manufacturer's recommendation for bleeding air out of the cooling system. When would air-locking in the cooling system be a concern?

_____

_____

11. Replace the radiator cap and check the coolant level in the recovery reservoir. Add coolant if needed.

12. Did the cooling system take the amount specified in the service information listed as cooling system capacity? Why not? _____

_____

Instructor's Response _____

_____

_____

JOB SHEET

Name _____ **Date** _____

LEAK TEST A COOLING SYSTEM

Upon completion of this job sheet, you should be able to leak test a cooling system.

NATEF Correlation

HEATING AND AIR CONDITIONING: Heating, Ventilation, and Engine Cooling Systems Diagnosis and Repair; *Perform cooling system, cap, and recovery system tests (pressure, combustion leakage, and temperature); determine necessary action.* **(P-1)**

Tools and Materials

Late-model vehicle

Shop manual

Safety glasses or goggles

Cooling system leak tester

Hand tools, as required

Describe the vehicle being worked on.

Year _____ Make _____ Model _____

VIN _____ Engine type and size _____

Procedure

Follow the procedures typically outlined in the Shop Manual. Write a brief summary of your procedure and findings following each step. Ensure that the engine is cold for this procedure. Wear eye protection while pressure testing the cooling system.

1. Note the cooling system operating pressure on the radiator cap. Verify the pressure by referring to the specifications in the manufacturer's service manual. Record radiator cap pressure: Does this match manufacturer recommendation Yes/No

2. Remove the radiator cap and adjust the coolant level to 0.5 in. (12.7 mm) below the bottom of the fill neck.

3. Attach the pressure tester to the filler neck of the radiator. Some radiators require an adapter.

4. While observing the gauge, pump the tester until it indicates the same pressure noted in the specifications. What is the specifications pressure? _____

   What gauge pressure could you pump in the system? _____

   Explain: _____

   _____

5. If the gauge pressure is not correct in step 4, proceed with step 6. If correct, proceed with step 8. Gauge pressure was: Correct _____ Not correct _____

6. Make a visual inspection for coolant leaks. List those areas where a leak is observed.

_____

_____

7. Repair leaks, as required, and repeat leak testing, beginning with step 4.
   Where any additional leaks defected? _____

_____

8. Let pressure stand for 5 minutes. If the pressure is the same as in step 4, proceed with step 9. If not, repeat leak testing beginning with step 4.

9. Release the pressure and remove the tester from the cooling system. Attach the tester to the pressure cap.

10. Pump the tester while observing the gauge. What is the maximum pressure? _____

    a. Does the pressure reach that noted on the radiator cap? _____

    b. Does the excess pressure release when rated pressure is reached? _____

    c. What are your conclusions regarding the pressure cap? _____

    _____

Instructor's Response _____

_____

_____

Name _____ **Date** _____

REPLACE THERMOSTAT

Upon completion of this job sheet, you should be able to replace a cooling system thermostat.

NATEF Correlation

HEATING AND AIR CONDITIONING: Heating, Ventilation, and Engine Cooling Systems Diagnosis and Repair; *Inspect, test, and replace thermostat and housing.* **(P-1)**

HEATING AND AIR CONDITIONING: Heating, Ventilation, and Engine Cooling Systems Diagnosis and Repair; *Flush system; refill system with recommended coolant; bleed system.* **(P-1)**

Tools and Materials

Late-model vehicle

Service manual

Safety glasses or goggles

Gasket scraper

Hand tools, as required

Thermostat and gasket, as required

Describe the vehicle being worked on.

Year _____ Make _____ Model _____

VIN _____ Engine type and size _____

Procedure

Follow the procedures outlined in the service manual. Give a brief description of your procedure, following each step. Ensure that engine is cold and that you wear eye protection.

1. Drain coolant to a level below the thermostat. Follow the appropriate procedures outlined in Job Sheet 7.
2. Is thermostat located near the upper or lower radiator hose? _____

   _____
3. Is this the outlet or inlet for the engine? _____

   _____
4. If required, remove components to gain access to thermostat housing. List components removed.
5. Remove retaining bolts or nuts and lift off thermostat housing.

6. Remove the thermostat. Using a gasket scraper, carefully remove any remaining seal or gasket material.

7. Install new thermostat and gasket. CAUTION: Make sure that the thermostat is not installed upside down. Have your instructor initial here at this time.

8. Replace the thermostat housing and retaining bolts or nuts.
   a. Torque specification _____ Procedures _____

   _____

9. Replace the components removed in step 2, if any.

10. Replace the coolant removed in step 1.

11. Bleed air out of the cooling system if required by the manufacturer.

12. Test for leaks.

Instructor's Response _____

_____

_____

Name _____ **Date** _____

TROUBLESHOOT AN ELECTRIC ENGINE COOLING FAN

Upon completion of this job sheet, you should be able to troubleshoot and repair electric engine cooling fans.

NATEF Correlation

HEATING AND AIR CONDITIONING: Heating, Ventilation, and Engine Cooling Systems Diagnosis and Repair; *Inspect and test electrical fan control system and circuits.* **(P-1)**

Tools and Materials

Late-model vehicle

Service manual

Fused jumper wire

Hand tools

Describe the vehicle being worked on.

Year _____ Make _____ Model _____

VIN _____ Engine type and size _____

Procedure

Refer to the appropriate service manual for the specific procedures for troubleshooting an electric engine cooling/condenser fan motor. Failure to do so can result in serious damage to the control system. The following may be used as a guide only. After each step write a brief report of your procedure.

1. Start the engine and allow the coolant to reach operating temperature. Did the fan turn on? _____ Explain _____

2. Turn on the air conditioner. Did the fan turn on? _____ Explain _____

   _____

   _____

3. Turn the air conditioner OFF and stop the engine. If the fan was running, does it continue to run? _____ Explain _____

   _____

   _____

4. Carefully disconnect the fan electrical connection. Is this a single-or double-lead connector?

   _____

   _____

5. Refer to the wiring diagram. What is the color of the wire that supplies power to the motor? _____ What is the color of the ground wire? _____

6. Using a fused jumper wire, connect between the fan motor lead and positive battery terminal. DO NOT connect to the motor ground lead. Does the motor operate? _____

   Did the fuse "blow?" _____ Explain: _____

   _____

   _____

7. If the motor ran in step 6, further testing of the electrical circuit is required. Follow the specific procedure outlined in the service manual. Give your step-by-step procedure and conclusions in the space below.

   a. _____

   _____

   b. _____

   _____

   c. _____

   _____

   d. _____

   _____

Instructor's Response _____

_____

_____

Name _____ **Date** _____

INSPECT ENGINE COOLING SYSTEM HOSES

Upon completion of this job sheet, you should be able to visually check the condition of cooling system hoses and perform the necessary action.

NATEF Correlation

HEATING AND AIR CONDITIONING: Heating, Ventilation, and Engine Cooling Systems Diagnosis and Repair; *Inspect engine cooling and heater system hoses and belts; perform necessary action.* **(P-1)**

Tools and Materials

Late-model vehicle

Service manual or information system

Safety glasses or goggles

Hand tools, as required

Hoses, as required

Describe the vehicle being worked on.

Year _____ Make _____ Model _____

VIN _____ Engine type and size _____

Procedure

Follow the procedures outlined in the service manual. Give a brief description of your procedure, following each step. Ensure that the engine is cold and that you wear eye protection.

1. Inspect the hoses as outlined on page 99 to determine which hoses require replacement. What were the results of the inspections? _____

   _____

2. Drain the coolant as outlined in Job Sheet 7 to a level below hose connections.

3. If required, remove components to gain access to the hoses and clamps. What components needed to be removed? _____

   _____

4. Loosen the hose clamps at both ends of the hose. Hoses with ferrules on the end may be removed by cutting the ferrules off and using standard hose clamps on replacement hoses. Was it necessary to remove ferrules from any nose ends? _____

   _____

5. Firmly but carefully twist and turn the hose to break it loose from the connection point. Use a box cutter to slice through the hose to facilitate its removal.

6. Remove the hose and save for customer inspection if requested (remember, some states require that old parts be saved for customer inspection).
Did customer request old parts? _____

_____

7. Compare the old hose to the new hose for proper fit and shape. Some hoses will need to be trimmed to size. Was it necessary to trim hose length? _____

_____

8. Apply a small amount of water or coolant to the ends of new hoses to aid in installation.

9. Install the hose clamps. If screw clamps are used, recheck the tension after the cooling system has been allowed to thermal cycle.

10. Refill the cooling system as outlined in Job Sheet 7.

11. Test the cooling system for leaks as outlined in Job Sheet 8. Were any leaks detected?

_____

Instructor's Response _____

_____

_____

Name _____ **Date** _____

INSPECT HEATER CONTROL VALVE

Upon completion of this job sheet, you should be able to inspect and test the heater control valve(s) and perform the necessary action.

NATEF Correlation

HEATING AND AIR CONDITIONING: Heating, Ventilation, and Engine Cooling Systems Diagnosis and Repair; *Inspect and test heater control valve(s); perform necessary action.* **(P-2)**

Tools and Materials

Late-model vehicle

Service manual or information system

Safety glasses or goggles

Hand tools, as required

Describe the vehicle being worked on.

Year _____ Make _____ Model _____

VIN _____ Engine type and size _____

Procedure

Follow the procedures outlined in the service manual. Give a brief description of your procedure, following each step. Ensure that the engine is cold and that you wear eye protection.

1. Check the operation of the heater control valve assembly. Proceed to step 2 if replacement is necessary. Is heater control value functioning as designed? _____

   _____

2. Drain the coolant as outlined in Job Sheet 7 to a level below the hose connections.
3. Remove the cable linkage, vacuum hose(s), or electrical connection from the control valve. How was heater control valve controlled? _____

   _____

4. Loosen the hose clamps and remove the inlet and outlet hoses from the control valve.
5. Inspect the hose ends prior to reinstalling on a new control valve. If the hose ends are hard or split, replace the hoses. Was it necessary to replace hoses? _____

   _____

6. Remove the heater control valve and save it for customer inspection, if requested (remember, some states require that old parts be saved for customer inspection).

7. Compare the old heater control valve to the new assembly for proper fit and shape and install the control valve assembly. Do the old and new assembly's match one another?

_____

_____

8. Apply a small amount of water or coolant to the ends of the new hoses to aid in installation.

9. Install the hose clamps. If screw clamps are used, recheck the tension after the cooling system has been allowed to thermal cycle.

10. Reinstall the cable linkage, vacuum hose(s), or electrical connection from the control valve.

11. Refill the cooling system as outlined in Job Sheet 7.

12. Test the cooling system for leaks as outlined in Job Sheet 8.
Were any leaks detected? _____

_____

Instructor's Response _____

_____

_____

Name _____ **Date** _____

INSPECT AND TEST FAN AND FAN CLUTCH

Upon completion of this job sheet, you should be able to inspect and test the fan clutch assembly and perform the necessary action.

NATEF Correlation

HEATING AND AIR CONDITIONING: Heating, Ventilation, and Engine Cooling Systems Diagnosis and Repair; *Inspect and test fan, fan clutch (electrical and mechanical), fan shroud, and air dams; perform necessary action.* **(P-1)**

Tools and Materials

Late-model vehicle

Service manual or information system

Safety glasses or goggles

Hand tools, as required

Thermometer

Timing light

Cardboard

Describe the vehicle being worked on.

Year _____ Make _____ Model _____

VIN _____ Engine type and size _____

Procedure

Follow the procedures outlined in the service manual. Give a brief description of your procedure, following each step. Ensure that the engine is cold and that you wear eye protection.

1. Inspect the fan shroud and air dams. Ensure that they are firmly mounted to the radiator support of the vehicle. Were any faults or failures detected?

   _____

   _____

2. Inspect the physical condition of the fan blades and the condition of the drive pulley for the fan. Note any faults detected.

   _____

   _____

3. If the vehicle is equipped with a fan clutch, inspect the assembly for lateral movement. Maximum allowable movement is ¼ in. as measured at the tip of the fan blades. Also inspect for signs of oil leaking from the clutch assembly. Note any faults or defects.

_____

_____

4. For the fan clutch assembly diagnosis, begin by spinning the fan blade by hand. If the fan blade revolves more than twice when spun by hand, replace the fan clutch. The fan clutch should have a slight drag when spun by hand. Note any faults or failures.

_____

_____

5. Place a thermometer between the fan and the radiator. It may be necessary to drill a small hole in the fan shroud for the thermometer placement.

6. Position a square of cardboard in front of the radiator to limit airflow and to raise the engine temperature.

7. Start the engine and turn on the air conditioning; raise the engine speed to 2,000 rpm.

8. Record the temperature when the fan clutch engages.
   a. Engagement temperature _____

   A 5–10°F (3–10°C) drop in air temperature should be noticed when the fan clutch engages, along with an increase in fan noise.
   If the fan clutch does not engage by 150–190°F (65–90°C), replace the assembly. Note any faults or failures.

   _____

   _____

9. Once the fan clutch has engaged, remove the cardboard blocking the radiator and reduce the engine speed to 1,500 rpm. There may be an increase in fan speed detected. Was there an increase in fan speed.

_____

_____

10. After several minutes of operation, the fan should disengage if it is operating properly. Is the fan clutch operating as designed?

_____

_____

11. If any of the above tests fail, replace the fan clutch assembly.
   Did the fan clutch fail any of the above tests? Which ones, if any?

_____

_____

Instructor's Response _____

_____

_____

Name _____ **Date** _____

DIAGNOSTIC CHECK LIST

Upon completion of this job sheet you should be able to visually check the condition of cooling system and properly fill out a cooling system diagnostic check list.

NATEF Correlation

P-1: V.C.3; HEATING AND AIR CONDITIONING: Heating, Ventilation, and Engine Cooling Systems Diagnosis and Repair; *Inspect engine cooling and heater system hoses and belts; perform necessary action.*

Tools and Materials

Late-model vehicle

Service manual or information system

Safety glasses or goggles

Hand tools, as required

Describe the vehicle being worked on.

Year _____ Make _____ Model

VIN _____ Engine type and size _____

Procedure

Follow procedures outlined in the service manual. Give a brief description of your procedure following each step. Ensure that the engine is cold and wear eye protection.

DIAGNOSTIC CHECK LIST

Inspection	OK	Marginal	Faulty
Coolant Level			
Coolant Color			
Coolant Freeze Point (–35°F)			
Radiator Cap (–1)			
No Sign of Coolant in Oil			
Radiator Condition (1)			
Expansion Tank			
Drive Belt(s)			
Water Pump (1)			
Hoses (1)			
Freeze Plugs (1)			
Head Gaskets (1)			
Other Leaks (1)			

Inspection	OK	Marginal	Faulty
Thermostat Opens at Set Temperature			
Upper Radiator Hose Hot			
Lower Radiator Hose Warm			
Fan Engages			
Heater Hoses Hot			
Passenger Compartment Heater Functions			
Radiator Flow (2)			
No sign of coolant from exhaust (steam)			
Comments:			

1. Test performed using pressure tester
2. To fully test radiator flow it should be removed from vehicle.

Instructor's Response _____

_____

_____

Name _____ Date _____

BENCH TEST THERMOSTAT

Upon completion of this job sheet you should be able to visually check the condition of and bench test a thermostat operation and diagnose outcomes of a faulty thermostat.

Tools and Materials

Hot plate

Thermometer

Mechanics wire

Glass beaker

Stirring rod or spoon

Safety glasses

Procedure

1. Record temperature rating marked on the bottom of thermostat.

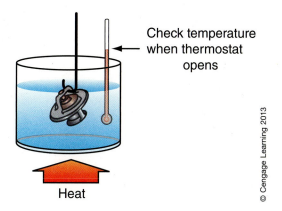

Check temperature when thermostat opens

Heat

© Cengage Learning 2013

⚠ **CAUTION: Do not rest the thermostat or the thermometer on the bottom of the container while heating the water. Contact with the bottom of the container causes the thermostat and the thermometer to be at a higher temperature than the solution.**

2. Place thermostat suspended by a wire in a beaker of warm water and place on hot plate with thermometer in water. Agitate the solution with a stirring rod or spoon in order to maintain a uniform temperature of the solution, the thermostat, and the thermometer.

3. Record temperature at which thermostat begins to open.
   The thermostat should begin to open within +/−3°F

4. Does this temperature match the rating on thermostat?
   1. Yes
   2. No
5. Record temperature when thermostat is fully open.
   The thermostat valve should start to open at the rated temperature. The thermostat should be fully open after the temperature has increased 25°F (11°C) above its rated temperature.
6. Is this a good thermostat? _____

7. If the thermostat opens too soon, what effect would this have on the cooling system? What would the customer complaint be? _____

_____

8. If the thermostat opens too late, what effect would this have on the cooling system? What would the customer complaint be? _____

_____

Instructor's Response _____

_____

_____

Chapter 4

THE MANIFOLD AND GAUGE SET

BASIC TOOLS
Manifold and gauge set with hoses
Safety glasses or goggles
Fender cover
Basic tool set
Thermometer

UPON COMPLETION AND REVIEW OF THIS CHAPTER, YOU SHOULD BE ABLE TO:

■ Describe the nomenclature and function of the manifold and gauge set.

■ Identify the scaling of the low- and high-side gauges in English and metric units.

■ Calibrate a gauge.

■ Connect a manifold and gauge set into an automotive air-conditioning system.

■ Hold a performance test on an automotive air-conditioning system.

THE MANIFOLD AND GAUGE SET

The **manifold and gauge set** is to the automotive air-conditioning service technician what the stethoscope and aneroid manometer is to the physician. Both are tools that are essential for the diagnosis of internal conditions that cannot otherwise be observed.

The manifold and gauge set is used to diagnose and troubleshoot various system malfunctions based on low- and high-side system pressures.

A basic tool for the air-conditioning service technician is the manifold and gauge set (Figure 4-1). The pressure measurement of an air-conditioning system is a means of determining system performance. The manifold and gauge set is an essential tool for making these measurements. The servicing of automotive air-conditioning systems requires the use of a two-gauge manifold set.

One gauge is used to observe pressure on the low (suction) side of the system. The second gauge is used to observe pressure on the high (discharge) side of the system. The service technician should have two gauge sets—one for **HFC-134a** and one for **CFC-12**. CFC-12 refrigerants are referred to as R-12 or Freon by service technicians, whereas HFC-134a refrigerants are referred to as R-134a. There is no significant difference between gauges designated for HFC-134a refrigerant and those designated for CFC-12 refrigerant. However, the two refrigerants are not compatible, and two sets should be used. The general description, however, is the same for both types.

> The **manifold and gauge set** is one of the primary service tools for the air-conditioning technician. It contains the gauges and service valve block along with high-side, low-side, and utility hoses and is specific for only one type of refrigerant.

> Manifold and gauge sets must be dedicated: one for R-12 (CFC-12) and one for R-134a (HFC-134a).

Low-Side Gauge

The **low-side gauge** (Figure 4-2), used to monitor low-side system pressure, is called a **compound gauge**. A compound gauge is designed to give both vacuum and pressure indications. This gauge is connected to the low side of the air-conditioning system through the manifold and low-side hose.

Calibration of the vacuum scale of a compound gauge is from 30 in. Hg to 0 psig. The pressure scale is calibrated to indicate pressures from 0 psig to 120 psig. The compound

> **Low-side gauge** is the left-side refrigerant gauge on the manifold used to read pressures on the low side of the system.

HFC-134a is hydrofluorocarbon refrigerant gas used as a refrigerant and is also known as R-134a. This refrigerant is not harmful to the ozone. It replaced CFC-12 (R-12) in the early to mid-1990s and is the only substitute refrigerant recognized by the major automotive manufacturers.

CFC-12 is dichloro-difluoromethane refrigerant gas (also referred to as R-12) used in automotive air conditioners until the early 1990s. The passage of the Clean Air Act in 1990 banned the production of CFC-12 and limited its further use.

A **compound gauge** registers both pressure and vacuum. The low-side refrigerant gauge is a compound gauge.

Low-side system operating pressures are generally 15–35 psig (103–241 kPa).

High-side system operating pressures are generally 160–220 psig (1,103–1,517 kPa).

FIGURE 4-1 A typical R-134a (HFC-134a) manifold.

© Cengage Learning 2013

FIGURE 4-2 A typical low-side gauge.

© Cengage Learning 2013

gauge is constructed in such a manner so as to prevent any damage to the gauge if the pressure should reach a value as high as 250–350 psig.

Low-side pressures above 80 psig are rarely experienced in an operating system. Such pressures may be noted, however, if the low-side manifold service hose is accidentally connected to the high-side fitting. Even experienced service technicians are known to make this type of error on R-12 systems.

FIGURE 4-3 A typical high-side gauge.

The metric gauge used on the low side of the system is scaled in absolute kiloPascal (kPa) units. For conversion, 1 psi is equivalent to 6.895 kPa.

High-Side Gauge

The high-side gauge (Figure 4-3) indicates pressure in the high-pressure side of the system. Under normal conditions, pressures in the high side seldom exceed 300 psig. As a safety factor, however it is recommended that the maximum indication of the high-side gauge be 500 psig. The high-side gauge, though not calibrated below 0 psig, is not damaged when pulled into a vacuum.

High-side metric pressure gauges are scaled in kPa.

The correct conversion is to kPa, whereby 1 psi equals 6.895 kPa. Atmospheric pressure at sea level (14.696 psia) is 101.32892 kPa, rounded off to 101 kPa.

$$14.696 \text{ psi} \times 6.895 \text{ kPa} = 101.32892$$

Gauge Calibration

Most quality gauges have a provision for a calibration adjustment. Generally, a gauge is accurate to about 2 percent of its total scale when calibrated so the needle rests on zero with atmospheric pressure applied.

To calibrate a gauge:

1. Remove the hose after recovering refrigerant (if applicable).
2. Remove the retaining ring (bezel) and plastic lens cover.
3. Locate the adjusting screw.
4. Use a small screwdriver (Figure 4-4) to turn the adjusting screw in either direction until the pointer is lined up with the zero mark.
5. Replace the plastic lens and bezel.

CAUTION:
Do not force the adjusting screw; to do so may damage the gauge or alter its accuracy.

MANIFOLD

The low- and high-side gauges are connected into the air-conditioning system through a manifold (Figure 4-5) and two high-pressure hoses. The manifold is a cast or machined block with two or more gauges attached and two to four shut-off valves incorporated

FIGURE 4-4 Use a small screwdriver to calibrate a gauge.

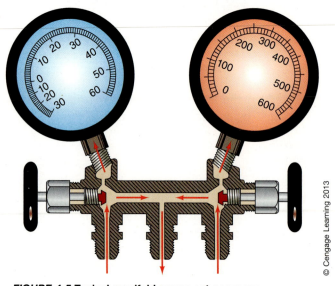

FIGURE 4-5 Typical manifold gauge set passages.

into it. Some more expensive manifolds also contain a visible sight glass, which allows the service technician to see the flow of liquid refrigerant during charging.

The manifold assembly has unique fittings for different refrigerants, and the hoses are not interchangeable between refrigerants of different types. This means that you will need a separate manifold and gauge assembly for each of the different refrigerants (R-134a and R-12) you service. The Society of Automotive Engineers (SAE) has set standards for automotive refrigerant service equipment. Standard J2197 states that all hose connections on a service manifold for R-134a refrigerant must be a ½ in. male ACME fitting and for R-12 refrigerants must be a $7/16$ in. × 20 thread pitch male fitting.

Most manifold sets contain two hand valves to provide flow control through the manifold, one for the low-side service hose and one for the high-side service hose, as was shown in Figure 4-2 and Figure 4-3. Turning the hand valve clockwise shuts off the common passage in the gauge block, and the gauge will read system pressure. Turning the hand valve counterclockwise opens the common passage for charging or recovery/evacuation. As was noted in Chapter 2, if both valves are closed, both high-side and low-side pressures will be read isolated from one another. However, if either the low- or high-side valve is open, that line pressure gauge circuit will be open to the center service hose port on the manifold set. If the center line is not connected to a charging or recovery station, refrigerant will be vented to the atmosphere. If the center service port hose is connected to a charge/recovery unit and both hand valves are open, both gauges will read the same pressure due to the common center passage in the gauge block giving inaccurate pressure readings.

Some manifolds have a third valve for connecting a vacuum service or refrigerant service hose through the manifold, yet other manifolds may have four valves with:

1. Low-side service hose
2. High-side service hose
3. Vacuum pump service hose
4. Refrigerant service hose

Hoses

Specific requirements for service hoses used in automotive air-conditioning service are given in the SAE standards J2196 and J2197. A refrigeration service hose (Figure 4-6) is constructed to withstand maximum working pressures of 500 psi (3,448 kPa). Some service hoses have a minimum burst pressure rating of up to 2,500 psi (17,238 kPa). The SAE has established different hose specifications for R-134a and R-12 refrigerants. This is due, in part, to the fact that R-134a has smaller molecules than R-12 refrigerants and tends to more readily leak through hoses and seals. Current service hoses have an impervious barrier to reduce the possibility of refrigerant leakage and are referred to as **barrier hoses**. The following is a brief overview of those standards as they apply to refrigerants R-134a and R-12. It must be noted that there are no less than nine other alternative refrigerants that have been approved for automotive use by the Environmental Protection Agency (EPA). Each of these alternate refrigerants requires its own unique fittings. The refrigerant manufacturer will supply any information about this requirement on request. The SAE standard J2197 specifies that R-134a service hoses have a ½ in.-16 ACME thread for connecting to manifold gauge sets or equipment. The service end can connect directly to a quick coupler (Figure 4-7), which has no external threads and a one-way check valve to avoid refrigerant being purged to the atmosphere when disconnected and connects to the vehicle service fitting. If necessary, a M14X1.5 fitting can be used between the hose and quick coupler adapter. The R-12 service hoses have ⁷/₁₆ in.-20 female refrigerant flare nuts on both ends of all service hoses. The unique service fittings are designed to prevent accidental mixing of refrigerants.

Barrier hoses have an impervious lining to prevent refrigerant leaking through the walls of the hose. Air-conditioning systems in vehicles have had barrier hoses since 1988.

FIGURE 4-6 A typical service hose.

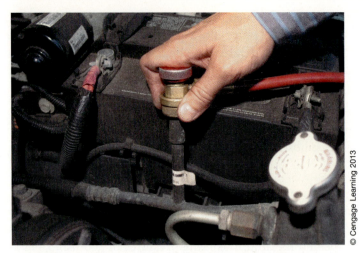

FIGURE 4-7 Quick coupler for R-134a manifold gauge set.

High-Side Service Hose. The high-side service hose is connected between the system high side and the manifold gauge set or service equipment. Hoses that are designed for HFC-134a will be labeled SAE J2196/R-134a and are solid red with a black stripe. The high-side service end has a quick disconnect coupler fitting with a 16 mm outside diameter (OD) as stipulated by SAE standard J639. Hoses that are designed for R-12 service are marked SAE J2196 and are available in either solid red or black with a red stripe. A shut-off valve must be placed within 1 ft. or 30 cm of the end connected to the system (Figure 4-8). This valve is intended to help reduce the amount of refrigerant that is vented to the atmosphere during a service procedure and must be on manifold and gauge sets as well as on recovery/recycling station high-/low-side service lines.

Low-Side Service Hose. The low-side service hose is connected between the system low side and the manifold gauge set or service equipment. Hoses that are designed for HFC-134a will be labeled SAE J2196/R-134a and are solid blue with a black stripe. The low-side service end has a quick disconnect coupler fitting with a 13 mm OD as stipulated by SAE standard J639. Hoses that are designed for R-12 service are marked SAE J2196 and are available in either solid blue or black with a blue stripe. A shut-off valve must be placed within 1 ft. or 30 cm of the end connected to the system on manifold and gauge sets as well as on recovery/recycling station high-/low-side service lines.

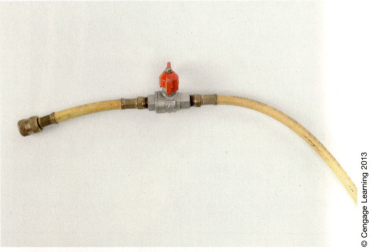

FIGURE 4-8 There must be a shut-off valve within 12 in. (30.5 cm) of the service hose end.

Utility Service Hose. The utility service hose connects to the center port on the manifold set and is used to connect to external service equipment such as recovery/charging stations, vacuum pumps, or disposable refrigerant tanks. Utility hoses that are designed for HFC-134a will be labeled SAE J2196/R-134a and are solid yellow with a black stripe. Utility hoses that are designed for R-12 service are marked SAE J2196, are available in either solid yellow or white or black with a yellow or white stripe, and have $^7/_{16}$ in.-20 female refrigerant flare nuts on both ends. A shut-off valve must be placed within 1 ft. or 30 cm of the end connected to service equipment for both R-134a and R-12 refrigerants.

CONNECTING THE MANIFOLD AND GAUGE SET

This procedure is used when connecting the manifold and gauge set on the automotive air-conditioning system to perform any one of the many operational tests and service procedures.

The transition from an ozone-depleting refrigerant (CFC-12) to an ozone-friendly refrigerant (HFC-134a) took place in some car lines during the 1993 model year. In other car lines the transition took place during the 1994 model year.

By the 1995 model year nearly all vehicles were equipped with the new refrigerant. Before proceeding with the installation of the manifold and gauge set, it is important to identify which type of refrigerant is in the air-conditioning system.

Manifold and gauge sets for R-12 and R-134a are not interchangeable. The two refrigerants and their oils are not compatible. Mixing refrigerants, oils, or components could result in serious damage to the air-conditioning system and poor system performance.

 WARNING: It is important to use extreme care and observe all service and safety precautions when working with refrigerants and refrigeration oil.

Procedure

This procedure is given in three parts that follow. Part I outlines the procedure used when connecting a manifold and gauge set into an R-134a system and is also shown in Photo Sequence 3. Part II outlines the procedure used when connecting a manifold and gauge set into an R-12 system equipped with Schrader-type service valves. Part III outlines the procedures to follow when connecting a manifold and gauge set into an R-12 system equipped with hand shut-off service valves located on or near compressor assemblies.

 WARNING: Safety glasses must be worn at all times while working with refrigerants. Remember, liquid refrigerant sprayed in the eyes can cause blindness.

 WARNING: The EPA requires positive shut-off provisions within 12 in. (30.5 cm) of the service end of each hose. There are various methods of accomplishing this; some are manual and some are semiautomatic.

Part I—HFC-134a System

1. Remove the protective caps from the high- and low-side service ports (Figure 4-9).

 WARNING: Remove the caps slowly to ensure that no refrigerant escapes past a defective service valve.

The utility hose is often referred to as the service hose.

SERVICE TIP: The quick disconnect coupler fittings for R-134a high-side and low-side hoses come in various lengths to allow access into tight locations. So if the couplings that came on your manifold set are too long, you can order a compact coupling. As a rule I always use the compact design on my gauge sets.

CAUTION: Place a fender cover on the vehicle to avoid damage to the finish.

FIGURE 4-9 Remove the protective caps slowly to ensure no refrigerant loss.

NOTE: If a leak is found in the service valve, it should be repaired or replaced.

2. Turn the low-side hose hand valve fully counterclockwise to retract the Schrader depressor, before attaching to the system.
3. Connect the low-side hose (Figure 4-10).
   a. Press a quick-joint-type hose fitting onto a matching system fitting.
   b. Push firmly until a clicking sound is heard, ensuring that it is locked in place.
   c. Turn the hand valve clockwise to extend the depressor.
4. Repeat steps 2 and 3 with the high-side hose.

Part II—CFC-12 System with Schrader-Type Valves

1. Remove the protective caps from the high- and low-side service ports.

 WARNING: Remove the caps slowly to ensure that no refrigerant is leaking past a defective Schrader valve.

NOTE: If a leak is found in the service valve, it should be repaired or replaced.

CAUTION:
Place a fender cover on the vehicle to avoid damage to the finish.

FIGURE 4-10 Connecting the hose fitting to an R-134a system.

TYPICAL PROCEDURE FOR CONNECTING A MANIFOLD AND GAUGE SET TO AN R-134A AIR-CONDITIONING SYSTEM

P3-1 Typical location of the low-side service valve on an R-134a air-conditioning system.

P3-2 Typical location of the high-side access fittings on an R-134a air-conditioning system.

P3-3 Ensure that the manifold and gauge set with hoses comply with SAEJ2211.

P3-4 Ensure that the manifold low-side hand valve is closed—turned fully clockwise.

P3-5 Ensure that the manifold high-side hand valve is closed—turned fully clockwise.

P3-6 Remove the protective cap from the low-side service valve fitting. Repeat this procedure with the high-side service valve fitting.

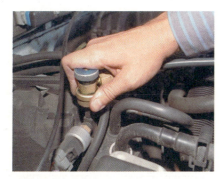

P3-7 Connect the low-side service hose quick connect adapter to the low-side service valve fitting. Repeat this procedure with the high-side service valve fitting.

P3-8 The manifold and gauge set is ready to be used for servicing the air-conditioning system.

FIGURE 4-11 A typical CFC-12 high-side access fitting.

Flexible adapter

45° adapter   90° adapter   Straight adapter

FIGURE 4-12 Adapters used to connect a high-side hose into an R-12 system.

2. Make sure that the manifold hand shut-off valves are closed before the next step.
3. Connect the low-side manifold hose, finger tight, to the suction side of the system.
4. Connect the high-side manifold hose, finger tight, to the discharge side of the system.

Part III—CFC-12 System with Hand Valves

The high-side and low-side compressor-mounted hand-operated manual service valves were used on some early air-conditioning systems—both factory and aftermarket installed—and had a cast iron or cast aluminum compressor manufactured by Tecumseh or York.

Though not used for years for automotive service, they are found in many over-the-road applications, such as Diamond Reo, Kenworth, Mack, and Peterbilt. They are also found in off-road applications such as Allis Chalmers, Caterpillar, International Harvester, and John Deere. The Tecumseh and York compressors, equipped with hand

FIGURE 4-13 **Remove the protective caps from the service valves.**

FIGURE 4-14 **Remove the protective caps from the service ports.**

shut-off service valves, are available new or rebuilt for CFC-12 or HFC-134a refrigerants having a lubricant charge of mineral oil, ester, or poly-alkaline glycol (PAG), as required.

1. Remove the protective caps from the service valve stems (Figure 4-13), if equipped.
2. Remove the protective caps from the service ports (Figure 4-14).

 WARNING: **Remove the caps slowly to ensure that no refrigerant is leaking past the service valve.**

> **NOTE:** If a leak is found in the service valve stem, it should be repaired or replaced. If there is a leak at the service port, the service valve should be replaced.

3. Connect the low-side manifold hose, finger tight, to the suction side of the system.
4. Connect the high-side manifold hose, finger tight, to the high side of the system.
5. Using a service valve wrench (Figure 4-15), rotate the suction-side service valve stem two or three turns clockwise.
6. Repeat step 5 with the discharge service valve stem.

SERVICE TIP:
Make certain that the hand shut-off valves are closed on the manifold set before the next step.

SPECIAL TOOL
Service valve wrench

Classroom Manual
Chapter 4, page 127

FIGURE 4-15 Use a service wrench to turn the service valve stem.

BASIC PERFORMANCE TESTING THE AIR-CONDITIONING SYSTEM

The following procedure serves as a guide for the service procedures required for basic **performance testing** of air-conditioning systems. Greater performance testing details are discussed in Chapter 6 of the Classroom Manual; this is meant only as an introduction to become familiar with normal system pressures and temperatures and is not meant to be a diagnostic test procedure. The service technician must always refer to the manufacturer's service information for specific data for any particular vehicle model.

> **CUSTOMER CARE:** When you are performing repairs or inspection services on a customer's vehicle, be sure to install disposable paper floor mats and plastic seat covers. The best repair in the world can go unrewarded if you leave a customer's car dirty.

Preparing and Stabilizing the System

1. Ensure that both manifold hand valves are in the closed position to prevent refrigerant venting.
2. Connect the manifold and gauge set into the system.
3. Start the engine; set the speed to about 1,500–1,700 rpm.
4. Place a fan in front of the radiator to assist the ram air flow (Figure 4-16).
5. Turn on the air conditioner; set all controls to maximum cooling or Recirculation mode; set the blower speed on HI.
6. Insert a thermometer in the air-conditioning duct as close as possible to the evaporator core (Figure 4-17).
7. Note the relative humidity levels and refer to Figure 4-19 to determine the approximate duct temperature. The atmospheric relative humidity has a dramatic effect on the effectiveness of the air-conditioning system.

Moisture in the air is known as relative humidity (RH).

Performance testing checks temperature and pressure readings under controlled operating conditions to determine if an air-conditioning system is operating at full efficiency.

SPECIAL TOOL

Large floor fan

FIGURE 4-16 A fan is placed in front of the vehicle to provide additional air while performance testing.

FIGURE 4-17 Need to be able to see low temperature scale of 0–100°F.

An evaporating pressure of 30 psig (207 kPa) corresponds to a temperature of 34.5°F (1.4°C) for R-134a and 32°F (0°C) for R-12.

A condensing pressure of 190 psig corresponds to a temperature of 127° F (52.8°C) for R-134a and 134°F (33.9°C) for R-12.

SERVICE TIP: When removing the plastic service caps from the R-134a refrigerant lines, inspect the condition of the O-rings inside the caps. These O-rings function as seals for the system and can be the source of refrigerant leaks.

Visual Check of the Air Conditioner

1. The average low-pressure gauge reading should be in the range of 20–30 psig (239–310 kPa absolute).

    NOTE: The term *average* must be considered: for example, 15–25 psig (103–172 kPa) "averages" 20 psig (239 kPa); 25–35 psig (172–241 kPa) "averages" 30 psig.

2. The high-side gauge should be within the specified range of 160–220 psig (1,103–1,517 kPa) depending on the ambient temperature and humidity.

3. The discharge air temperature should be within the specified range of 40–50°F (4.4–10°C).

Inspect the High and Low Sides for Even Temperatures

1. Feel the hoses and components in the high side of the system to determine if the components are evenly heated.

Air-conditioning temperature control is by thermostat or low-pressure control.

 WARNING: Certain system malfunctions cause the high-side components to become superheated to the point that a serious burn can result if care is not taken when handling these components.

2. Note the inlet and outlet temperatures of the receiver/drier assembly. A change in the temperature is an indication of a clogged or defective receiver/drier.
3. All lines and components on the high side should be warm to the touch (see the Warning following step 1).
4. All lines and components on the low side of the system should be cool to the touch.
5. Note the condition of the thermostatic expansion valve (TXV) or fixed orifice tube (FOT). The high pressure inlet should be warmer than the low pressure outlet.

Test the Thermostats and Control Devices

1. Refer to the service manual for the performance testing of the particular type of control device used.
2. Determine that the thermostat or low-pressure switch engages and disengages the clutch. There should be about a 12°F (6.7°C) temperature rise between the cut-out (off) and cut-in (on) point. Many compressors today are of the variable displacement design and do not cycle the compressor clutch, but instead vary the amount of refrigerant flow in the system to regulate evaporator temperature. This is discussed in greater detail in Chapter 8. Guides for determining the proper gauge readings and temperatures are shown in Figure 4-18. The RH at any particular temperature is a factor in the quality of the air (Figure 4-19A and Figure 4-19B). These figures should be regarded as guides only.
3. Complete performance testing. Refer to the manufacturer's service manual for specific requirements.

Ambient Air Temperature, °F	70	80	90	100	110
Average Compressor Head Pressure, psig	150–190	170–220	190–250	220–300	270–370
Average Evaporator Temperature, °F	38–45	39–47	40–50	42–55	45–60

(A) English

Ambient Air Temperature, °C	21	27	32	38	43
Average Compressor Head Pressure, kPa	1,034–1,310	1,172–1,517	1,310–1,724	1,517–2,069	1,862–2,551
Average Evaporator Temperature, °C	3.3–7.2	3.9–8.3	4.4–10	5.5–12.8	7.2–15.6

(B) Metric

FIGURE 4-18 Head pressure performance charts: (A) English; (B) metric.

Ambient Temperature, °F	70			80			90			100		
Relative Humidity, %	50	60	90	50	60	90	40	50	60	20	40	50
Discharge Air Temperature, °F	40	41	42	42	43	47	41	44	49	43	49	55

(A) English

Ambient Temperature, °C	21			27			32			38		
Relative Humidity, %	50	60	90	50	60	90	40	50	60	20	40	50
Discharge Air Temperature, °C	4.4	5	5.5	5.5	6.1	8.3	5	6.6	9.4	6.1	8.3	12.7

(B) Metric

FIGURE 4-19 Relative humidity performance charts: (A) English; (B) metric.

Return the System to Service

1. Return the engine speed to normal idle.
2. Back seat the high- and low-side compressor service valves, if equipped.
3. Close the service hose valves.
4. Remove the service hoses.
5. Replace the protective caps to prevent a loss of refrigerant in case the service valve leaks.
6. Turn off the air-conditioning controls.
7. Stop the engine.

CAUTION:
If service hoses are equipped with manual shut-off, be sure to close them before disconnecting from the system.

CASE STUDY

A customer brings a car, recently purchased from a reputable used car dealer, in for service with a repair authorization to "fix a water leak." The complaint is that water spills out of the air conditioner onto her feet when she makes a right turn. She also notes that the floor mat on the passenger side is damp most of the time.

The technician verifies the damp floor mat and suspects that the heater core is leaking. However, there is no evidence of antifreeze solution on the floor mat. Further discussion with the customer verifies that the cooling system is apparently "sound." While talking with the customer, however, the technician notices that there are no familiar drips on the shop floor usually experienced when the air conditioner is operating.

Inspection of the drain tube of the evaporator reveals that a recently applied undercoating material has sealed the opening. This causes the water to back up into the evaporator and spill out. Cleaning the drain tube solved the problem.

TERMS TO KNOW

Barrier hoses
CFC-12 (R-12)
Compound gauge
HFC-134a (R-134a)
Low-side gauge
Manifold and gauge set
Performance testing

ASE-STYLE REVIEW QUESTIONS

1. *Technician A* says that the performance test determines if system pressures are proper.
   Technician B says that the test determines if the system temperatures are proper.
   Who is correct?
   A. A only
   B. B only
   C. Both A and B
   D. Neither A nor B

2. *Technician A* says that humidity has an effect on system performance.
   Technician B says that poor airflow has an effect on system performance.
   Who is correct?
   A. A only
   B. B only
   C. Both A and B
   D. Neither A nor B

3. *Technician A* says that the blower should be run on high speed for the performance test.
   Technician B says it does not matter at what speed the fan is run.
   Who is correct?
   A. A only
   B. B only
   C. Both A and B
   D. Neither A nor B

4. *Technician A* says that a slight temperature change at the inlet and outlet of the receiver-drier indicates a restriction.
   Technician B says that a slight temperature change at the inlet and outlet of an accumulator indicates a restriction.
   Who is correct?
   A. A only
   B. B only
   C. Both A and B
   D. Neither A nor B

5. *Technician A* says that a change in temperature from the inlet to the outlet of the thermostatic expansion valve (TXV) is not acceptable.

Technician B says that a change in temperature from the inlet to the outlet of a fixed orifice tube (FOT) is acceptable. Who is correct?

A. A only C. Both A and B
B. B only D. Neither A nor B

6. *Technician A* says that an R-12 manifold and gauge set may be used on an R-134a system.

Technician B says that an R-134a manifold and gauge set may be used on an R-12 system. Who is correct?

A. A only C. Both A and B
B. B only D. Neither A nor B

7. *Technician A* says that the kPa scale is used to denote pressure on the metric gauges.

Technician B says that the Hg/psig scale is used to denote pressure on the English scale.

Who is correct?

A. A only C. Both A and B
B. B only D. Neither A nor B

8. *Technician A* says that the maximum working pressure of service hoses should be 500 psig (3,448 kPa).

Technician B says that 500 psig (3,448 kPa) is also the burst pressure.

Who is correct?

A. A only C. Both A and B
B. B only D. Neither A nor B

9. *Technician A* says that the low-side gauge may be used on the high side, if necessary.

Technician B says that the high-side gauge may be used on the low side, if necessary.

Who is correct?

A. A only C. Both A and B
B. B only D. Neither A nor B

10. *Technician A* says both hand valves must be closed to read service pressures.

Technician B says there is a common passage on a manifold and gauge set.

Who is correct?

A. A only C. Both A and B
B. B only D. Neither A nor B

ASE CHALLENGE QUESTIONS

1. All of the following statements about a Schrader-type service valve are correct, *except:*
A. The valve may be front seated.
B. The valve may be back seated.
C. The valve may be midpositioned.
D. The valve may be opened.

2. The service valve adapter shown right is turned in the _____ direction to allow access to the air-conditioning system.
A. Clockwise (cw) C. Either A or B
B. Counterclockwise (ccw) D. Neither A nor B

© Cengage Learning 2013

3. *Technician A* says that a manifold and gauge set must have service shut-off valves within 1 ft. (30 cm) of the hose service end.

   Technician B says that a recovery/recycling station must have service shut-off valves within 1 ft. (30 cm) of the hose service end.

   Who is correct?

   A. A only
   B. B only
   C. Both A and B
   D. Neither A nor B

4. Each type of refrigerant uses a unique set of service fittings that:

   A. Are designed to prevent accidental loss of refrigerant.
   B. Are designed to prevent accidental mixing of refrigerants.
   C. Are more cost effective for manufacturers.
   D. Are easy to hook up to any system.

5. *Technician A* says that a manifold and gauge set is one of the primary service tools for the air-conditioning service technician.

   Technician B says that a manifold and gauge set must be dedicated to the type of refrigerant being serviced.

   Who is correct?

   A. A only
   B. B only
   C. Both A and B
   D. Neither A nor B

Name _____ Date _____

INTERPRETING GAUGE PRESSURE

Upon completion of this job sheet, you should be able to troubleshoot an air-conditioning system based on gauge pressure readings.

NATEF Correlation

HEATING AND AIR CONDITIONING: A/C System Diagnosis and Repair; *Identify and interpret heating and air-conditioning concern; determine necessary action.* **(P-1)**

HEATING AND AIR CONDITIONING A/C System Diagnosis and Repair; *Identify refrigerant type; conduct a performance test of the A/C system; determine necessary action.* **(P-1)**

Tools and Materials

None required

Procedure

The gauge pressure reading is given. Identify the problem and suggest the remedy.

High-side gauge reading too high:

PROBLEM	REMEDY
1. _____	_____
2. _____	_____
3. _____	_____

High-side gauge reading is too low:

PROBLEM	REMEDY
4. _____	_____
5. _____	_____

Low-side gauge reading is too high:

PROBLEM	REMEDY
6. _____	_____
7. _____	_____
8. _____	_____

Low-side gauge reading is too low:

PROBLEM	REMEDY
9. _____	_____
10. _____	_____

Instructor's Response _____

_____

_____

Name _____ **Date** _____

INTERPRETING SYSTEM CONDITIONS

Upon completion of this job sheet, you should be able to troubleshoot an air-conditioning system based on visual, smell, and touch.

NATEF Correlation

HEATING AND AIR CONDITIONING: A/C System Diagnosis and Repair; *Identify and interpret heating and air-conditioning concern; determine necessary action.* **(P-1)**

Tools and Materials

None required

Procedure

The condition is given. Identify the problem and suggest the remedy.

Sight glass is clear:

PROBLEM	REMEDY
1. _____	_____
2. _____	_____

Sight glass is cloudy:

PROBLEM	REMEDY
3. _____	_____
4. _____	_____

Oily hose or fittings; worn or abraded hose:

PROBLEM	REMEDY
5. _____	_____
6. _____	_____

Icing condition or temperature change at a component:

PROBLEM	REMEDY
7. _____	_____
8. _____	_____

Musty smell from evaporator:

PROBLEM	REMEDY
9. _____	_____
10. _____	_____

Instructor's Response _____

_____

_____

Name _____ **Date** _____

IDENTIFYING REFRIGERATION SYSTEM TYPE

Upon completion of this job sheet, you should be able to determine the type of refrigerant in the system.

NATEF Correlation

HEATING AND AIR CONDITIONING: A/C System Diagnosis and Repair; *Identify refrigerant type; conduct a performance test of the A/C system; determine necessary action.* **(P-1)**

Tools and Materials

None required

Procedure

The refrigerant type is not known. The system condition is given. Identify the probable cause and suggest the remedy.

System pressure is higher than expected:

PROBLEM	REMEDY
1. _____	_____
2. _____	_____
3. _____	_____

System pressure is lower than expected:

PROBLEM	REMEDY
4. _____	_____
5. _____	_____
6. _____	_____

Expected low-side pressure range for a properly operating system:

	R-12		R-134a	
7. _____ to _____ psig			_____ to _____ psig	
8. _____ to _____ kPa			_____ to _____ kPa	

Expected high-side pressure range for a properly operating system:

	R-12		R-134a	
9. _____ to _____ psig			_____ to _____ psig	
10. _____ to _____ kPa			_____ to _____ kPa	

Instructor's Response _____

_____

_____

Chapter 5

SERVICING SYSTEM COMPONENTS

BASIC TOOLS

Mechanic's basic tool set (English and metric)

Fender cover

Flare nut wrench set

Torque wrench

Hacksaw with 32 tpi blade

Spring lock coupling tool set

Single-edge razor blade

FOT remove-replace tool

Tube cutter

Calibrated container

Internal-external snapring pliers

Pressure/temperature switch socket

UPON COMPLETION AND REVIEW OF THIS CHAPTER, YOU SHOULD BE ABLE TO:

- Identify and compare the differences between English and metric fasteners.

- State the purpose of good safety practices when servicing an automotive air-conditioning system.

- Diagnose air-conditioning system malfunctions based on customer complaints.

- Identify the different types of automotive air-conditioning systems.

- Remove the refrigerant (R-134a or R-12) from the vehicle air-conditioning system using an approved refrigerant recovery/recycling unit.

- Remove and replace the refrigerant hoses and fittings.

- Remove and replace the thermostatic expansion valve (TXV).

- Remove and replace the fixed orifice tube (FOT).

- Remove and replace the accumulator assembly.

- Remove and replace the condenser.

- Remove and replace the receiver-drier.

- Remove and replace the superheat or pressure switch.

ENGLISH AND METRIC FASTENERS

The servicing of automotive air-conditioning systems seems to become more and more complex each year. Although basic theories do not change, refrigeration and electrical control are redesigned or modified year to year. To add to the confusion, domestic automotive manufacturers use metric nuts and bolts on many components and accessories. Both English and **metric fasteners** can be found on the same automobile.

Some metric fasteners closely resemble **English fasteners** in size and appearance. The automotive service technician must be very careful to avoid mixing these fasteners. English and metric fasteners are not interchangeable. For example, a metric 6.3 (6.3 mm) capscrew may replace by design an English ¼-28 (¼ in. by 28 threads per in.) capscrew. Note in Figure 5-1 that the diameters of the two fasteners differ by only 0.002 in. (0.05 mm). The threads differ by only 2.6 per in. (1 per cm). There are 28 threads per in. (11 threads per cm) for a ¼-28 capscrew, and 30.6 threads per in. (12 threads per cm) for a 6.3 mm capscrew.

While the differences are minor, an English ¼-28 nut will not hold on a metric 6.3 capscrew. Mismatching of fasteners can cause component damage or early failure. Such component failure can result in personal injury.

> Both English and metric fasteners may be found on an assembly.

> **Metric fasteners** are any type of fastener with metric size designations, numbers, or millimeters.

> **English fasteners** are any type of fastener with English size designations, numbers, decimals, or fractions of an inch.

English Series				Metric Series			
Size	Diameter		Threads Per Inch	Size	Diameter		Threads Per Inch (prox)
	in	mm			in	mm	
#8	0.164	4.165	32 or 36				
#10	0.190	4.636	24 or 32				
1/4	0.250	6.350	20 or 28	M6.3	0.248	6.299	25
				M7	0.275	6.985	25
5/16	0.312	7.924	18 or 24	M8	0.315	8.001	20 or 25
3/8	0.375	9.525	16 or 24				
				M10	0.393	9.982	17 or 20
7/16	0.437	11.099	14 or 20				
				M12	0.472	11.988	14.5 or 20
1/2	0.500	12.700	13 or 20				
9/16	0.562	14.274	12 or 18	M14	0.551	13.995	12.5 or 17
5/8	0.625	15.875	11 or 18				
				M16	0.630	16.002	12.5 or 17
				M18	0.700	17.780	10 or 17
3/4	0.750	19.050	10 or 16				
				M20	0.787	19.989	10 or 17
				M22	0.866	21.996	10 or 17
7/8	0.875	24.765	9 or 14				
				M24	0.945	24.003	8.5 or 12.5
1	1.000	25.400	8 or 14				
				M27	1.063	27.000	8.5 or 12.5

FIGURE 5-1 A comparison of English and metric fasteners.

SAFETY

It must be recognized that the skills and procedures of those performing **service procedures** vary greatly. It is not possible to anticipate all of the conceivable ways or conditions under which service procedures may be performed. It is, therefore, not possible to provide precautions for every possible hazard that may result.

The following precautions are basic and apply to any type of automotive service:

- Wear safety glasses or goggles for eye protection when working with refrigerant (Figure 5-2). This is important while working under the hood of the vehicle.
- If the engine is to be operated, set the parking brake.
  a. Place the gear selector in PARK if the vehicle is equipped with an automatic transmission.
  b. Place the transmission in NEUTRAL if the vehicle is equipped with a manual transmission.

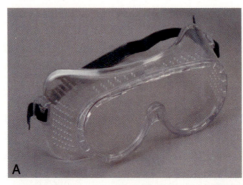
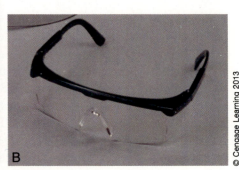

FIGURE 5-2 Wear safety (A) goggles or (B) glasses when servicing air conditioning components.

- Unless required otherwise for the service procedure, be certain that the ignition switch is turned to the OFF position.
- If the engine is to be operated, be certain that the vehicle is in a well-ventilated area or that provisions are made to vent the exhaust gases.
- Avoid loose clothing. Roll up long shirt sleeves. Tie long hair securely behind the head. Remove rings, watches, and loose-hanging jewelry.
- Keep clear of all moving parts when the engine is running. Engine-driven cooling fans have been known to separate. A loose fan blade can cause serious injury.
- Keep hands, clothing, tools, and test leads away from the engine cooling fan. Electric cooling fans may start without warning even when the ignition switch is in the OFF position.
- Avoid personal contact with hot parts such as the radiator, exhaust manifold, and high-side refrigerant lines.
- Disconnect the battery when required to do so (Figure 5-3). Follow the recommendations of the manufacturer; Chrysler, for example, requires that the negative (−) cable be disconnected to disable the air bag. General Motors requires the positive (+) cable to be disconnected, and Ford requires both cables be disconnected, first the negative (−), then the positive (+).
- Batteries normally produce explosive gases. DO NOT allow flames, sparks, or any lighted substances to come near the battery. Always shield your face and protect your eyes when working near a battery.

  NOTE: When the battery has been reconnected after being disconnected, volatile memory information—such as radio station presets, clock, seat, mirror, and window memory as well as customer input keyless entry codes—will be lost. Also, when the battery is reconnected, some abnormal drive symptoms may occur for the first 10 miles (18 kilometers) or so.

- If in doubt, ASK; do not take a chance. If there is no one to ask, consult an appropriate service manual. Again, do not take chances.

 WARNING: The technician must exercise extreme caution and pay heed to every established safety practice when performing these or any automotive air conditioning service procedures.

© Cengage Learning 2013

FIGURE 5-3 Carefully disconnect the battery cable.

Diagnostic Techniques

Before attempting to service an automotive air-conditioning system, be certain that the diagnosis is based on sound reasoning. Consider the following:

- Did you listen carefully to the customer's complaint?
- Does your diagnosis of the problem have merit based on the customer complaint?
- What type of system are you working on?
  - **a.** Cycling or noncycling clutch?
  - **b.** CFC-12, HFC-134a, or unknown refrigerant?
  - **c.** Fixed orifice tube or thermal expansion valve?
- Do you have the proper tools, equipment, and parts to service the air-conditioning system?
  - **a.** What tools are required?
  - **b.** What equipment is required?
  - **c.** What parts are required?

Listen carefully to the customer's complaint of the problem. If you do not understand, ask questions. Suppose, for example, that the customer complains of a moaning sound. The word *moaning* means different things to different people, so ask questions. Take a test drive with your customer so the noise can be identified.

Assume that you test drive the vehicle and hear nothing. When you return to the shop and tell the customer you heard nothing, the response may be, "Well, you did not start off very fast at the traffic light." Further discussion will reveal that the noise is only heard when pulling away from a traffic light after stopping.

The diagnosis to this problem is a relatively simple one. You are looking for a problem that only occurs during heavy acceleration. The problem can be a defective vacuum check valve or split vacuum hose (Figure 5-4), anything that may cause a vacuum loss during heavy acceleration.

NOTE: The loss of a vacuum signal at the control head will generally cause domestic vehicles to "fail safe" in either the heat or defrost mode.

Proper Tools, Equipment, and Parts

Having and using the proper tools and equipment is an essential part of performing a successful repair procedure. A screwdriver, for example, makes a very poor chisel; an adjustable wrench is a very poor hammer.

© Cengage Learning 2013

FIGURE 5-4 A split vacuum hose can cause system malfunction.

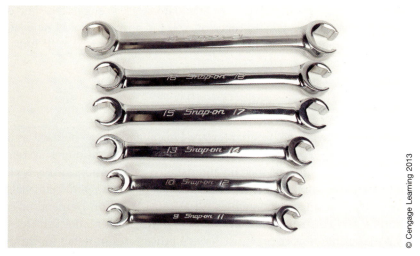

FIGURE 5-5 **A flare nut wrench set.**

Tools

It is important that tools be used for the purpose for which they are designed. For example, a flare nut wrench, not an open-end wrench, should be used on flare nuts (Figure 5-5). A flare nut wrench should not be used to remove a bolt or nut; this service requires either an open, box, combination, or socket wrench.

Equipment

It is equally important that the proper equipment be used for a particular service procedure. A vacuum pump, for example, can be constructed using an old refrigerator compressor (Figure 5-6). It is neither as attractive nor as efficient as a commercially available vacuum pump, however.

Parts

It is not practical to stock all of the parts that may be required in the course of doing business. The local parts distributors are responsible for that. If, however, a particular customer

FIGURE 5-6 **A typical refrigerator compressor.**

relies on your shop for a certain service, an adequate stock is suggested. For example, if a local off-road equipment repair facility relies on you for hoses, it would be wise to stock an adequate supply of various types of hose fittings and the several sizes of bulk hose necessary to supply the customer's needs.

SERVICE PROCEDURES

The service procedures given in this chapter are typical and are to be used as a guide only. Due to the great number of variations in automotive air-conditioning system configurations, it is impossible to include all specific and detailed information in this text. When specific and more detailed information is required, the service technician must consult a computer-based information system such as All Data/Mitchell on Demand or the appropriate manufacturer's service manual for any particular year and model vehicle. General service manuals are available that cover most service procedures in detail for automobiles of a specific year, make, and model.

The information is given only as a guide for the student technician to perform basic service procedures that are normally required. Proper service and repair procedures are vital to the safe, reliable operation of the system. Most important, proper service procedures and techniques are essential to providing personal safety to those performing the repair service and to the safety of those for whom the service is provided.

Be sure to follow the manufacturer's recommendations. To **disarm** the air bag restraint system, for example, Chrysler had suggested disconnecting the battery ground (−) cable; General Motors had suggested disconnecting the positive (+) battery cable; and Ford had suggested disconnecting both cables—first the ground (−) cable, then the positive (+) cable.

Since the computers are disabled and often must "relearn" a program, disconnecting the battery is now discouraged. General Motors suggest the following typical procedures:

1. Turn off the ignition switch.
2. Remove the restraint system (air bag) fuse from the fuse panel.
3. Remove the left-side sound insulator.
4. Disconnect the connector position assurance (CPA) and yellow two-way connector found at the base of the steering column.

There is no set universal procedure. The best approach is to refer to the specific service manual appropriate for the make and model vehicle being serviced to ensure that proper procedures are followed.

PREPARATION

There are certain basic procedural steps that must be taken before attempting to perform any service procedure. Whenever applicable, this procedure will be referenced in the various service procedures that follow.

SPECIAL TOOLS

Battery pliers
Recovery equipment

1. Place the ignition switch in the OFF position.
2. Place a fender cover on the car to protect the finish.
3. Disconnect the battery cable. Follow procedures as outlined in the appropriate manufacturer's service manual.
4. Recover the system refrigerant. Use the proper recovery equipment and follow the instructions outlined in the next section.
5. Locate the component that is to be removed for repair or replacement.
6. Remove access panel(s) or other hardware necessary to gain access to the component.

REFRIGERANT RECOVERY

To **purge** an air-conditioning system, in general terms, is to remove all of its contents, primarily refrigerant. While the term is generally understood to refer to refrigerant, it may also include air and moisture. Purging a system of refrigerant is usually necessary when a component is to be serviced or replaced. Until recently, to "purge" was to vent refrigerant into the atmosphere. The **Federal Clean Air Act** Amendments of 1990, however, required that after July 1, 1992, no refrigerants could be intentionally vented. The EPA requires that service equipment hoses have shut-off valves within 12 inches (30 cm) of the service end coupler. These valves must be closed before removing the coupler from the air-conditioning system service fitting. R-134a manifold and gauge sets as well as Recovery/Recycling/Recharging equipment generally have service hoses with an automatic check valve in the coupler to avoid the venting of refrigerant into the atmosphere.

It is required that a refrigerant **recovery system** (Figure 5-7) be used to purge an air-conditioning system of refrigerant. Manufacturers' specifications and procedures should be followed to ensure safe and adequate performance. The following procedure is typical and should be used only as a guide.

 WARNING: Adequate ventilation must be maintained during this procedure.

Procedure

1. Determine what type of refrigerant is in the system being serviced (i.e., R-134a or R-12) in order to select the correct recovery/recycling/recharge unit to be used.
   a. On some vehicles it may be advisable to test the purity of the refrigerant in the vehicle's air-conditioning system with a refrigerant identifier and to test the system for sealant contamination prior to attaching the refrigerant recovery unit. The procedures for using both these devices are outlined in Chapter 6 of this manual.

NOTE: Some units are dual refrigerant recovery/recycling stations and have separate gauges and lines for both R-134a and R-12 refrigerants.

© Cengage Learning 2013

FIGURE 5-7 Typical refrigerant Recovery/Recycling/Recharging machine.

To **purge** is to remove refrigerant, moisture, and air from a system or by flushing with a dry gas such as nitrogen (N) to remove all moisture from a system.

The **Federal Clean Air Act**, Title 6, is an amendment signed into law in 1990 that established national policy relative to the reduction and elimination of ozone-depleting substances.

To recover refrigerant is to remove it, in any condition, from the system.

A **recovery system** refers to the circuit inside the recovery unit used to recycle and transfer refrigerant from the air-conditioning system to the recovery cylinder.

The term *purge* is used to refer to the removal of refrigerant.

SERVICE TIP: Any time a major air conditioning component is replaced or if the refrigerant system has been open to the atmosphere (moisture contamination) or empty for an extended period it is advisable that the receiver-drier/accumulator be replaced, along with other necessary repairs. The desiccant contained inside acts to absorb unwanted moisture from the refrigerant system, and heavy contamination particles may settle to the bottom of the container. Most compressor manufacturers will not honor the warranty on a new or rebuilt compressor if the receiver-drier/accumulator is not also replaced at the time of service.

Classroom Manual

Chapter 5, page 153

2. Connect the recovery unit high- and low-side lines to the vehicle's refrigerant service fittings and note the system pressures.

   a. If you are using a unit that is not equipped with gauges on the control console, it will be necessary to connect a manifold and gauge set to the vehicle system and attach the yellow service hose of the manifold and gauge set to the stand-alone recovery/recycling unit.

3. Start the engine and set the air-conditioning system controls to the MAX cold position with the blower on HI.

   NOTE: Some system malfunctions, such as a defective compressor or low system refrigerant charge level, may make the next four steps impossible to perform, in which case proceed with step 8.

4. Raise the engine speed to 1000–1200 rpm and operate for approximately 10 minutes to allow the system to stabilize and to allow most of the refrigerant oil to return to the compressor assembly.

5. Return the engine speed to normal idle.

6. Turn off the air conditioning controls.

7. Shut of the engine.

8. Start the recovery unit, following the instructions for the recovery/recycling/recharge unit being used, and begin the recovery process.

   a. In general this requires the opening of the valves on the refrigerant recovery tank and turning on the recovery switch located on the recovery unit console.

   b. If a manifold and gauge set is being used it will be necessary to open the hand valves on the manifold set high side and low side.

9. When the recovery process is complete, note the gauge readings; the low-side gauge should read 4 in. Hg (102 mm) of vacuum or lower.

10. Wait 2 minutes and observe the gauge readings again. If a positive pressure is noted, the recovery process will need to be repeated again. If a vacuum or 0 psig is noted, the recovery process is complete.

See Photo Sequence 4 for a typical procedure for recovering (purging) refrigerant from the system.

SERVICING REFRIGERANT HOSES AND FITTINGS

There are many types of fittings used to join refrigerant hoses to the various components of the air-conditioning system. Some of these fittings are (Figure 5-8):

- Male and female **SAE** flare
- Male and female upset flange, commonly called **O-ring**
- Male and female **spring lock**
- Male barb
- Peanut
- Beadlock

This procedure will be given in three parts:

1. Repairing a hose using a beadlock **insert fitting**
2. Servicing spring lock fittings
3. O-ring service

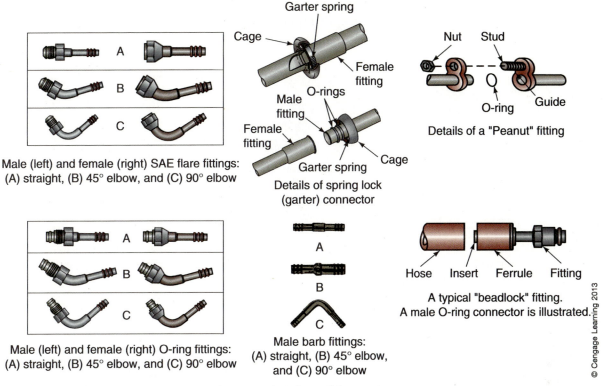

Male (left) and female (right) SAE flare fittings: (A) straight, (B) 45° elbow, and (C) 90° elbow

Details of spring lock (garter) connector

Details of a "Peanut" fitting

Male (left) and female (right) O-ring fittings: (A) straight, (B) 45° elbow, and (C) 90° elbow

Male barb fittings: (A) straight, (B) 45° elbow, and (C) 90° elbow

A typical "beadlock" fitting. A male O-ring connector is illustrated.

© Cengage Learning 2013

FIGURE 5-8 Various types of fittings used in automotive air condition service.

Part 1. Hose Repair Using a Beadlock Fitting

In order to meet new SAE refrigerant leakage specifications, a beadlock ferrule fitting with a captive metal shell (Figure 5-9) must be used for R-134a hose repairs.

BEADLOCK FITTING CRIMP STYLE

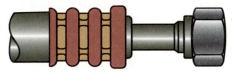

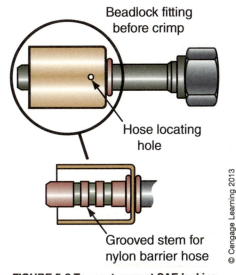

Beadlock fitting before crimp

Hose locating hole

Grooved stem for nylon barrier hose

© Cengage Learning 2013

FIGURE 5-9 To meet current SAE leaking standards, a beadlock fitting with a captive sheel must be used or R-134a line or fitting repairs.

SAE is the Society of Automotive Engineers.

An **O-ring** is a synthetic rubber or plastic gasket with a round- or square-shaped cross section used as seals on line fittings.

A **spring lock** is a special fitting used to form a leak-proof joint.

An **insert fitting** is a fitting designed to fit inside a hose, such as a barb fitting or beadlock fitting.

Use caution when using a razor blade. Always cut away from the body.

TYPICAL PROCEDURE FOR RECOVERING (PURGING) REFRIGERANT FROM THE SYSTEM

P4-1 Attach the service hoses of the refrigerant recovery/recycling/charging system to the vehicle high- and low-side service fittings.

P4-2 Open all hose shut-off valves.

P4-3 Connect the refrigerant recovery system to an approved electrical power supply.

P4-4 Turn on the main switch.

P4-5 Turn on the recovery (compressor) switch.

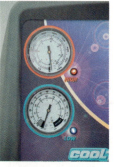

P4-6 Operate until a vacuum pressure is indicated. The recovery system will automatically shut off. If it is not equipped with an automatic shut-off, turn the compressor switch to OFF after achieving a vacuum pressure.

P4-7 Observe the gauges for at least 5 minutes. If the vacuum does not rise, complete refrigerant recovery. If the vacuum rises but remains at 0 psig (0 kPa) or below, a leaking system is indicated. Complete refrigerant recovery and repair the system.

P4-8 If the vacuum rises to a positive pressure, above 0 psig (0 kPa), the refrigerant was not completely removed from the system. Repeat the recovery procedure, starting with step P4-5 of this procedure.

P4-9 Repeat step P4-6 until the system holds a stable vacuum for at least 2 minutes.

P4-10 After all of that refrigerant has been recovered from the system, close all valves. Close the service hose shut-off valves, the low- and high-side manifold valves, and the recovery system inlet valve. Disconnect all the hoses from the system service valves or fittings. Cap all fittings and hoses prevent dirt, foreign matter, or moisture from entering the system. This is most important for an R-134a system. The lubricant used in this system is very hygroscopic.

A **barb fitting** is a fitting that slips inside a hose and is held in place with a gear-type clamp. Ridges (barbs) on the fitting prevent the hose from slipping off.

CAUTION:
Use the proper oil for the refrigerant used in the system; that is, mineral oil for R-12 and PAG oil for R-134a (Figure 5-11). Some service information indicates that mineral oil may be used for all applications because it aids in forming a natural barrier to R-134a leaks.

Barb fittings and screw (worm gear) style hose clamps are not acceptable repair methods for barrier-type hoses and do not meet current SAE standards. To repair lines using a beadlock fitting, follow the steps below and refer to Photo Sequence 5 for beadlock hose assembly repair.

1. Measure and mark the required length of replacement hose, or determine how much hose must be cut ahead of the damaged fitting.
2. Use a single-edge razor blade knife (utility knife) to cut the hose.
3. Trim the end of the hose to be used to ensure that the cut is at a right angle (square).
4. Select the correct fitting outside diameter (OD) for the hose inside diameter (ID). See Figure 5-10 for beadlock fitting sizes.
5. Apply clean refrigerant oil to the inside of the hose to be used.
6. Be sure that the beadlock ferrule fitting is free of nicks and burrs, and coat the fitting end to be inserted with clean refrigerant oil.
7. Slip the beadlock ferrule fitting into the refrigerant line in one constant, deliberate twisting motion.
8. Position the hose and beadlock ferrule fitting into the crimping tool clamp, verifying that proper crimping die was selected.
9. Apply clamping pressure as directed by the tool manufacturer to achieve proper crimp, generally until the die halves almost touch.
10. Reinstall the hose assembly on the vehicle.

Part 2. Servicing Spring Lock Fittings

For simplicity, this procedure is given in two parts: To Separate, steps 1 through 4, and To Join, steps 5 and 6.

SPECIAL TOOLS

Flare nut wrench set

¼ in. drive socket set

Torque wrench

Hacksaw with 32 tpi blade

Pliers

Beadlock crimp set

Spring lock coupling tool set

Single-edge razor blade

ASSEMBLE BEADLOCK HOSE ASSEMBLY

P5-1 Measure the hose and mark the proper length.

P5-2 Cut the hose, ensuring that the end is "squared."

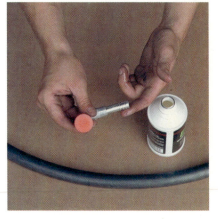

P5-3 Liberally lubricate the beadlock fitting with mineral oil.

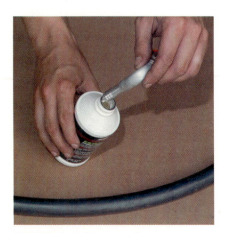

P5-4 Shake off excess oil.

P5-5 Insert the hose with the ferrule into the fitting.

P5-6 Crimp the ferrule.

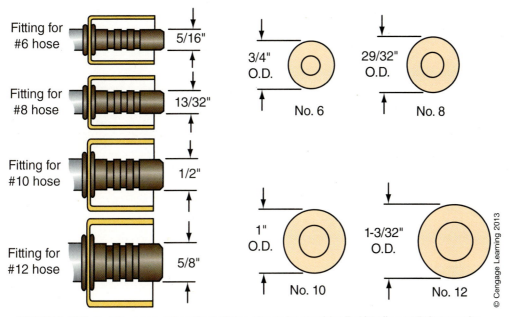

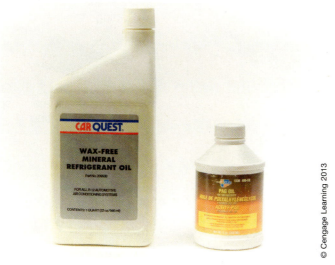

FIGURE 5-10 Select the correct beadlock fitting size to be used for R-134a line or fitting repairs.

FIGURE 5-11 Containers of mineral and PAG oil.

To Separate

1. Install the special tool onto the coupling so it can enter the cage to release the garter spring (Figure 5-12).
2. Close the tool and push it into the cage to release the female fitting from the garter spring (Figure 5-13).
3. Pull the male and female coupling fittings apart (Figure 5-14).
4. Remove the tool from the disconnected spring lock coupling.

To Rejoin

5. Lubricate two new O-rings with clean refrigeration oil and install them on the male fitting.

   NOTE: The O-ring material is of a special composition and size. To avoid leaks, use the proper O-rings. Also, see the caution following step 4 of Part 1. Always use new O-rings.

6. Insert the male fitting into the female fitting and push them together to join.

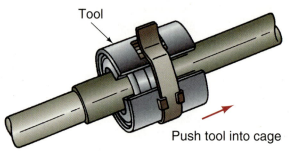

Tool

Push tool into cage

FIGURE 5-13 **Close the tool and push it into the cage to release the coupling.**

© Cengage Learning 2013

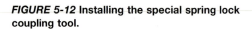

Tool

Cage

FIGURE 5-12 **Installing the special spring lock coupling tool.**

© Cengage Learning 2013

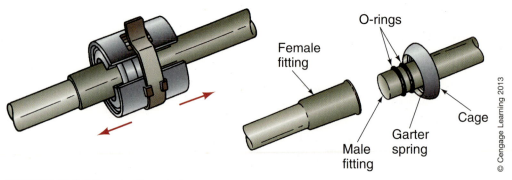

FIGURE 5-14 **Pull the coupling apart.**

O-rings

Female fitting

Male fitting

Garter spring

Cage

© Cengage Learning 2013

Part 3. Servicing O-rings

O-rings must be replaced whenever a component fitting is removed for any reason. They do not usually leak if not disturbed. On occasion, however, an O-ring may be found to be leaking and must be replaced.

If it becomes necessary to replace an O-ring, be aware that there are several different types available for different applications. For example, in addition to spring lock fittings, there are two different types of O-ring fittings used on systems (Figure 5-15):

1. Captive
2. Standard

Although R-134a O-rings are similar to R-12 O-rings, they are made of a different material. Most R-12 and R-134a O-rings are not compatible. When replacing them, it is important to use the proper O-ring for the fitting.

The ID as well as the OD are also important considerations to ensure a leak-free connection.

Use the proper O-ring.

SERVICE TIP:
Coat O-ring seals with mineral oil to form an improved barrier seal.

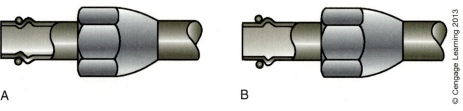

A   B

FIGURE 5-15 **Captive (A) and standard (B) O-ring fittings.**

© Cengage Learning 2013

REPLACING AIR-CONDITIONING COMPONENTS

The following procedures may be considered typical for step-by-step instructions for the replacement of air-conditioning system components. For specific replacement details, however, refer to the manufacturer's shop service manual for the particular year, model, and make of the vehicle.

REMOVING AND REPLACING THE THERMOSTATIC EXPANSION VALVE (TXV)

This procedure will be given in two parts:

1. Servicing the Standard TXV
2. Servicing the H-Block TXV

Part 1. Servicing the Standard TXV

1. Follow the procedures outlined in "Preparation."
2. Remove the insulation tape from the remote bulb.
3. Loosen the clamp to free the remote bulb.
4. Disconnect the external equalizer, if the TXV is so equipped.
5. Remove the liquid line from the inlet of the TXV.
6. Remove and discard the O-ring, if equipped.
7. Inspect the inlet screen (Figure 5-16).
   a. If it is clogged, clean and replace the screen. Skip to step 14.
   b. If it is not clogged, proceed with step 8.
8. Remove the evaporator inlet fitting from the outlet of the TXV.
9. Remove and discard the O-ring, if equipped.
10. Remove the holding clamp (if provided on the TXV), and carefully lift the TXV from the evaporator.
11. Carefully locate the new TXV in the evaporator.
12. Insert new O-ring(s) on the evaporator inlet, if equipped.
13. Attach the evaporator inlet to TXV outlet. Tighten to the proper torque.
14. Install the new O-rings on the liquid line fitting, if so equipped.
15. Attach the liquid line to the TXV inlet. Tighten to the proper torque.
16. Reconnect the external equalizer tube, if equipped.

SPECIAL TOOL
Flare nut wrench set

Classroom Manual
Chapter 5, page 160

Take care not to damage the capillary tube or remote bulb.

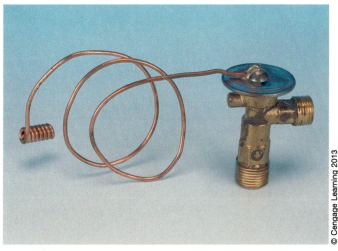

FIGURE 5-16 **Inspect the inlet screen of the thermostatic expansion valve.**

© Cengage Learning 2013

17. Position the remote bulb and secure it with a clamp.
18. Tape the remote bulb to prevent it from sensing ambient air.
19. Proceed with step 14 of Part 2.

Part 2. H-Valve TXV

1. Follow the procedures outlined in "Preparation."
2. Disconnect the wire connected to the pressure cut-out or pressure differential switch, as applicable.
3. Remove the bolt from the line sealing plate found between the suction and liquid lines.
4. Carefully pull the plate from the H-valve.
5. Cover the line openings to prevent the intrusion of foreign matter.
6. Remove and discard the plate to the H-valve gasket.
7. Remove the two screws from the H-valve.
8. Remove the H-valve from the evaporator plate.
9. Remove and discard the H-valve to the evaporator plate gasket.
10. Install a new H-valve with the gasket (Figure 5-17).
11. Replace the two screws. Torque to 170–230 in.-lb (20–26 N·m).
12. Replace the line plate to the H-valve gasket.
13. Hold the line assembly in place and install the bolt. Torque to 170–230 in.-lb (20–26 N·m).
14. Replace the access panels and any hardware previously removed.
15. Reconnect the battery following the instructions given in the manufacturer's shop manual.
16. Leak-test, evacuate, and charge the system with refrigerant as outlined in Chapter 6 of this manual.
17. Complete the system performance test job sheet.

Be careful not to mar or nick the mating surface(s).

Classroom Manual
Chapter 5, page 165

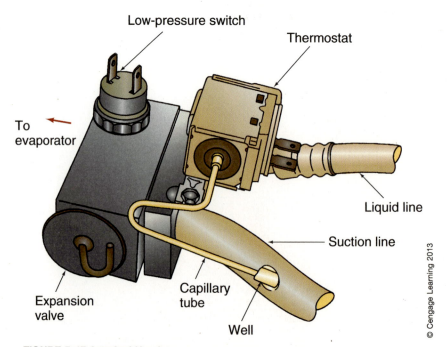

FIGURE 5-17 A typical H-valve.

REMOVING AND REPLACING THE FIXED ORIFICE TUBE (FOT)

The fixed orifice tube (FOT) is also known as:

- Cycling clutch orifice tube (CCOT)
- Fixed orifice tube/cycling clutch (FOTCC)
- Variable displacement orifice tube (VDOT)

It should be noted that orifice tubes are not interchangeable. An orifice tube used on Ford car lines may not be used on GM car lines. The same service tool may be used, however, to remove and replace any of them. When replacing the orifice tube, it is most important that the correct replacement part be used.

Some car lines have a non-accessible orifice tube in the liquid line. Its exact location, anywhere between the condenser outlet and the evaporator inlet, is determined by a circular depression or three indented notches in the metal portion of the liquid line. An orifice tube replacement kit is used to replace this type orifice tube and 2.5 in. (63.5 mm) of the metal liquid line.

This service procedure is given in two parts:

1. Servicing the Accessible FOT
2. Replacing the Non-accessible FOT

   NOTE: On some models a service kit is not available for replacement of the orifice tube and the liquid line will have to be replaced. In addition a manufacturer may require the replacement of the liquid line that houses the orifice tube if service is indicated.

Part 1. Servicing the Accessible FOT

The procedure (Photo Sequence 6) is given in two parts: Removing the FOT, steps 1 through 9, and Replacing the FOT, steps 10 through 14.

Removing the FOT

1. Perform the procedures outlined in "Preparation."

   NOTE: It may not be necessary to disconnect the battery for this procedure.

2. Using the proper flare nut wrenches, remove the liquid line connection at the inlet of the evaporator to expose the FOT.
3. Remove and discard the O-ring(s) from the liquid line fitting, if equipped.
4. Pour a very small quantity of clean refrigeration oil into the FOT well to lubricate the seals.
5. Insert the FOT removal tool onto the FOT (Figure 5-18).
6. Turn the T-handle of the tool slightly clockwise (cw) only enough to engage the tool onto the tabs of the FOT.
7. Hold the T-handle and turn the outer sleeve or spool of the tool clockwise to remove the FOT. Do not turn the T-handle.

   NOTE: If the FOT breaks during removal, proceed with step 8. If it does not break, proceed with step 10.

8. Insert the extractor into the well and turn the T-handle cw until the threaded portion of the tool is securely inserted into the brass portion of the broken FOT (Figure 5-19).
9. Pull the tool. The broken FOT should slide out.

   NOTE: The brass tube may pull out of the plastic body. If this happens, remove the brass tube from the puller and reinsert the puller into the plastic body. Repeat steps 8 and 9.

SPECIAL TOOLS

Flare nut wrench set

FOT removal tool

Extractor tool

Torque wrench

Tube cutter

> Although orifice tubes may look alike, they are not interchangeable.

CAUTION:
Use an oil that is proper for use with the system refrigerant; that is, mineral oil for CFC-12 and PAG for HFC-134a.

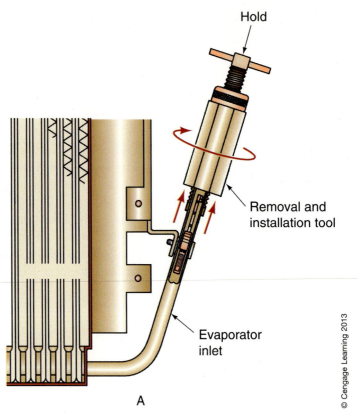

FIGURE 5-18 Insert the FOT removal tool.

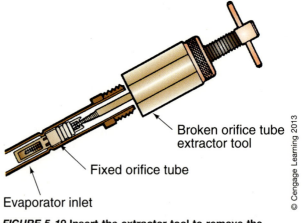

FIGURE 5-19 Insert the extractor tool to remove the broken FOT.

Installing the FOT

10. Liberally coat the new FOT with clean refrigeration oil.
11. Place the FOT into the evaporator cavity and push it in until it stops against the evaporator tube inlet dimples.
12. Install a new O-ring, if equipped.
13. Replace the liquid line and tighten it to the recommended torque.
14. Replace the accumulator as outlined in this chapter.

TYPICAL PROCEDURE FOR REPLACING A FIXED ORIFICE TUBE

This service procedure presumes that all of the refrigerant has been properly removed from the air-conditioning system.

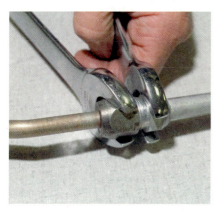

P6-1 Using the proper open end or flare nut wrenches, remove the liquid line from the evaporator inlet fitting.

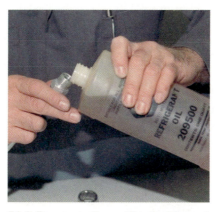

P6-2 Pour a small quantity of refrigerant lubricant into the orifice tube well to lubricate the O-rings.

P6-3 Insert an orifice tube removal tool and turn the T-handle slightly clockwise to engage the orifice tube.

P6-4 Hold the T-handle and turn the outer sleeve clockwise to remove the orifice tube. Do not turn the T-handle. **NOTE:** If the orifice tub broke, proceed with P6-5. If not broken, proceed with P6-6.

P6-5 Insert the broken orifice tube extractor into the orifice tube well and turn the T-handle clockwise several turns until the tool has been threaded into the orifice tube.

P6-6 Pull the tool. The orifice tube should slide out.

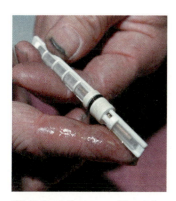

P6-7 Coat the new orifice tube liberally with clean refrigeration lubricant.

P6-8 Slide the new orifice tube into the evaporator until it stops against the tube inlet dimples.

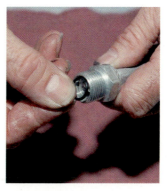

P6-9 Slide a new O-ring onto the evaporator or liquid, as applicable.

P6-10 Connect the liquid line to the evaporator and tighten the nut with the proper wrenches.

Part 2. Servicing the Non-accessible FOT

1. Follow the procedures outlined in "Preparation."

   NOTE: It may not be necessary to disconnect the battery.

2. Remove the liquid line from the evaporator inlet. Remove and discard the O-rings, if equipped.

3. Remove the liquid line from the condenser outlet. Remove and discard the O-rings, if equipped.

4. Locate the orifice tube. The outlet side of the orifice tube can be identified by a circular depression or three notches (Figure 5-20).

5. Use a tube cutter to remove a 2.5 in. (63.5 mm) section of the liquid line (Figure 5-21).

6. Slide a compression nut onto each section of the liquid line.

7. Slide a compression ring onto each section of the liquid line with the taper portion toward the compression nut.

8. Lubricate the two O-rings with clean refrigeration oil of the proper type and slide one onto each section of the liquid line. See the warning following step 4 of part 1.

9. Attach the orifice tube housing, with the orifice tube inside, to the two sections of the liquid line (Figure 5-22).

10. Hand tighten both compression nuts. Note the flow direction indicated by the arrows. The flow should be toward the evaporator.

11. Hold the orifice tube housing in a vise or other suitable fixture to tighten the compression nuts.

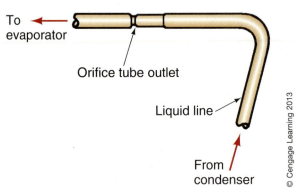

FIGURE 5-20 Locate the FOT.

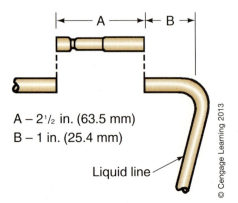

A – 2½ in. (63.5 mm)
B – 1 in. (25.4 mm)

FIGURE 5-21 After locating the FOT, remove 2.5 in. (63.5 mm) of the liquid line.

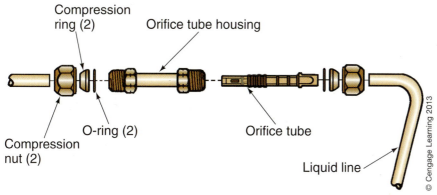

Compression ring (2)

Orifice tube housing

Compression nut (2)

O-ring (2)

Orifice tube

Liquid line

© Cengage Learning 2013

FIGURE 5-22 The replacement orifice tube housing assembly.

12. Tighten each compression nut to 65–70 ft.-lb (87–94 N·m) torque.
13. Insert new O-rings on both ends of the liquid line, if equipped.
14. Install the liquid line:
    a. Attach the condenser end of the liquid line to the condenser and tighten to the proper torque.
    b. Repeat step 14a with the evaporator end of the liquid line.
15. Leak-test, evacuate, and charge the system as outlined in Chapter 6 of this manual.
16. Repeat or continue performance testing.

REMOVING AND REPLACING THE ACCUMULATOR

If the refrigerant system is suspected of being contaminated with moisture it is recommended that the accumulator/receiver dryer be replaced since it contains moisture absorbing desiccant material. It is also recommended that the accumulator/receiver dryer be replaced when any major component in the refrigerant system is replaced (i.e., compressor, metering device, evaporator, condenser) since debris and contaminants may be trapped in the container. In addition, most compressor manufacturers/rebuilders will not honor their warranty if the accumulator/receiver is not replaced at the same time.

1. Follow the procedures outlined in "Preparation."
2. Disconnect the electrical connection on the pressure control switch.
3. Remove the accumulator inlet fitting.
4. Remove and discard the O-ring.
5. Remove the accumulator outlet fitting.
6. Remove and discard the O-ring.
7. Remove the bracket attaching screw and bracket.
8. Remove the accumulator from the vehicle (Figure 5-23).
9. Remove the pressure switch.
10. Remove and discard the O-ring from the pressure switch.
11. Pour the oil from the accumulator into a calibrated container.
12. With a new O-ring, install the pressure switch on the new accumulator.
13. Add a like amount of oil, as removed in step 11, to a new accumulator or 0.2 fl. oz. (5 mL) to 0.5 fl. oz. (15 mL) if no fluid was recovered from the old accumulator.
14. Position the new accumulator.
15. Using new O-rings, attach the evaporator outlet line to the accumulator inlet and finger tighten.
16. Using new O-rings, attach the suction line to the accumulator outlet and finger tighten.
17. Replace the retainer bracket, removed in step 7.

SERVICE TIP:
Be sure that the hose bends are in the same position as when removed for ease in replacing the liquid line.

SPECIAL TOOLS
Flare nut wrench set
Calibrated container
Torque wrench

The accumulator contains the desiccant in a fixed orifice tube system.

CAUTION:
Use the proper oil for the refrigerant in the system; for example, mineral oil for R-12 and PAG for R-134a.

Do not reuse oil that was removed from the system.

FIGURE 5-23 Removing the accumulator from the vehicle.

18. Using appropriate flare nut wrenches, torque the inlet and outlet fittings, steps 15 and 16, as specified.
19. Reattach the electrical connector to the pressure switch.
20. Connect the battery, if removed in step 1, according to the manufacturer's recommendations.

NOTE: For steps 20 through 22, refer to Chapter 6.

21. Leak-test the system.
22. Evacuate the system.
23. Charge the system.
24. Complete the system performance test job sheet.

REMOVING AND REPLACING THE CONDENSER

1. Follow the procedures outlined in "Preparation."

NOTE: It is not generally necessary to disconnect the battery for this procedure.

2. Remove the hood hold-down mechanism and any other cables or hardware that inhibit access to the condenser.
3. Remove the hot-gas line at the top of the condenser.
4. Remove and discard any O-rings.
5. Remove the liquid line at the bottom of the condenser.
6. Remove and discard any O-rings.
7. Remove and retain any attaching bolts or nuts holding the condenser in place.
8. Lift the condenser from the car.
9. Install the new condenser by reversing the procedure given in steps 2, 3, 5, and 7 and 1.0 fl. oz. (30 mL) of clean refrigerant oil.

NOTE: Be sure to install new O-rings, if equipped, on the hot-gas and liquid lines.

10. Leak-test the system.
11. Evacuate the system.
12. Charge the system with refrigerant.

REMOVING AND REPLACING THE RECEIVER-DRIER

1. Recover the air-conditioning system refrigerant as outlined earlier in this chapter.
2. Remove the low- or high-pressure switch wire, if applicable.
3. Remove the inlet and outlet hoses (liquid lines) from the receiver-drier.

Classroom Manual
Chapter 5, page 173

SPECIAL TOOLS
Flare nut wrench set
Torque wrench

Do not reuse O-rings.

Classroom Manual
Chapter 5, page 173

SPECIAL TOOLS
Flare nut wrench set
Torque wrench

The receiver-drier contains the desiccant in a TXV system.

FIGURE 5-24 Remove the receiver-drier from the vehicle.

Classroom Manual
Chapter 5, page 158

4. Remove and discard O-rings or gaskets, if applicable.
5. Loosen and remove the mounting hardware.
6. Remove the receiver-drier from the vehicle (Figure 5-24).
7. Drain and measure the amount of refrigerant oil that was removed from the receiver-drier. Add the same amount of new manufacturer recommended refrigerant oil to the replacement receiver-drier to be installed in step 9. If no oil is removed, add 0.2 fl. oz. (5 mL) to 0.5 fl. oz. (15 mL) of new refrigerant oil to the new receiver-drier.
8. Remove the pressure switch from the drier, if applicable. Discard the gasket or O-ring.
9. Install a new receiver-drier. Reverse the order of removal, steps 2, 3, and 5–7.

NOTE: Most receiver-driers are marked with an arrow (→) or the word(s) "IN" and "OUT" to denote the direction of refrigerant flow. Remember, flow is away from the condenser and toward the metering device.

10. Leak-test, evacuate, and charge the air-conditioning system.

CUSTOMER CARE: It is becoming a more common practice today for technicians to leave a small personalized business card with the technician's name on it thanking the customer for the opportunity to serve them.

SUPERHEAT OR PRESSURE SWITCH

Determine the location of the switch to be replaced. If it is the superheat switch on the rear head of a Harrison compressor, follow the applicable steps in the procedure for replacing the compressor.

If it is a pressure switch on the accumulator, follow the appropriate steps in the procedure for replacing the accumulator. If it is a drier pressure switch, follow the appropriate

SPECIAL TOOLS

Internal snaping pliers, if applicable

Appropriate wrench for the type pressure/temperature switch

steps in the procedure for replacing the drier. If it is anywhere else in the system, the following general procedures may apply:

1. Recover the refrigerant as outlined earlier in this chapter.
2. Using the proper tool, remove the defective component.
3. Remove and discard the gasket or O-ring.
4. Place a new gasket or O-ring on the new component.
5. Install the new component, again using the proper tool.
6. Leak-test, evacuate, and charge the system with refrigerant.
7. Hold the performance test, if indicated.

IN CONCLUSION

- Note the amount of refrigerant and oil removed from the system during refrigerant recovery process.
- Use only components and parts designated for a particular system: R-12 or R-134a.
- Use new gaskets or O-rings when replacing a component or removal of a line or fitting.
- Liberally coat all components with clean refrigeration oil before reassembly.
- For reassembly, reverse the removal procedure.
- Do not overfill or underfill the A/C system with refrigerant oil. When replacing refrigerant components refer to manufacturers recommendation on how must refrigerant oil needs to be added back into the refrigerant system and include this amount to the amount that was removed during the recovery and evacuation process. The pie chart in Figure 5-25 gives an example of refrigerant system oil distribution. Only use fresh clean oil recommended by manufacturer.
- Recharge the system with the specified amount of refrigerant.

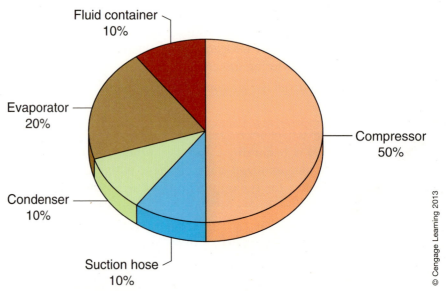

FIGURE 5-25 This pie chart roughly represents the distribution of oil throughout the refrigerant system. When replacing refrigerant components refer to manufacturer's recommendation on how much refrigerant oil needs to be added back into the refrigerant system.

© Cengage Learning 2013

CASE STUDY

A customer complains of a chattering noise that occurs only when driving home.

Technician: "The noise occurs only on the way home, never on the way to work?"

Customer: "That's right, only on the way home."

Technician: "What time do you go home?"

Customer: "Usually seven or eight o'clock. I've even tried to take different routes. The chatter always happens about 5 miles from work."

Technician: "Do you operate your air conditioner in the evening?"

Customer: "I seldom turn it off. I like to avoid the road fumes whenever I can."

Technician: "Are you sure that it is the air conditioner?"

Customer: "Yes. When it makes a noise, I turn the air conditioner off. The noise stops."

Technician: "Only on the way home from work. Does it ever happen when you're on the way home from a movie or the grocery store?"

Customer: "When we go out later, I take my wife's car."

Technician: "Why?"

Customer: "I have a bad generator (alternator) and use my little battery charger to keep the battery up. An overnight charge is just enough to get me to work and back."

Based on what you have learned, you know that low voltage will cause the clutch to chatter. By the time the customer drove approximately 5 miles toward home with his headlamps on, the battery voltage was reduced to a level that caused this chatter. The problem is a defective alternator; it is not an air conditioning problem at all.

Get to know your customers. The more you communicate, the more effective you are when performing troubleshooting and diagnosis of needed service. Be it a groan, chatter, squeak, or bang, the customer will eventually reveal the problem. Often, the customer will hear a noise that you may not identify as being a problem. The customer hears it as a noise that is different than those experienced since owning the vehicle. Questioning the customer will often reveal when, why, and how.

Other problems may cause an air-conditioning system to malfunction.

TERMS TO KNOW

Barb fitting
Disarm
English fasteners
Federal Clean Air Act
Insert fitting
Metric fasteners
O-ring
Purge
Recovery system
SAE
Service procedures
Spring lock

ASE-STYLE REVIEW QUESTIONS

1. Before installation:

   Technician A says refrigerant system O-rings can be coated with a gasket sealant.

   Technician B says that refrigerant system O-rings should be coated with new refrigerant oil.

   Who is correct?

   A. A only
   B. B only
   C. Both A and B
   D. Neither A nor B

2. Before attempting to service an automotive air-conditioning system consider all of the following except:

   A. Listen to customer complaint.
   B. Does your diagnosis make sense based on customer complaint?
   C. Can customer afford the repair?
   D. Do you have the proper tools and equipment to service the system?

3. *Technician A* says that it is permissible, according to the EPA, to release refrigerant from the hoses of a manifold and gauge set because it is such a small amount.

   Technician B says that, to minimize refrigerant loss, shut-off valves are required by the EPA within 12 in. (30 cm) of the service end of the manifold and gauge set hoses. Who is correct?

   A. A only
   B. B only
   C. Both A and B
   D. Neither A nor B

4. *Technician A* says that desiccant is found in the receiver-drier.

   Technician B says that desiccant is found in the accumulator.

   Who is correct?

   A. A only
   B. B only
   C. Both A and B
   D. Neither A nor B

5. *Technician A* says that fixed orifice tubes are interchangeable in order to tailor system performance.

   Technician B says that fixed orifice tubes may be located anywhere between the evaporator outlet and the compressor inlet.

   Who is correct?

   A. A only   C. Both A and B

   B. B only   D. Neither A nor B

6. The term purge an air-conditioning system in general terms means to:

   A. Remove the refrigerant from the system.

   B. Flush the refrigerant system.

   C. Recharge the refrigerant system.

   D. Add sealant to the refrigerant system.

7. *Technician A* says that the accumulator should be replaced if there is moisture in the system.

   Technician B says that the accumulator must be replaced if the fixed orifice tube (FOT) is replaced.

   Who is correct?

   A. A only   C. Both A and B

   B. B only   D. Neither A nor B

8. When removing the accessible orifice tube:

   Technician A says that both the T-handle and the outer sleeve are turned.

   Technician B says that either the T-handle or the outer sleeve may be turned.

   Who is correct?

   A. A only   C. Both A and B

   B. B only   D. Neither A nor B

9. *Technician A* says that all orifice tubes are the same size, but they are not interchangeable.

   Technician B says that all orifice tubes are not the same size, but they are interchangeable.

   Who is correct?

   A. A only   C. Both A and B

   B. B only   D. Neither A nor B

10. *Technician A* says that mineral oil may be used to lubricate O-rings used on an R-134a system.

    Technician B says that PAG lubricants may be used on O-rings for an R-134a system. Who is correct?

    A. A only   C. Both A and B

    B. B only   D. Neither A nor B

ASE CHALLENGE QUESTIONS

1. A leaking evaporator is being replaced on a vehicle.

   Technician A says that some additional refrigerant oil should be added to the air-conditioning system prior to recharging it.

   Technician B says that the new O-rings or seals should be coated with clean refrigerant oil.

   Who is correct?

   A. A only   C. Both A and B

   B. B only   D. Neither A nor B

2. After the refrigerant has been recovered from an air-conditioning system, which of the following should a technician do?

   A. Change the recovery unit filters.

   B. Record the amount of refrigerant oil removed from the system.

   C. Check the purity of the refrigerant in the recovery tank.

   D. Measure the acidity level of the refrigerant in the recovery tank.

3. The refrigerant is being recovered from an air-conditioning system.

   Technician A says that both the vapor and liquid tank valves on the recovery unit storage tank valves must be opened.

   Technician B says that the high- and low-pressure valves on the manifold and gauge set must be opened.

   Who is correct?

   A. A only   C. Both A and B

   B. B only   D. Neither A nor B

4. Most vendors will not honor the warranty on a new or rebuilt compressor if the _____ is not also replaced at the time of service.
   A. Accumulator or receiver-drier
   B. Expansion valve or orifice tube
   C. O-ring seals or gaskets
   D. Oil or lubricant

5. The illustration below shows:
   A. A spring lock coupling being removed.
   B. An O-ring seal or gasket being removed.
   C. Both A and B
   D. Neither A nor B

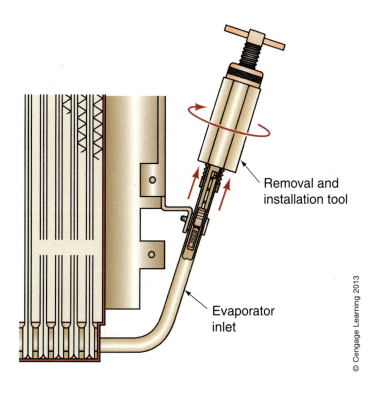

Removal and installation tool

Evaporator inlet

© Cengage Learning 2013

Name _____ **Date** _____

DETERMINING THE TYPE OF AIR-CONDITIONING SYSTEM

Upon completion of this job sheet, you should be able to identify the type of air-conditioning system: cycling clutch or noncycling clutch.

NATEF Correlation

A/C System Diagnosis and Repair; *Identify refrigerant type; conduct a performance test of the A/C system; determine necessary repairs.* **(P-1)**

Tools and Materials

An air conditioned vehicle

Manufacturer's service manual

Describe the vehicle being worked on.

Year _____ Make _____ Model _____

VIN _____ Engine type and size _____

Procedure

1. Visually inspect the air-conditioning system and describe its overall condition.

    _____

    _____

2. In a well-ventilated area, start the engine, place the transmission in PARK, and turn on the air-conditioning system to MAX cooling. If a standard transmission, place in NEUTRAL and chock the wheels. Describe your procedure for accomplishing this step.

    _____

    _____

3. Move the cold control from one extreme to the other.
    a. Does the compressor cycle off and on? Describe your findings.

    _____

    _____

    b. Does the heater control valve change positions? Describe your findings.

    _____

    _____

4. What is the type of system? Cycling clutch or noncycling clutch?

    _____

    _____

5. How did you make this determination?

_____

_____

6. Locate the temperature control information in the shop manual. Where is this information located?

_____

_____

7. Does the information presented in step 6 verify your determination of step 5?

_____

_____

Instructor's Response _____

_____

_____

Name _____ **Date** _____

DETERMINING THE REFRIGERANT TYPE

Upon completion of this job sheet, you should be able to identify the type of refrigerant used in the air-conditioning system.

NATEF Correlation

HEATING AND AIR CONDITIONING: A/C System Diagnosis and Repair; *Identify refrigerant type; conduct a performance test of the A/C system; determine necessary repairs.* **(P-1)**

Tools and Materials

An air conditioned vehicle

Manufacturer's shop manual

Describe the vehicle being worked on.

Year _____ Make _____ Model _____

VIN _____ Engine type and size _____

Procedure

1. Visually inspect the air-conditioning system and describe its overall condition.
   _____
   _____

2. Inspect the low-side service fitting. Is it a R-12, R-134a, or other type of fitting? Describe.
   _____
   _____

3. Inspect the high-side service fitting. Is it a R-12, R-134a, or other type of fitting? Describe.
   _____
   _____

4. Are there any decals under the hood to identify the refrigerant type? Describe your findings.
   _____
   _____

5. What type of refrigerant is in the system?
   _____
   _____

6. Does the shop manual verify your findings?

_____

_____

7. What special precautions would you take when recovering this refrigerant?

_____

_____

Instructor's Response _____

_____

_____

Name _____ Date _____

COMPONENT TEMPERATURE TESTING

Upon completion of this job sheet, you should be able to determine the normal external temperature of various air-conditioning system components and determine if repairs are required.

NATEF Correlation

HEATING AND AIR CONDITIONING: A/C System Diagnosis and Repair; *Identify and interpret heating and air conditioning concerns; determine necessary action.* **(P-1)**

Tools and Materials

Late-model vehicle

Service manual or information system

Safety glasses or goggles

Hand tools, as required

Thermocouple on a digital multimeter or an infrared noncontact thermometer

Describe the vehicle being worked on.

Year _____ Make _____ Model _____

VIN _____ Engine type and size _____

Procedure

Give a brief description of your procedure following each step. Always wear eye protection when working on or around refrigerant system components.

1. Start the engine and turn the air-conditioning system on. Allow the vehicle to run for several minutes.

   _____

   _____

2. Using a contact thermocouple on a digital multimeter or an infrared noncontact thermometer, measure the temperature of each component listed below.

3. If the temperature is in a safe range, you may touch the hoses and components on the high side of the system to determine if the components' temperature is consistent throughout. A refrigerant pressure of 199 psig corresponds to a temperature of 130°F for R-134a. Measure or touch the entire length of the hose or component and note any temperature differential. All high-side system components should be warm.

   a. High-side discharge refrigerant line from the compressor to the condenser: Temperature recorded _____

> ⚠️ **CAUTION:**
> Be aware that system malfunctions can cause the high-side components to become superheated to the point that a serious burn may result. Care must be taken when handling any high-side system components.

b. On the graph below, record the condenser surface temperature at various grid locations to show how heat flows through the condenser.

c. High-side liquid line leaving the condenser to the receiver-drier (if equipped): Condenser outlet temperature _____

d. If equipped, record the receiver-drier inlet and outlet temperatures:

Receiver-drier inlet temperature _____

Receiver-drier outlet temperature _____

4. Record the inlet and outlet temperatures of the expansion valve or fixed orifice tube (FOT).

Expansion valve (FOT) inlet temperature _____

Expansion valve (FOT) outlet temperature _____

Frost on the outlet side of the metering device is acceptable. But frost on the inlet side of the metering device may be an indication of a defective metering device or a sign of excessive moisture in the refrigerant system.

5. If the temperature is in a safe range, you may touch the hoses and components on the low side of the system to determine if the components' temperature is consistent throughout. A refrigerant pressure of 27 psig corresponds to a temperature of 31°F for R-134a. Measure or touch the entire length of the hose or component and note any temperature differential. All low-side system components should be cool.

a. Record the evaporator inlet and outlet temperatures.
Evaporator inlet temperature _____

Evaporator outlet temperature _____

b. Record the evaporator suction line-to-compressor inlet temperature. _____

c. If equipped, record the accumulator inlet and outlet temperatures.
Accumulator inlet temperature _____

Accumulator outlet temperature _____

Instructor's Response _____

_____

_____

Name _____ **Date** _____

IDENTIFYING HOSE FITTINGS

Upon completion of this job sheet, you should be able to identify the type of hose fittings used in the air-conditioning system.

NATEF Correlation

HEATING AND AIR CONDITIONING: Refrigeration System Component Diagnosis and Repair; Evaporator, Condenser, and Related Components; *Remove and inspect A/C system mufflers, hoses, lines, fittings, O-rings, seals, and service valves; perform necessary action.* **(P-2)**

Tools and Materials

An air conditioned vehicle

Manufacturer's shop manual

Describe the vehicle being worked on.

Year _____ Make _____ Model _____

VIN _____ Engine type and size _____

Procedure

1. Visually inspect the air-conditioning system and describe its overall condition.

   _____

   _____

2. Inspect the hose-to-condenser inlet fitting. Describe its type.

   _____

   _____

3. Inspect the high-side liquid line-to-evaporator inlet fitting. Describe its type.

   _____

   _____

4. Are the two hose fittings (steps 2 and 3) interchangeable? Explain.

   _____

   _____

5. Why do you think that a barb-type fitting should not be used with a barrier hose?

   _____

   _____

6. Locate hose fittings in the shop manual. Does the shop manual verify your findings?

_____

_____

7. What special precautions would you take when servicing hoses and fittings?

_____

_____

Instructor's Response _____

_____

_____

Name _____ Date _____

RECOVER AND RECYCLE REFRIGERANT

Upon completion of this job sheet, you should be able to recover and recycle refrigerant.

NATEF Correlation

HEATING AND AIR CONDITIONING: Refrigerant Recovery, Recycling, and Handling; *Identify (by label application or use of a refrigerant identifier) and recover A/C system refrigerant.* **(P-1)**

HEATING AND AIR CONDITIONING: Refrigerant Recovery, Recycling, and Handling; *Recycle refrigerant.* **(P-1)**

Tools and Materials

Vehicle with air-conditioning system in need of service

Selected air-conditioning system service tools

Manifold and gauge set with hoses

Refrigerant recovery/recycle machine

Recovery tank

Describe the vehicle being worked on.

Year _____ Make _____ Model _____

VIN _____ Engine type and size _____

Procedure

After each of the following procedures, briefly explain how you performed the task:

1. What type of refrigerant does the under-hood label identify in the system? _____

2. Do the vehicle refrigerant service fittings match the refrigerant listed on the identification label? _____

3. Attach the refrigerant purity identifier to the system.
   a. Type of refrigerant identified _____

   b. Does refrigerant identified match refrigerant label on vehicle? _____

   c. Purity of refrigerant tested _____

   d. Was the presence of contamination detected? _____

4. Connect the gauge manifold hoses to the air-conditioning system service ports.
   _____

   _____

5. What type connectors are found:

On the low-side service port? _____

On the high-side service port? _____

6. Observe the gauges.

The low-side gauge reads _____

The high-side gauge reads _____

NOTE: If zero (0 psig or 0 kPa) or below pressure is observed in step 6, it may be assumed that there is no refrigerant in the system. If it is presumed that there are residual traces of refrigerant in the lubricant, proceed with step 7.

7. Connect the manifold gauge hose to the recovery unit.

8. What type connector is found on the recovery unit?

_____

_____

9. Start the recovery unit, open the appropriate hand valves, and recover air-conditioning system refrigerant following instructions provided by the recovery equipment manufacturer.

_____

_____

10. Close the hand valves (opened in step 9) and turn off the recovery unit. Observe the gauges. What is the reading in psig or kPa?

	Now	After 5 Min.	After 10 Min.	After 15 Min.
Low Side	_____	_____	_____	_____
High Side	_____	_____	_____	_____

11. Explain the conclusion of the results of step 10.

_____

_____

12. Carefully remove the manifold and gauge set hoses ensuring that no ambient air enters the system.

Instructor's Response _____

_____

_____

Name _____ **Date** _____

REPLACING THE RECEIVER-DRIER/ACCUMULATOR

Upon completion of this job sheet, you should be able to identify a receiver-drier/accumulator and describe how to replace it.

NATEF Correlation

HEATING AND AIR CONDITIONING: Refrigeration System Component Diagnosis and Repair, Evaporator, Condenser, and Related Components; *Remove and reinstall receiver-drier or accumulator-drier; measure oil quantity; determine necessary action.* **(P-1)**

Tools and Materials

An air conditioned vehicle with a defective receiver-drier/accumulator

Refrigerant recovery station

Set of mechanic's hand tools

Manufacturer's shop manual

Describe the vehicle being worked on.

Year _____ Make _____ Model _____

VIN _____ Engine type and size _____

Procedure

1. Visually inspect the air-conditioning system and describe its overall condition.
   _____
   _____

2. Locate the receiver-drier/accumulator. Describe its location.
   _____
   _____

3. Troubleshoot the receiver-drier/accumulator following procedures outlined in the shop manual. Is it defective? Explain your findings.
   _____
   _____

4. Are there decals under the hood to identify the refrigerant type? Describe your findings. What type of refrigerant is in the system?
   _____
   _____

5. Does the shop manual verify your findings?

_____

_____

6. What procedure would you follow to replace the receiver-drier/accumulator?

_____

_____

Instructor's Response _____

_____

_____

Name _____ **Date** _____

REPLACING THE SUPERHEAT OR PRESSURE SWITCH

Upon completion of this job sheet, you should be able to troubleshoot and replace a high-pressure release device.

NATEF Correlation

HEATING AND AIR CONDITIONING: Refrigeration System Component Diagnosis and Repair; *Diagnose A/C system conditions that cause the protection devices (pressure, thermal, and PCM) to interrupt system operation; determine necessary action.* **(P-2)**

HEATING AND AIR CONDITIONING Operating Systems and Related Controls Diagnosis and Repair; *Test and diagnose A/C compressor clutch control systems; determine necessary action.* **(P-1)**

Tools and Materials

An air conditioned vehicle with pressure switch

Refrigerant recovery station

Set of mechanic's hand tools

Manufacturer's shop manual

Describe the vehicle being worked on.

Year _____ Make _____ Model _____

VIN _____ Engine type and size _____

Procedure

1. Visually inspect the air-conditioning system and describe its overall condition.

   _____

   _____

2. Locate the pressure switch(es). Describe its/their location.

   _____

   _____

3. Troubleshoot the low-/high-pressure switch following procedures outlined in the shop manual. Is it defective? Explain your findings.

   _____

   _____

4. Can the pressure switch be replaced without recovering the refrigerant? Explain your answer.

   _____

   _____

5. Are there decals under the hood to identify the refrigerant type? Describe your findings. What type of refrigerant is in the system?

_____

_____

6. Does the shop manual verify your findings?

_____

_____

7. What procedure would you follow to replace the low-/high-pressure switch?

_____

_____

Instructor's Response _____

_____

_____

Name _____ **Date** _____

AIR-CONDITIONING SYSTEM PERFORMANCE TEST

Upon completion of this job sheet, you should be able inspect and test the air-conditioning system for normal operation.

NATEF Correlation

HEATING AND AIR CONDITIONING: A/C System Diagnosis and Repair; *Performance test A/C system; diagnose A/C system malfunctions using principles of refrigeration.* **(P-1)**

HEATING AND AIR CONDITIONING A/C System Diagnosis and Repair; *Identify refrigerant type; conduct a performance test of the A/C system; determine necessary action.* **(P-1)**

Tools and Materials

Late-model vehicle

Service manual or information system

Safety glasses or goggles

Hand tools, as required

Manifold and gauge set

Thermometer

Describe the vehicle being worked on.

Year _____ Make _____ Model _____

VIN _____ Engine type and size _____

Procedure

Following the procedures outlined in the service manual, give a brief description of your procedure following each step. Ensure that the engine is cold and wear eye protection.

NOTE: The following procedures outlined are for an R-134a refrigerant system.

Task Completed

1. Place a fender cover on the vehicle to protect the finish. ☐
2. Remove the protective caps from the high- and low-side service ports. ☐

   NOTE: Remove the caps slowly to ensure that no refrigerant escapes past a defective service valve, and inspect the O-ring seal on the cap.

3. Ensure that all manifold valves are closed (clockwise) to prevent refrigerant venting from service ports. ☐
4. Turn the low-side hose hand valve fully counterclockwise to retract the Schrader depressor into the service port. ☐
5. Clip on the low-side hose; push firmly until it clicks in place. ☐
   a. Turn the low-side hose hand valve fully clockwise to extend the Schrader depressor. ☐
6. Clip on the high-side hose; push firmly until it clicks on. ☐
   b. Turn the high-side hose hand valve fully clockwise to extend the Schrader depressor. ☐

☐   7. Place a fan in front of the vehicle to assist in cooling the A/C condenser.

☐   8. Start the engine.

☐   9. Turn on the air conditioner; set all controls to control settings listed in table and note the blower motor operation.

Control Setting	OK/NOT OK
LO	
LO-MED	
HI-MED	
HI	

☐   10. Turn on the air conditioner; set all controls to maximum cooling; set the blower speed on HI.

☐   11. Insert a thermometer in the air conditioning duct as close to the evaporator core as possible and note which duct this is. Also record the following duct temperatures.

12.

	Right Duct	Center Duct	Left Duct
Duct Outlet Temperature			

13. If equipped with a sight glass on the receiver-drier, note the refrigerant flow.

     a. Clear _____

     b. Bubbles _____

     c. Foggy _____

14. Record pressure levels on gauges.

	Low Side	High Side
System Pressure		
Outside air temperature		
System expected pressure based on Chart 4-21 Class Manual		

Instructor's Response _____

_____

_____

Name _____ **Date** _____

REPLACE AC EXPANSION VALVE

Upon completion of this job sheet, you should be able to recover refrigerant from an A/C system, remove and replace the expansion valve assembly, evacuate and recharge the system, as well as perform a leak test of the system.

NATEF Correlation

Inspect engine cooling and heater system hoses and belts; perform necessary action. **(P-2)**

Tools and Materials

Test vehicle

Refrigerant recovery and recycling unit

Manifold and gauge set

Basic hand tools

Safety glasses

Describe the vehicle being worked on.

Year _____ Make _____ Model _____

VIN _____ Engine type and size _____

Procedure

Task Completed

1. Determine the type of refrigerant and expansion valve used in the system.
   a. Refrigerant _____
   _____

   b. Expansion valve _____
   _____

2. Connect the gauge set to the system. Note the high-side and low-side pressures with the system off.
   a. High-side pressure _____
   _____

   b. Low-side pressure _____
   _____

3. Connect the manifold gauge set-to-recovery unit. Start the unit following unit manufacturer instructions. ☐

4. Wait 5 minutes and note if pressure is detected (if pressure is present rerun the recovery process). System should hold a vacuum for 2 minutes. Did it? _____ ☐

5. Follow manufacturer's recommended procedure for removal and replacement of component. ☐

Standard Valve

6. Remove the bulb from the insulation. Describe the location and the method of holding the bulb and external equalizer, if equipped, in place. _____

_____

_____

_____

_____

_____

_____

7. Disconnect the inlet side from the line. Which line is this one and which tool is required to disconnect? _____

_____

_____

_____

_____

_____

_____

8. Inspect the screen. Describe its condition and your actions to correct any problems. ___

_____

_____

_____

_____

_____

_____

9. Disconnect the valve from the evaporator. Which tool is required to disconnect? ____

_____

_____

_____

_____

_____

10. Remove the valve from the evaporator. Explain the procedures and the tools used. ___

_____

_____

_____

_____

_____

☐ 11. Remove all O-rings. Install all new O-rings. Lube the rings with the proper lubricant.

☐ 12. Position the valve in the evaporator and tighten the clamp, if equipped.

13. Connect the valve to the evaporator. Torque _____

_____

14. Connect the line to the valve inlet. Torque _____

_____

15. Position and secure the equalizer and bulb. How is the bulb secured? _____

_____

_____

_____

_____

_____

_____

H-block Expansion Valve   ☐

16. Complete steps 1 and 2.

17. Are there any electrical connections? If so, what circuit(s) is (are) involved? Disconnect as needed. _____

_____

_____

_____

_____

_____

_____

18. Disconnect the line-sealing plate from the valve. Explain the procedures and the tools used. _____

_____

_____

_____

_____

_____

_____

19. Remove the valve from the evaporator. Explain the procedures and the tools used. ___

_____

_____

_____

_____

_____

_____

20. Remove all gaskets and clean the mating surfaces as needed.   ☐
21. Install the valve/evaporator gasket.   ☐

☐

22. Mount the valve to the evaporator. Torque _____

_____

_____

☐

23. Install the line plate/valve gasket.

24. Connect the line-sealing plate to the valve. Torque _____

_____

25. Evacuate the system. Did the system hold a vacuum?

26. Recharge the system.

    a. Record the amount of refrigerant added._____

_____

☐

27. Check the system for leaks using a halon leak detector.

28. Record the system running pressure.

    a. High-side pressure_____

_____

    b. Low-side pressure_____

_____

Instructor's Response _____

_____

_____

Name _____ **Date** _____

REPLACE A/C ORIFICE (EXPANSION) TUBE

Upon completion of this job sheet, you should be able to recover refrigerant from an A/C system, remove and replace the orifice (expansion) tube assembly, evacuate and recharge the system, as well as perform a leak test of the system.

NATEF Correlation

HEATING AND AIR CONDITIONING: Refrigeration System Component Diagnosis and Repair; *Remove and install expansion valve or orifice (expansion) tube.* **(P-2)**

Tools and Materials

Test vehicle

Refrigerant recovery/recycling/recharge unit

Manifold and gauge set

Basic hand tools

Safety glasses

FOT and O-ring seal

Describe the vehicle being worked on.

Year _____ Make _____ Model _____

VIN _____ Engine type and size _____

Procedure

1. Determine the type of refrigerant used in the system.
   a. Refrigerant _____

2. Connect both the high- and low-side gauges to the vehicle refrigerant system. Note the high-side and low-side pressures with the system off.
   a. High-side pressure _____

   b. Low-side pressure _____

3. Begin the refrigerant recovery process as outlined in Job Sheet 21, Recover and Recycle Refrigerant, by initializing the recovery unit. Start the unit following unit manufacturer's instructions.

4. Wait 5 minutes and note if pressure is detected (if pressure is present, rerun the recovery process). System should hold a vacuum for 2 minutes. Did it? _____

5. Record the amount of refrigerant recovered, if the service unit you are using has this feature. _____

6. Locate the fixed orifice tube (FOT) in the high-side liquid refrigerant line between the condenser outlet and the evaporator inlet. There is usually a visible dimple in the liquid line where the FOT is located and a line connection point.

   a. Where was the FOT in relation to the condenser outlet and the evaporator inlet? _____

   _____

7. Disconnect the line fitting where the FOT is located. Which line is this one and which tool is required to disconnect? _____

8. Remove and discard the O-ring(s) from the liquid line fitting, if equipped.

9. Insert the FOT removal tool onto the FOT.

10. Turn the T-handle and turn the outer sleeve or spool of the tool clockwise to remove the FOT. Do not turn the T-handle.

    NOTE: If the FOT breaks during removal, proceed with step 11. If it does not break, proceed to step 13.

11. Insert the extractor tool into the line where the remaining piece of the FOT is located and turn the T-handle clockwise until the threaded portion of the tool is securely inserted into the brass portion of the broken FOT.

12. Pull the tool. The broken FOT should slide out.

    NOTE: If the FOT brass tube pulls out of the plastic body, remove the brass tube and repeat steps 11 and 12.

13. Inspect the screen on the FOT. Describe its condition and your actions to correct any problems. _____

    _____

14. Verify that the replacement FOT is the same color as the one that was removed.

15. Liberally coat the new FOT with clean refrigerant oil.

16. Place the FOT into the refrigerant line and push it until it stops against the dimples on the line.

17. Install all new O-rings. Lube the rings with the proper lubricant.

18. Connect the line connection and torque to specifications. Torque _____

19. Evacuate the system. Did the system hold a vacuum? _____

20. Recharge the system.

    a. Record the amount of refrigerant added. _____

21. Check the system for leaks using a halon leak detector.

22. Record the system running pressure.

    a. High-side pressure _____

    b. Low-side pressure _____

Instructor's Response _____

_____

_____

Name _____ **Date** _____

REPLACE AIR-CONDITIONING CONDENSER

Upon completion of this job sheet, you should be able to recover refrigerant from an A/C system, remove and replace the condenser assembly, evacuate and recharge the system, as well as perform a leak test of the system.

NATEF Correlation

Evaporator, Condenser, and Related Components; *Inspect A/C condenser for airflow restrictions; perform necessary action.* **(P-1)**

Tools and Materials

Test vehicle

Refrigerant recovery and recycling unit

Manifold and gauge set

Basic hand tools

Safety glasses

Describe the vehicle being worked on.

Year _____ Make _____ Model _____

VIN _____ Engine type and size _____

Procedure

_____

1. Connect the gauge set to the system. Note the high-side and low-side pressures with the system off.

   a. High-side pressure _____

   _____

   b. Low-side pressure _____

   _____

Task Completed

2. Connect the manifold gauge set to the recovery unit. Start the unit following the unit manufacturer's instructions. ☐

3. Wait 5 minutes and note if pressure is detected (if pressure is present, rerun the recovery process). The system should hold a vacuum for 2 minutes. Did it? _____

4. Follow the procedures outlined earlier in "Removing and Replacing the Condenser" and refer to the manufacturer's recommended procedure for removal and replacement of the component. ☐

5. Evacuate the system. Did the system hold a vacuum? _____

6. Follow the manufacturer's recommended procedure for the removal and replacement of the component. ☐

7. Recharge the system.

   a. Record the amount of refrigerant added. _____

☐ 8. Check the system for leaks using a halon leak detector.

9. Record the system running pressure.

   a. High-side pressure _____

_____

   b. Low-side pressure _____

_____

Instructor's Response _____

_____

_____

Name _____ **Date** _____

REPLACE THE A/C REFRIGERANT LINE

Upon completion of this job sheet, you should be able to recover refrigerant from an A/C system, remove and replace a refrigerant line assembly, evacuate and recharge the system, as well as perform a leak test of system.

NATEF Correlation

HEATING AND AIR CONDITIONING: Refrigeration System Component Diagnosis and Repair; *Remove and inspect A/C system mufflers, hoses, lines, fittings, O-rings, seals, and service valves; perform necessary action.* **(P-2)**

Tools and Materials

Test vehicle

Refrigerant recovery/recycling/recharge unit

Manifold and gauge set

Basic hand tools

Safety glasses

Refrigerant line and O-ring seals

Describe the vehicle being worked on.

Year _____ Make _____ Model _____

VIN _____ Engine type and size _____

Procedure

1. Determine the type of refrigerant used in the system.
   a. Refrigerant _____

2. Connect both the high- and low-side gauges to the vehicle refrigerant system. Note the high-side and low-side pressures with the system off.
   a. High-side pressure _____

   b. Low-side pressure _____

3. Begin the refrigerant recovery process as outlined in Job Sheet 21, Recover and Recycle Refrigerant, by initializing the recovery unit. Start the unit following the unit manufacturer's instructions.

4. Wait 5 minutes and note if pressure is detected (if pressure is present rerun the recovery process). System should hold a vacuum for 2 minutes. Did it? _____

5. Record the amount of refrigerant recovered, if the service unit you are using has this feature. _____

6. Locate the refrigerant line to be replaced.
   a. What other components will have to be removed to gain access to the refrigerant line requiring replacement? _____

   _____

   _____

   b. What is the name of the line being replaced and is it located on the high- or low-side of the refrigerant system? _____

   _____

   _____

   c. Does the line normally transport refrigerant in the liquid or gaseous state? _____

   _____

7. Disconnect the line fittings using the appropriate service tools. Which tool is required to disconnect the line and fittings? _____

   _____

8. Remove the old line from the vehicle.
9. Remove and discard the O-ring(s) from the line fittings, if equipped.
10. Position the new line into mounting brackets, if equipped.
11. Install all new O-rings. Lube rings with the proper lubricant.
12. Connect the line connection and torque to specifications. Torque _____

13. Evacuate system. Did system hold a vacuum? _____

14. Recharge system.
    a. Record the amount of refrigerant added. _____

15. Check the system for leaks using a halon leak detector.
16. Record the system running pressure.
    a. High-side pressure _____

    b. Low-side pressure _____

17. Record the system vent temperature.

Instructor's Response _____

_____

_____

Air-Conditioning System Servicing and Testing

BASIC TOOLS

Manifold and gauge set (R-12 or R-134a, as applicable)

Refrigerant recovery system (R-12 or R-134a, as applicable)

Vacuum pump (oil-less, R-12 or R-134a, as applicable)

Charging cylinder (R-12 or R-134a, as applicable)

Scales (if using refrigerant cylinders)

Can tap (if using small cans of refrigerant)

Dye injector

Halogen leak detector

Safety glasses

Fender cover

Small brush

Upon Completion and Review of this Chapter, you should be able to:

- Determine refrigerant system contamination by performing a refrigerant purity test with a refrigerant analyzer.

- Perform an unloaded refrigerant system performance test.

- Analyze performance test results.

- Determine if the refrigerant system is contaminated with refrigerant sealant.

- Leak-test an air-conditioning system using a soap solution.

- Leak-test an air-conditioning system using a halogen leak detector.

- Leak-test an automotive air-conditioning system using a dye solution.

- Evacuate an air-conditioning system using the single evacuation method.

- Evacuate an air-conditioning system using the triple evacuation method.

- Charge the system with refrigerant CFC-12 (R-12).

- Charge the system with refrigerant HFC-134a (R-134a).

- Test the refrigerant for noncondensable gas contamination.

Generally, the first piece of equipment that a service technician reaches for is the manifold and gauge set (Figure 6-1). The manifold and gauge set to the technician is much the same as a blood pressure test kit is to a physician. It provides a means to "see" what is happening inside the system. Pressures inside an air-conditioning system are as important to the system as pressures inside the body are to the human.

Procedures for the proper use of the manifold and gauge set are given in Chapter 4 of this manual. Review of Chapter 4 would be a good idea at this time.

Some systems also have a sight glass.

 WARNING: Safety glasses must be worn while working with refrigerants. Remember, liquid refrigerant splashed in the eyes can cause blindness.

Refrigeration Contamination

Before servicing an air-conditioning system, it should first be determined what type of refrigerant is in the system and, perhaps more important, if the refrigerant is contaminated. Recent studies indicate that, on average, 23 out of every 1000 motor vehicles tested contained some form of **contaminated refrigerant**. That amounts to over 450,000

Contaminated refrigerant refers to any refrigerant that is not 98 percent pure. Refrigerant may be contaminated if it contains excess air or another type of refrigerant.

FIGURE 6-1 Manifold and gauge set.

© Cengage Learning 2013

contaminated systems out of the 20 million vehicles serviced each year. A system is considered to be contaminated if it contains more than 2 percent of a "foreign" substance. Since the average system contains less than 3 lb. (1.42 L) of refrigerant, contaminants may not exceed 0.96 oz. (28.4 mL). There are actually several types of contamination to be found in the automotive air-conditioning system. In order of occurence, these contaminants include air, moisture, mixed refrigerant types, and illegal refrigerants.

A refrigerant identifier should be used to determine the purity of the refrigerant before servicing an air-conditioning system. Failure to do so may result in personal injury if it contains a flammable substance, or in a contaminated recovery cylinder if it contains other types of refrigerant.

The refrigerant gas analyzer measures and displays the purity of the refrigerant being tested as a percentage along with the percentage of air (noncondensable gas) and the presence of hydrocarbons (HC) in the sample being tested.

The following procedure is for the GA500-Plus Gas Analyzer (Figure 6-2). Procedures may vary for other models. Always follow the operating instructions that accompany the equipment being used:

Classroom Manual
Chapter 6, page 181

1. Connect the purity test unit power cable to the vehicle's battery.
2. Attach the sample hose to the vehicle's refrigerant system service fitting.
   a. Select the R-134a (or the R-12) hose and push the quick-connect coupler onto the fitting on the bottom of the analyzer. Turn the fitting knob to the left to retract the valve activating pin.
   b. Connect the sample hose service port coupler to the low side (vapor port) of the air-conditioning system and turn the knob to the full right position to open the valve port.
3. Fully depress the purity test unit pump button four times to start the test cycles.
4. Evaluate the sample using light-emitting diode (LED) indicators and liquid crystal display (LCD).
   a. Record the refrigerant purity:
   _____ R-134a _____ R-12 _____ HC _____ R-22 _____ Air
   b. Percent of air contamination

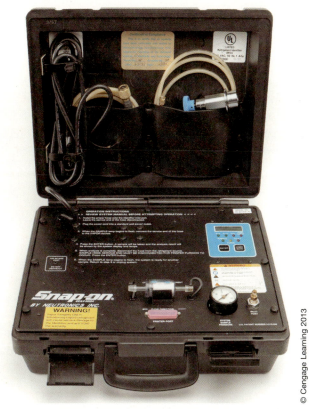

FIGURE 6-2 **The refrigerant gas analyzer measures and displays the purity of the refrigerant being tested.**

NOTE: Because air is a noncondensable gas, the analyzer ignores its presence in calculating gas concentrations. Consequently the total of all displays may be greater than 100 percent.

5. To repeat the test, pump the button four times to restart the test cycles.
6. Disconnect the sample hose attached to the vehicle's R-134a system; push the pump button to purge the refrigerant in the analyzer.

Refrigerant systems should be tested with a refrigerant identifier because cross-contamination of refrigerant will also affect the temperature-pressure relationship of the refrigerant gas. Cross-contamination is when a system designed and containing one refrigerant is partially charged with another refrigerant on top of the existing refrigerant. An example of this can occur when a system containing R-12 is not properly recovered, evacuated, and recharged with R-134a with proper service fittings and labels installed, but instead is partially charged with another refrigerant such as R-134a on top of the R-12 that remains in the original system. Unfortunately with retrofit kits for R-12 to R-134a readily available at all parts and discount retailers, any consumer can purchase and incorrectly service his or her pre-1994 R-12 vehicle. Be wary of retrofitted R-12 vehicles that do not have a retrofit label installed by a reputable service facility.

Noncondensable (air) or cross-contaminated refrigerant can cause reduced system performance, lack of lubrication issues, and chemical breakdown. If recovery/recycling equipment is connected to a cross-contaminated system the unit will have to be cleaned out and all the filters and dehydrator element will have to be replaced. If the cross-contaminated equipment is unknowingly connected to other vehicle air-conditioning systems being serviced, the resulting performance issues can spread like a virus, creating a service nightmare for the repair shop as well as extremely high repair costs to fix the issues created.

If a system or portable recovery container is cross-contaminated or contains an unknown refrigerant, it must be disposed of properly. The Environmental Protection Agency (EPA) does not allow venting of any automotive refrigerant into the atmosphere no matter what is in the system. The acceptable method is to recover the refrigerant into a dedicated recovery-only unit for unknown or cross-contaminated refrigerants. The contaminated refrigerant should be recovered into a Department of Transportation (DOT) approved gray-with-yellow-top recovery tank. When full it must be disposed of with a local recycler in accordance with local, state, and federal laws. In addition, check with local ordinances that govern the storage of combustible or hazardous materials. Because of the many requirements and additional service equipment required, many shops will not service contaminated refrigerant systems.

UNLOADED SYSTEM PERFORMANCE TESTING

When a customer complains of poor air conditioning performance an air-conditioning system unloaded performance test should be the initial test to determine if the refrigerant system is operating as designed. As a service technician you should always follow the manufacturer's service and diagnostic information to determine if the air-conditioning system is operating as designed.

A thorough diagnosis of the air-conditioning system begins with a comparison of the temperature-pressure relationship (Figure 6-3) data of system performance. The role of this

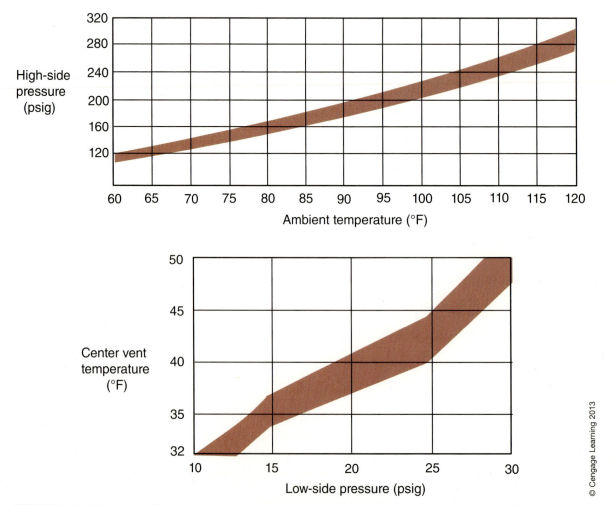

FIGURE 6-3 R-134a air-conditioning system unloaded performance test data chart requires that the vehicle's windows and doors be closed and that the recirculation vent control mode be selected for accurate test results.

© Cengage Learning 2013

type of chart is to provide a performance test platform that is reliable and easy to understand. The relative humidity of any given day has no effect on the unloaded performance test. It will not affect the high-side system pressure unless the Fresh Air mode is selected instead of the Recirculation mode, which should be selected. The unloaded performance test takes into account system pressures and temperatures in analyzing system condition.

During the unloaded performance test you are required to select the Recirculation mode, which dries the air as it is recirculating during system stabilization. The unloaded performance test's use of recirculated dry air allows for consistent achievement of the lowest possible vent temperatures and provides a consistent performance comparison of similar vehicles regardless of ambient air temperatures or humidity levels on a particular day. Thus preconditioning the vehicle in the pretest procedure that follows is critical to accurate system testing.

Pretest Procedure

It is important to prepare and inspect the vehicle prior to beginning the test:

1. Move the vehicle to a shaded area not in direct sunlight and allow it to cool down. The performance test should not be performed until the vehicle has reached ambient air temperature.
2. Inspect the condenser fins and clean them with a soft brush and non-pressurized running water. A condenser that restricts airflow can give the appearance, based on pressure readings, that the system is fully charged when it is actually low on refrigerant.
3. Close all windows and doors.
4. Open the vehicle hood.
5. Turn the vent fan on high and activate the air-conditioning system in the Recirculation mode.
6. Next, run the engine at 1,000 rpm and allow the system to stabilize for a minimum of 15 minutes. Measure the air temperature leaving the center vents. The unloaded test strips the air in the passenger compartment of heat and humidity as it is recirculated through the evaporator, allowing the lowest possible vent temperatures to be achieved. The performance chart in Figure 6-3 for the unloaded performance test requires that the engine speed be 1,000 rpm ($\pm$50 rpm) to be accurate.

Test Procedure

1. Connect an R-134a recovery/recycling station to the air-conditioning system high-side and low-side pressure test fittings.
2. Record the ambient air temperature approximately 12 in. in front of the vehicle's condenser.
3. With the engine running at 1000 rpm check both the high- and low-side system pressures. If the vehicle is equipped with a rear air-conditioning system ensure that the rear blower is off and record the result.
4. Place a thermometer in the center dashboard vent and record the outlet air temperature.
5. If the vehicle is equipped with a rear air-conditioning system, turn the rear blower on high and place a thermometer in the rear outlet vent and record the outlet air temperature. With the engine running at 1000 rpm check both the high- and low-side system pressures and record the result.
6. Increase the engine speed to 3000 rpm and measure the low-side system pressure as well as the dashboard center vent outlet air temperature and record the results.

   NOTE: The pressure gauge readings at 1000 rpm are not comparable to the observations at 3000 rpm.
7. Compare the readings recorded to the temperature pressure graph in Figure 6-3.

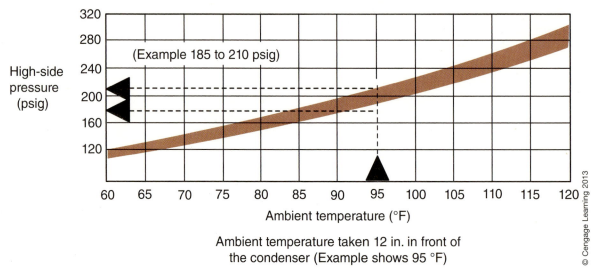

FIGURE 6-4 The high-side performance chart for R-134a allows the technician to determine if the vehicle's air-conditioning system high-side pressure is in the normal range for current ambient air temperature.

To interpret the high-side performance chart, first determine the shop temperature (step 2) and draw a vertical line on the graph (Figure 6-4). Next, draw horizontal lines from both the top and bottom of the black pressure band on the chart across to the high-side pressures on the left of the graph. The high-side pressure reading (step 3) should be between the two horizontal pressure marks for a normally operating system.

Next, interpret the low-side performance chart. First, determine the low-side system pressure (step 3) and draw a vertical line on the graph (Figure 6-5). Next, draw a horizontal line from the point where the vertical line meets the top of the black pressure band on the chart across to the center vent temperature on the left of the graph. The center vent temperature in the passenger compartment (step 4) should be at or below the temperature indicated on the graph for a normally operating system. As long as the evaporator is not icing, the temperature may be colder than the chart indicates but the temperature should never be lower than 33°F (0.6°C). In general, to obtain 40°F (4.4°C) center vent temperature air, the air conditioning low-side temperature generally needs to be 20 psig (137.9 kPa).

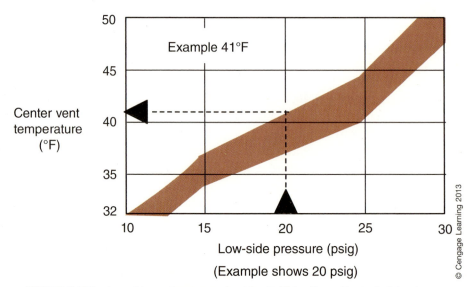

FIGURE 6-5 The low-side performance chart for R-134a allows the technician to determine if the vehicle's air-conditioning system vent temperature is correct for the current low-side operating pressure.

Test Results

This table shows the possible causes for test results being out of specifications.

Test Results	Possible Causes
Outlet air temperature from the center dashboard vent is higher than normal.	• Air mix door is out of adjustment.
Outlet air temperature from the corner dashboard vents varies by more than 10°F.	• Air mix door is out of adjustment.
Refrigerant system low-side pressure is too low, and the outlet air temperature from the right corner dashboard vent is at least 10°F cooler than the left corner dashboard vent.	• Low refrigerant charge. • Restricted refrigerant flow through the evaporator.
Pressure on the high side of the refrigerant system is too high.	• Restricted airflow through the condenser. • Condenser/radiator fan is inoperative. • Restricted refrigerant flow through the system.
Pressure on the high side of the refrigerant system is too low.	• System is low on refrigerant charge.

SEALANT CONTAMINATION DETECTION

Refrigerant systems contaminated with air-conditioning system sealant additives can cause irreparable damage to air-conditioning service equipment and void manufacturer warranties of both service equipment and vehicle. These sealants are carried through the system by the refrigerant and as the refrigerant and sealant escape through a leak point (hole) the sealant reacts with moisture and air, solidifying to form a seal. These sealants will also enter the air-conditioning service equipment and can cause serious damage to the solenoids, valves, and tubing in there cover/recycle/recharge units and manifold and gauge sets. A sealant detection kit like the "Neutronics QuickDetect A/C Sealant Detection Kit" (Figure 6-6) is an

Classroom Manual
Chapter 6, page 182

© Cengage Learning 2013

FIGURE 6-6 A simple-to-use Neutronics Inc. refrigerant system sealant identifier will determine if an air-conditioning system has been contaminated by sealant.

invaluable tool for today's service technician and shop owner. It is important to gather information from the customer about past vehicle service history to determine if you should test the system for sealant contamination. If the system was ever serviced by a noncertified person (backyard mechanic) or shop, always test refrigerant purity and test the system for sealant contamination before proceeding with any air-conditioning system service or repair.

Procedure

1. Prepare the sensing plug by injecting water into both ends with the syringe. After injecting water, shake the sensing plug once to remove excess water.
2. Insert the non-ribbed end of the sensing plug into the quick-disconnect adapter coupling for the type of refrigerant being tested.
3. Install the safety cap (steel washer) over the sensing plug to prevent accidental release of the plug from the coupler during testing.
4. Insert one end of the rubber hose over the ribbed end of the sensing plug, ensuring that the hose is fully seated onto the plug.
5. Attach the other end of the rubber hose onto the graduated flowmeter, ensuring that the hose is fully seated.
6. Start the vehicle and turn the air-conditioning system on, selecting MAX cooling, lowest temperature setting, and high blower speed.
7. Allow the air-conditioning system to run and circulate refrigerant for a minimum of 2 minutes.
8. Shut off the air-conditioning system and turn off the engine. Allow the refrigerant system 3 minutes to stabilize and equalize pressures.
9. Suspend the flowmeter in a vertical position (Figure 6-7).
10. Attach the flowmeter assembly refrigerant system service coupler to the vehicle's air-conditioning system high-side service port.

FIGURE 6-7 The flowmeter of the sealant identifier must be in a vertical position in order to obtain accurate results if the air-conditioning system has been contaminated by sealant.

11. Measure the flow rate coming from the air conditioning high side on the graduated flowmeter scale and slide a rubber O-ring on the graduated flowmeter scale to the reading observed. If the flow rate is above 1.5 proceed to step 12. If the flow rate is below 1.5 repeat steps 1 through 11. If the flow rate is still below 1.5 see "Troubleshooting low flow" in step 11a.

   a. Troubleshooting low flow during initial installation. Low flow may be caused by:
      i. An inoperable service port
      ii. Reusing a sensing plug
      iii. A clogged tube
      iv. Low refrigerant system charge. If the first three steps were checked and deemed okay and a low refrigerant charge is suspected, add a small amount of refrigerant to the system and retest.

12. Monitor the reading on the flowmeter for 3 minutes. During the first 30–60 seconds the flow rate may rise as water is pushed through the sensing bulb.

13. Note the highest reading recorded during the first 60 seconds and compare it to the reading after 3 minutes. If the final flow rate drops by more than 30 percent of the highest initial flow rate observed, then refrigerant system sealant is present. If the final reading is within 30 percent of the initial reading, then sealant is not present.

14. After the test is complete, disconnect the tester from the vehicle and replace the vehicle service cap on the service fitting. Throw away the used sensing plug and return the rest of the components to the kit.

LEAK TESTING THE SYSTEM

Before undertaking any leak-detection procedures, perform a visual inspection of all system components, fittings, and hoses for signs of lubricant leakage, damage, wear, or corrosion.

Note that R-134a refrigerant polyalkaline glycol (PAG) lubricant may evaporate and therefore not be visible, whereas R-12 refrigerant mineral oil, on the other hand, does not evaporate and will leave a visible stain. To prevent an inaccurate or false reading with an electronic leak detector, make sure there is no refrigerant vapor or tobacco smoke in the vicinity of the vehicle being checked. Also, because there could be more than one leak in the air-conditioning system, when a leak has been found, continue to check for additional leaks. Perform the leak test in a relatively calm area so the leaking refrigerant is not dispersed in air movement. With the engine not running:

1. Connect a proper manifold and gauge set (R-12 or R-134a) to the air-conditioning system's low- and high-side service ports.

2. Ensure that the refrigerant pressure in the air-conditioning system is at least 50 psig (345 kPa) and that the ambient temperature is 60°F (15.6°C) or above.

   NOTE: If less than specified, recover, evacuate, and recharge the system with enough refrigerant to perform the leak test.

   NOTE: At temperatures below 60°F (15.6°C), leaks may not be detected since the system pressure may not reach 50 psig (345 kPa).

3. Conduct the leak test from the high side to the low side at points shown in Figure 6-8, as follows.

Compressor

Check the high- and low-side hose fittings, relief valve, gaskets, and shaft seal. Check the service valves, if compressor-mounted, with the protective covers removed.

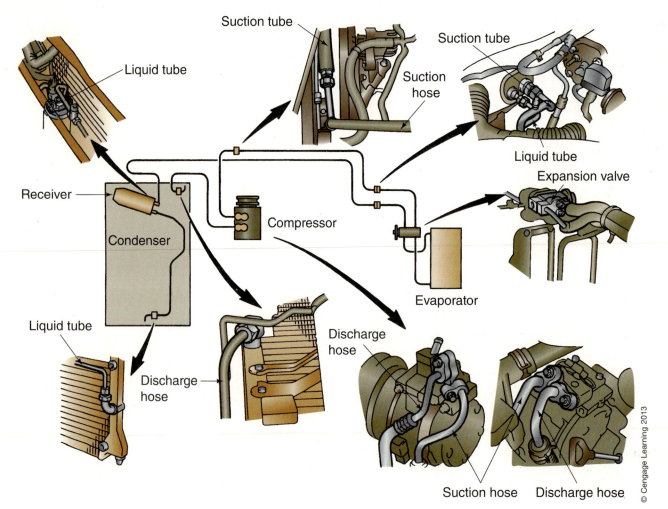

Suction tube

Liquid tube

Suction tube

Suction hose

Receiver

Condenser

Liquid tube

Compressor

Liquid tube

Expansion valve

Evaporator

Discharge hose

Discharge hose

Suction hose   **Discharge hose**

© Cengage Learning 2013

FIGURE 6-8 **Test points for leak detection.**

Accumulator

Check the inlet and outlet fittings, pressure switch, weld seams, and low-side service fitting (with cap removed), if equipped.

Receiver-Drier

Check the inlet and outlet hose fittings, pressure switch, weld seams and the fusible plug, and sight glass, if equipped.

Service Valves

Check all around the low- and high-side service valves with the caps removed. Ensure that the service valve caps are secured on the service valves after testing to prevent future leaks.

NOTE: After removing the manifold gauge hoses from the service valves, wipe away residue to prevent any false readings by the leak detector. Blowing low-pressure air across the service valve should clear any refrigerant residue vapor. If a service valve port is found to be leaking, repair it as required before replacing the valve cap.

Evaporator

In some systems, the blower motor resistor block may be removed to gain access to the evaporator core for testing. Since refrigerant is heavier than air, however, the leak may be best detected at the evaporator condensate drain hose. If using an electronic leak detector, place the probe near the drain hose for 10 to 15 seconds immediately after stopping the engine. Take care not to contaminate the end of the test probe. If it is a dual air-conditioning system, do not forget to check both evaporators, front and rear.

Condenser

Check all around the discharge (inlet) line and liquid (outlet) line connections. Check the front (face) of the condenser for any leaks that may be due to road damage. If the air-conditioning system has an auxiliary condenser, check it for leaks as well.

Metering Device

Carefully check all connections, inlet and outlet, of the metering device if a thermostatic expansion valve (TXV) or the spring lock coupling if an orifice tube. In a dual air-conditioning system, check both front and rear metering devices.

Hoses

Although barrier-type refrigeration hoses, now required by the EPA, are relatively leak proof, they may on occasion develop a pinhole leak. Visually inspect all hoses carefully for telltale traces of lubricant or dye that indicate a leak. Leaks are not so easily detected visually since PAG lubricant may evaporate from the surface of the hose. Any of the leak test methods may be employed if a leak in a hose is suspected and is not visually detected. If it is a dual air-conditioning system, check all hoses, front to back, of the vehicle.

Pressure Controls

Check all around a pressure control. Though rare, a pressure control has been known to leak past the seal of the electrical connector. Remove the connection and thoroughly check the control.

Methods

The three most popular methods of leak detection today use soap bubbles, electronic halogen, and ultraviolet dye. They are covered in the following sections. The once popular halide (gas) leak detector is not recommended. It is only effective for detecting CFC refrigerants and is considered hazardous in view of the danger of encountering flammable refrigerants.

SOAP SOLUTION

The soap solution is used as a method of leak detection when it is impractical or impossible to pinpoint the exact location of a leak by using the halogen leak detector methods. A commercial soap solution is available that is generally more effective than a homemade solution. A good grade of sudsing liquid dishwashing detergent may be used, however, if a prepared commercial solution is not available.

Preparing the System

1. Connect the manifold and gauge set to the system.
2. Make certain that the high- and low-side manifold hand valves are in the closed (cw) position.

The least expensive method is by soap bubbles.

Classroom Manual
Chapter 6, page 192

Some detergents may be used undiluted.

221

FIGURE 6-9 Leaks are detected when a bubble forms.

3. If the system is equipped with manual service valves, place the high- and low-side valves in the cracked position.
4. Open the low- and high-side hose shut-off valves.
5. Determine the presence of refrigerant in the system. A minimum value of 50 psig (348 kPa) is needed.
6. If there is an insufficient **charge** of refrigerant in the system, continue with the next step, "Adding Refrigerant for Leak Testing." If the charge is sufficient, omit the next step and proceed with leak testing.

Adding Refrigerant for Leak Testing

1. Attach the center manifold service hose to a source of refrigerant.
2. Open the refrigerant container service valve.
3. Open the high-side manifold hand valve until a pressure of 50 psig (348 kPa) is reached on the low-side gauge. Then close the high-side hand valve.
4. Close the refrigerant container service valve.
5. Close the service hose shut-off valve.
6. Remove the hose from the refrigerant container.

Procedure

1. Apply soap solution to all joints and suspected areas by using the dauber supplied with the commercial solution or by using a small brush with household solution.
2. Leaks are exposed when a bubble forms (Figure 6-9).
3. Repair any leaks found.

TRACER DYE LEAK DETECTION

As a rule of thumb, the presence of oil at a fitting or connection generally indicates a refrigerant leak. This is not always the case, however, because oil is used on fittings as an aid in assembly procedures. If a leak is suspected, the area should be wiped clean and the leak verified. This may be accomplished by either of the several methods discussed in this chapter.

A popular method of refrigerant leak detection is with the use of a fluorescent tracer dye that is easily detected with an ultraviolet (UV) lamp (Figure 6-10). The fluorescent

Charge refers to a specific amount of refrigerant or oil by volume or weight.

SPECIAL TOOL

Refrigerant identifier

CAUTION:
Use a refrigerant identifier to determine that the source refrigerant is the same type— R-12 or R-134a— that is used in the air-conditioning system.

A minimum of 50 psig (348 kPa) is required for leak testing.

The hose shut-off valve is provided to reduce emissions.

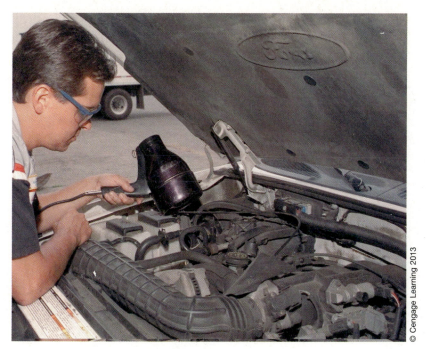

FIGURE 6-10 Ultraviolet (UV) lamp used for leak testing.

© Cengage Learning 2013

tracer dye lasts for about 500 hours of air-conditioning system use. When expended, another injection of tracer dye is required. To inject dye into the system, the system pressure must be above 80 psig (551.6 kPa). For proper use and accurate results, always follow the instructions included with the tracer dye. The refrigerant dye must be approved for automotive air-conditioning systems. A specific dye for an electric motor-driven air-conditioning compressor that is non-conductible is required. The dye is generally available in either yellow or red and after a few days of being added to a system the point of the leak should be visible.

To pinpoint the leak, scan all the air-conditioning system components, fittings, and hoses with the UV lamp. The exact location of a leak will be revealed by a bright yellow-green glow of the tracer dye. Leaks are best detected in low ambient-light conditions. In areas where the lamp cannot be used, such as where the ambient light is high, a mechanic's mirror may be used. The technician may also wipe the suspected area with a disposable, non-fluorescent towel, which is then examined with the lamp for traces of the dye.

After the leak has been repaired, the dye can be removed from the exterior of the leaking area by using an oil solvent. To verify that the repair has been made, operate the air-conditioning system for about 5 minutes and reinspect the area with the UV lamp. Since more than one leak may occur, it is wise to check the entire system.

Many technicians prefer to add a dye trace solution to the refrigerant any time the air-conditioning system is opened for service. With the high cost of refrigerant and technician labor, this practice is well worth the additional cost at the time of repairs.

HALOGEN (ELECTRONIC) LEAK DETECTION

The halogen (electronic) leak detector (Figure 6-11) is the most sensitive of all types of leak detectors. These leak detectors, must comply with SAE standard SAE J-2791 standard for R134a leak detecting equipment. This SAE standard was created to improve leak detection equipment sensitivity requirements to a sensitivity level of 0.15 oz./year (4 g/y). Leak detectors that meet this standard will have at least three sensitivity scales that can be selected manually: (A) 0.15 oz./year (4 g/y), (B) 0.25 oz./year (7 g/y), (C) 0.5 oz./year (14 g/y).

Classroom Manual
Chapter 6, page 193

Classroom Manual
Chapter 6, page 194

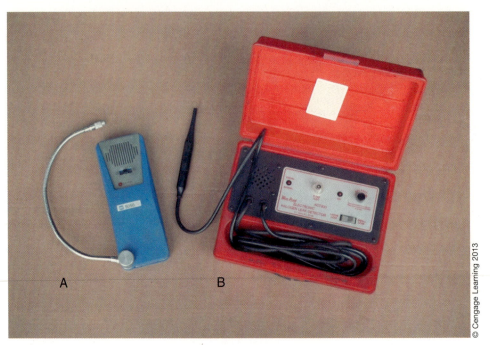

FIGURE 6-11 Electronic leak detectors: (A) cordless; (B) corded.

© Cengage Learning 2013

It must be calibrated to detect a refrigerant leak within 2 seconds from a distance of 3/8 in. (9.5 mm) moving at a rate of 3 in. (75 mm) per second and must be able to self-clear itself within 2 seconds once moved away from the leak. This type of leak detector can be of great value in detecting the "impossible" leak.

When using an electronic leak detector, ensure that the instrument, if required, is calibrated and set properly according to the operating instructions provided by the manufacturer. In order to use the leak detector properly, read the operating instructions and perform any specified maintenance.

Other vapors in the service area or any substances on the components—such as antifreeze, windshield washing fluid, or solvents and cleaners—may falsely trigger the leak detector. Make sure that all surfaces to be checked are clean. Do not permit the detector sensor tip to come into contact with any substance; a false reading can result, and the leak detector can be damaged.

 WARNING: **A halogen electronic leak detector must be used in well-ventilated areas only. It must never be used in spaces where explosive gases may be present.**

The following list and Photo Sequence 7 illustrate typical procedures for using an electronic leak detector:

1. Hold the probe in position about 3/16 in. (5 mm) from the area to be checked (Figure 6-12).
2. When testing, circle each fitting with the probe (Figure 6-13).
3. Move the probe along the component about 1–2 in. (25–50 mm) per second (Figure 6-14).
4. If a leak is detected, verify it by fanning or blowing compressed air into the area of the suspected leak, then repeat the leak check.

TYPICAL PROCEDURE FOR CHECKING FOR LEAKS

P7-1 Turn the control or sensitivity knobs to OFF or ZERO. If the leak detector is corded, connect it to an approved voltage source. Skip this step if it is cordless.

P7-2 Turn on the switch. Allow a warm-up period of about 5 minutes. There is usually no warm-up period required for cordless models.

P7-3 Place the probe at the reference leak. Adjust the control or sensitivity knobs until the detector reacts.

P7-4 Remove the probe. The reaction should stop.

P7-5 If the reaction continues, the sensitivity control is adjusted too high. Repeat the procedure of step P7-3. If the reaction stops, the sensitivity adjustment is adequate.

P7-6 Slowly move the search hose under and around all of the joints and connections.

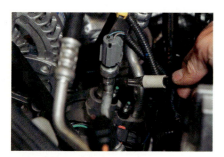

P7-7 Check all seals and screw-in pressure control devices.

P7-8 Check the service fittings. It will be necessary to remove the service cap for this test.

P7-9 Check the evaporator at the outlet duct drain, or in some cases the blower motor resistor block may be removed to gain access.

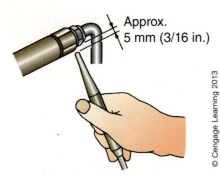

Approx.
5 mm (3/16 in.)

© Cengage Learning 2013

FIGURE 6-12 Hold the probe about ³⁄₁₆ in. (5 mm) away.

© Cengage Learning 2013

FIGURE 6-13 Circle each fitting.

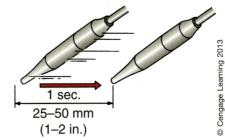

1 sec.
25–50 mm
(1–2 in.)

© Cengage Learning 2013

FIGURE 6-14 Move the probe 1–2 in. (25–50 mm) per second.

To **evacuate** is the process of creating a vacuum within a system to remove all traces of air and moisture.

Be sure to use the proper refrigerant, only R-12 or R-134a, as appropriate.

Classroom Manual
Chapter 6, page 197

A **standing vacuum test** is a leak test performed on an air-conditioning system by pulling a vacuum and then determining, by observation, if the vacuum holds for a predetermined period of time to ensure that there are no leaks.

Only enough refrigerant to increase the system pressure to 50 psi (345 kPa) is required.

Repair System

1. After the leak is located, recover the system refrigerant as outlined in this chapter.
2. Repair the leak as indicated.
3. Add or change oil, if required.
4. **Evacuate** the system.
5. Charge the system with refrigerant.
6. Recheck the system for leaks; verify the repair.

EVACUATING THE SYSTEM

An important step in the repair of an automotive air-conditioning system is proper evacuation. The air-conditioning system must be evacuated whenever it is serviced to the extent that the refrigerant was recovered. Proper evacuation rids the system of all air and most moisture that may have entered during repair service. Photo Sequence 8 illustrates the typical procedure for evacuating the system. The following procedure assumes that the system has been serviced and does not contain refrigerant.

Checking for Leaks

A **standing vacuum test** may be made to leak test the system. Proceed as follows:

1. Evacuate the systems outlined in Photo Sequence 8.
2. Close the manifold hand valves and turn off the vacuum pump.
3. Note the low-side gauge reading; it should be 29 in. Hg. (3.4 kPa absolute) or lower.
4. Allow the system to "rest" for 5 minutes, then again note the low-side gauge reading.

   NOTE: The low-side gauge needle should not raise faster than 1 in. Hg. (3.4 kPa absolute) in 5 minutes.

   If the system does not meet the requirement of step 4, a leak is indicated. A partial charge of refrigerant must be installed, and the system must be leak checked. After the leak is detected, the refrigerant must be recovered. After the leak is repaired, again perform the standing vacuum test, starting with step 1 above.

TRIPLE EVACUATION METHOD

The basic steps in the **triple evacuation** method are given here. The procedures assume that the system is sound after the refrigerant has been removed and repairs, if any, have been made.

Procedure

As previously detailed, connect the manifold and gauge set into the system. Be sure that all hoses and connections are tight and sound, and that all appropriate valves are in the closed position.

TYPICAL PROCEDURE FOR EVACUATING THE SYSTEM

P8-1 Connect the manifold and gauge set low- and high-side service hoses to the system.

P8-2 Make sure that the high- and low-side manifold hand valves are in the closed position and both gauges read zero or less.

P8-3 Connect the service hoses to the vehicle refrigerant system high- and low-side fittings.

P8-4 Remove the protective caps from the inlet and exhaust of the vacuum pump.

P8-5 Connect the center manifold service hose to the inlet of the vacuum pump.

P8-6 Open the shut-off valve of the three service hoses.

P8-7 Connect the power cord of the vacuum pump to an approved power source.

P8-8 Turn on the vacuum pump.

P8-9 Open the low-side manifold hand valve and observe the low-side gauge needle. The needle should be immediately pulled down to indicate a slight vacuum.

P8-10 After about 5 minutes, the low-side gauge should indicate 20 in. Hg (33.8 kPa absolute) or less. The high-side gauge needle should be slightly below the zero index of the gauge.

P8-11 If the high-side needle does not drop below zero, unless restricted by a stop, a blockage in the system is indicated. If the system is blocked, discontinue the evacuation. Repair or remove the obstruction. If the system is clear, continue the evacuation.

P8-12 Operate the pump for another 15 minutes and observe the gauges. The system should be at a vacuum of 24–26 in. Hg (20.3–13.5 kPa absolute). If it is not, close the low-side hand valve.

P8-13 Observe the compound (low-side) gauge. If the needle rises, indicating a loss of vacuum, there is a leak that must be repaired before the evacuation is continued. If no leak is evident, continue the evacuation.

P8-14 Reopen the low-side manifold hand valve.

P8-15 Open the high-side manifold hand valve.

P8-16 Allow the vacuum pump to operate for a minimum of 30 minutes, longer if time permits. After pump-down, close the high- and low-side manifold hand valves. Turn off the vacuum pump and close the service hose shut-off valves. Turn off the vacuum pump valve, if equipped. Then, disconnect the manifold service hose from the vacuum pump. Replace the protective caps, if any.

First Stage

1. Pump a vacuum to the highest efficiency for 25–30 minutes.
2. Close the manifold hand valves.
3. Close the service hose shut-off valves.
4. Close the vacuum pump shut-off valve, if equipped. If the system is not equipped with a shut-off valve, turn off the pump.
5. Disconnect the service hose from the vacuum pump.
6. Connect the service hose to a **dry nitrogen** (Figure 6-15) source.
7. Open the service hose shut-off valve.
8. Open the nitrogen supply valve.
9. Open the low- and high-side service hose shut-off valves.
10. Open the low-side manifold hand valve to break the vacuum:
    a. Slowly increase the pressure to 1–2 psig (6.8–13.7 kPa).
    b. Close all valves: the manifold low-side hand valve, service hose shut-off valve, low- and high-side hose shut-off valves, and the nitrogen supply valve.
11. Disconnect the service hose from the nitrogen supply.

Second Stage

1. Allow one-half hour, which is sufficient time for the dry nitrogen to "**stratify**" the system.
2. Reconnect the service hose to the vacuum pump.
3. Open the vacuum pump shut-off, if equipped. If not equipped, turn on the pump.
4. Open the service hose shut-off valve.
5. Open the manifold low- and high-side hand valves.
6. Pump a vacuum to the highest efficiency for 25-30 minutes.
7. Repeat steps 2 through 11 as outlined in "First Stage" procedures.

Third Stage

1. Follow steps 1 through 6 as outlined in "Second Stage" procedures.
2. Close all valves: the service hose shut-off valve, vacuum pump shut-off valve, if equipped (if not equipped, turn off the pump), low- and high-side service hose shut-off valves, and manifold low- and high-side hand valves.
3. Turn off the vacuum pump (if not previously done).
4. Remove the service hose from the vacuum pump.
5. The system is now ready for charging. Follow the appropriate procedure outlined in this chapter.

FIGURE 6-15 Typical dry nitrogen setup.

Triple evacuation is the process of evacuation that involves three pump downs and two system purges with an inert gas such as dry nitrogen (N).

Dry nitrogen is the element nitrogen (N) that has been processed to ensure that it is free of moisture.

Several different types of service hose shut-off valves are available.

CAUTION: Make sure that the nitrogen supply has proper regulators and that supply pressure does not exceed 75 psig (517 kPa).

Be sure the nitrogen pressure is regulated before opening the valve.

To **stratify** is the process of arranging or forming into layers to fully blend.

Air contains moisture in the form of humidity.

CUSTOMER CARE: While under the hood servicing a vehicle, make a point of checking all fluid levels and topping them off as needed. Customers may never say anything, but when they push the washer button and there is always fluid, they appreciate it.

CHARGING THE SYSTEM

The typical methods of charging an automotive air-conditioning system refrigerant are given in this service procedure: from a recovery/recycling/recharge unit, from **pound cans** with the system off, from pound cans with the system operating, and from a bulk source.

The following additional safety precautions must also be followed when handling refrigerant.

- Do not deliberately inhale refrigerant.
- Do not apply a direct flame to a refrigerant container.
- Do not place an electric resistance heater close to a refrigerant container.
- Do not abuse a refrigerant container.
- Do not use pliers or vice grips to open and close refrigerant valves. Use only approved wrenches.
- Do not handle refrigerant without suitable eye protection.
- Do not discharge refrigerant into an enclosed area.
- Do not expose refrigerant to an open flame.
- Do not lay cylinders flat. Store containers in an upright position only. Secure large cylinders with a chain to prevent them from tipping over.

Preparing the System

This procedure assumes that the system has been evacuated. If it has not been evacuated, refer to the proper procedure for evacuation *before* attempting to charge the system.

If charging from pound cans, set up the refrigerant recovery system according to the equipment manufacturer's instructions to be used to recover any residual refrigerant that remains in a can while charging the system.

Recharging System with a Recovery/Recycling/Recharge Unit

A service facility must have a recovery/recycling/recharge unit (Figure 6-16) for each type of refrigerant serviced (i.e., R-134a and R-12). This unit is used to charge refrigerant systems after the repair has been completed in most instances, though some shops also use manifold and gauge sets and bulk refrigerant containers and separate scales for the charging process. Some units are dual system and designed to service both R-134a and R-12 refrigerants.

Charging the refrigerant system with the correct amount of refrigerant is one of the most critical aspects of air-conditioning service. Close is not good enough, especially with today's small capacity systems. A 1.2 lb. (0.54 kg) system is overcharged or undercharged by almost 10 percent with an error of only ±2 oz. (0.06 kg). Efficient system operation depends on proper charge level of refrigerant. Too little or too much refrigerant can lead to poor cooling complaints or system failure. Partial charging a refrigerant system is no longer an acceptable practice except for the purpose of adding refrigerant to determine the location of a leak during the diagnostic process. In addition, with the speed and accuracy of today's recovery/recycling/recharge units there is no need for the guesswork involved in a partial charge. If the charge level of a refrigerant system is in doubt, then recover, evacuate, and recharge the system before proceeding with refrigerant system diagnosis.

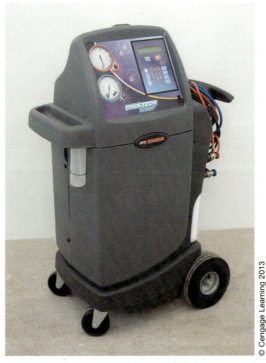

FIGURE 6-16 A typical air-conditioning system recovery/recycling/recharge unit.

An undercharged system will result in poor or inadequate cooling under high heat loads. Slightly undercharged systems may provide satisfactory cooling under moderate heat loads, but cooling performance will be poor under high ambient temperatures and heat loads due to lack of reserve refrigerant. This can lead to improperly diagnosed systems and either unneeded repairs or no repair at all. In addition, a low refrigerant charge level will cause the clutch cycling switch to cycle the compressor more often than normal.

An overcharged system will also result in poor or inadequate cooling. The higher than designed level of refrigerant can also lead to compressor failure. An overcharged system will generally have higher than normal system pressures. In addition, the refrigerant system may also exhibit noise coming from the compressor assembly due to liquid refrigerant entering the compressor and higher system-operating pressures.

The following is an outline of the basic charging procedures for a recovery/recycling/recharge unit. Always follow the specific instructions that are provided by the equipment manufacturer when using a charging station. Though systems are similar in operation each unit has specific steps that must be followed.

Procedure

1. Connect both the high- and low-side service lines from the recovery/recycling/recharge unit to the vehicle refrigerant system high- and low-side service fittings, respectively.
2. Ensure that the vehicle refrigerant system has been properly evacuated prior to proceeding with the recharging of the refrigerant system.
3. Open the refrigerant tank valve(s) on the recharge station (Figure 6-17).
4. Set the unit to deliver the required amount of either new or recycled refrigerant for the vehicle refrigerant system being charged (Figure 6-18). Remember, refrigerant charge level is system specific. Refer to the refrigerant label located in the engine compartment or refer to the appropriate service information for system refrigerant capacity.

FIGURE 6-17 Open the refrigerant tank valve(s) on the recovery/recycling/recharge unit.

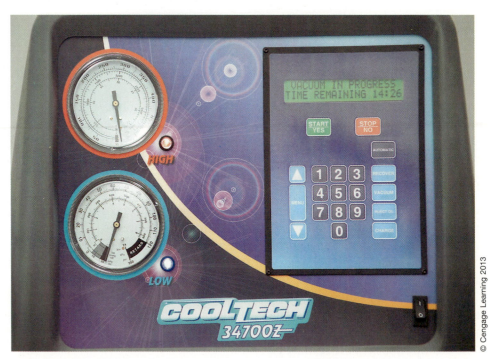

FIGURE 6-18 A typical air-conditioning system recovery/recycling/recharge console control panel.

5. Start the recharge process; most systems are capable of recharging the refrigerant system with the vehicle off.
6. Once the charging process is complete the unit may be disconnected from the vehicle and the vehicle service fitting caps may be reinstalled on the vehicle high- and low-side access fittings.
7. After the system has been recharged, perform an unloaded air conditioning performance test to verify proper function of the system.

Photo Sequence 9 illustrates a typical procedure for system charging.

TYPICAL PROCEDURE FOR COMPLETING THE SYSTEM CHARGE

P9-1 Connect both the high- and low-side service lines from the recovery/recycling/recharge unit to the vehicle refrigerant system high- and low-side service fittings, respectively.

P9-2 Open the service hose shut-off valve on the quick-disconnect coupling.

P9-3 Open the refrigerant tank valve(s) on the charge station.

P9-4 Set the unit to deliver the required amount of either new or recycled refrigerant for the vehicle refrigerant system being charged.

P9-5 Start the recharge process; most systems are capable of recharging the refrigerant system with the vehicle off.

P9-6 Once the charging process has been completed, start the vehicle engine.

P9-7 Turn the blower motor control on high speed.

P9-8 Activate the air-conditioning system in the Recirculation mode and set the temperature control to the coldest setting.

P9-9 Place a fan in front of vehicle to assist in cooling the air-conditioning condenser.

P9-10 With the engine running at 1,000 rpm check both the high- and low-side gauge pressures.

P9-11 Place a thermometer in the center dashboard vent and record the outlet air temperature.

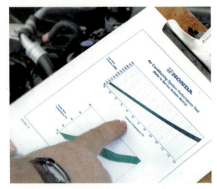

P9-12 Compare both gauge and thermometer readings to the temperature/pressure chart for the vehicle being serviced.

P9-13 Disconnect the recovery/recycling/recharge unit from the vehicle and shut the refrigerant tank valve(s) on the charge station.

P9-14 Replace the vehicle service fitting caps on the high- and low-side access fittings.

FIGURE 6-19 Replace the recovery/recycling/recharge unit filters at the recommended maintenance interval to provide optimum system performance.

© Cengage Learning 2013

Operation and Maintenance of Refrigerant Recover/Recycling Equipment

It should be a routine practice to verify the correct operation and to determine the required maintenance of refrigerant handling equipment in every shop. In some shops one individual is assigned the task of maintaining equipment, whereas in many other shops this task is not assigned but rather becomes the responsibility of all the technicians using the equipment. It is important to remember that refrigerant system operation after the repair depends on the accuracy of the service equipment being used. Improperly maintained equipment creates diagnostic and performance issues that lead to lost time and productivity for the vehicle owner, the repair shop, and the individual technician.

Procedure. Always follow the procedures outlined in the service manual for the specific refrigerant recovery/recycling unit being used.

1. Based on the service information for the unit, follow the replacement service interval for the filter(s) and desiccant (Figure 6-19). Often this is every 20 hours of unit operation. Also, make note if the unit is equipped with a filter change indicator.
2. Does the vacuum pump require routine oil change maintenance or is the unit equipped with a oil-less pump?
3. Locate and record the recovery tank date code stamped on the tank's collar (Figure 6-20).
4. Refrigerant recovery tanks require recertification every 5 years. Verify that the recovery tank is the correct color for the refrigerant it contains: sky blue for R-134a, white for R-12, or gray with a yellow top for unknown refrigerants. The U.S. DOT, under Title 49 as well as SAE standard J2296, requires that every 5 years a recovery tank be both externally and internally inspected and a hydrostatic pressure test be performed. The date code indicates when the tank was produced or recertified and must not be older than 5 years. Using a tank that is out of date can subject the owner to a $25,000 fine. Use only DOT-certified Title 49 or UL-approved containers with a stamp of DOT-4BA or DOT-4BW on the tank as proof.

INSTALL THE CAN TAP VALVE ON A "POUND" CONTAINER

This procedure may be followed prior to servicing a system using pound cans.

1. Be sure that the **can tap valve** stem is in the fully counterclockwise (ccw) position.
2. Attach the valve to the can (Figure 6-21). Secure the valve with the locking nut, if so equipped. Secure the valve with the clamping lever, if so equipped.

A **can tap valve** is a valve found on a can tap used to control the flow of refrigerant.

FIGURE 6-20 Note and record the recovery tank date code.

FIGURE 6-21 Attach the can tap valve.

There are two basic types of can tap valve: locking nut and clamping ring.

3. Make sure that the manifold service hose shut-off valve is closed.
4. Connect the manifold service hose to the can tap port.
5. Pierce the can by turning the valve stem all the way in the clockwise (cw) direction.
6. Back the can tap valve out, turning in a counterclockwise (ccw) direction.

7. The center (service) hose is charged with refrigerant to the shut-off valve and is under a vacuum between the manifold and shut-off valve.

 WARNING: **Do not open the high- or low-side hand valves at this time.**

Checking the System for Blockage

This procedure may be followed when charging the system from pound cans.

1. Open the service hose shut-off valve.
2. Open the low- and high-side service hose shut-off valves.
3. Open the high-side gauge manifold hand valve. Observe the low-side gauge pressure. Close the high-side hand valve.
4. Close the hose shut-off valves. Close the service hose and the low- and high-side hoses.

USING POUND CANS (SYSTEM OFF)

1. Open the service hose shut-off valve.
2. If not previously done, fully open the can tap valve.
3. Open the high-side gauge manifold hand valve.
4. Observe the low-side pressure gauge. If the gauge indication does not move from the vacuum range to the pressure range, a system blockage is indicated. If the system is not blocked, proceed with step 5. If the system is blocked, correct the condition and evacuate the system before continuing with step 5.
5. Invert the container (Figure 6-22) and allow the liquid refrigerant to enter the system.
6. Tap the refrigerant container on the bottom. An empty can produces a hollow ringing sound.

    The system generally requires 10 percent less R-134a than was required for R-12 when retrofitting.

7. Use the recovery system to remove any residual refrigerant from the empty can.
8. Repeat steps 5, 6, and 7 with additional cans of refrigerant as required to charge the air conditioner. Refer to the manufacturer's specifications for system capacity.

FIGURE 6-22 **Invert the container to allow liquid refrigerant to flow.**

If the discharge valve and plate are in good condition, refrigerant must circulate through the system to impress pressure on the low-side gauge.

SERVICE TIP: If the low-side gauge does not move from the vacuum range into the pressure range, a system blockage is indicated. Correct the blockage, then evacuate and continue with the appropriate service procedure.

CAUTION: Charging liquid refrigerant into a compressor while it is running may cause damage to the compressor and injury to the technician.

The ¼ in. (6.4 mm) service hose acts like a capillary tube. At 40 psig (276 kPa), the refrigerant will vaporize by the time it travels the 6 ft. (1.8 m) of hose.

Using Pound Cans (System Operational)

1. Start the engine and adjust the speed to about 1250 rpm by turning the idle screw or the setting on the high cam.
2. Make sure that both the manifold hand valves are closed.
3. Adjust the air conditioning controls for MAX cooling with the blower on HI.
4. Open the service hose shut-off valve.
5. Open the low- and high-side hose shut-off valves.
6. If not previously done, open the can tap valve.
7. With the can in an upright position, open the low-side manifold hand valve.
8. After the pressure on the low side drops below 40 psig (377 kPa absolute), the can may be inverted to allow more rapid removal of the refrigerant.
9. Tap the can on the bottom to determine if it is empty. An empty refrigerant can will give a hollow ringing sound. One may also shake the can to determine if refrigerant is "sloshing around" inside.
10. Repeat steps 7, 8, and 9 with additional cans of refrigerant as required to charge the system completely. Refer to the manufacturer's specifications regarding system capacity.
11. Close all valves: the can tap valve (if refrigerant remains in the can), service hose shut-off valve, low- and high-side service hose shut-off valves, low- and high-side manifold hand valves.
12. Remove the refrigerant container. If refrigerant remains in the can, close the charging hose shut-off valve and remove the can tap from the center service hose. If the can is empty, use the recovery system to remove any residual refrigerant from the manifold, hoses, and can.
13. Remove the manifold and gauge set.
14. Replace all protective caps and covers.

CHARGING FROM A BULK SOURCE WITH A MANIFOLD AND GAUGE SET

Shops that perform a large volume of air-conditioning service can obtain bulk refrigerant in 10-, 15-, 25-, 30-, 50-, and 145-pound (4.5-, 6.8-, 11.3-, 13.6-, 22.7-, and 65.8-kilogram) cylinders. The use of bulk containers requires a set of scales or other approved measuring device to determine when the proper system charge is obtained.

The high cost of refrigerant, particularly R-12, has prompted many service facilities to charge for it by the ounce. An ounce (29.6 mL) of R-12 now generally costs more than 10 times what a pound (473 mL) did before the federal restrictions on refrigerants.

Regulations by the Bureau of Weights and Measures, in most states, require that certified scales be used to ensure that the costs of refrigerant to the customer are proper. Two types of scales easily meet this requirement: electronic and beam.

An electronic scale (Figure 6-23) is more properly referred to as an electronic charging meter. Some charging meters display refrigerant weight as it is being charged from the tank; others may be programmed to automatically charge a selected amount of refrigerant and will shut off when the programmed amount has been reached.

Some charging meters measure refrigerant in increments of 0.5 oz. (14.8 mL), while other; are accurate to and measured in 0.25 oz. (7.4 mL) increments. Charging meters can generally accept cylinders having a gross weight of 100 lbs. (45.4 kg) or more. When buying refrigerant in any size container, it is suggested that the cylinder be checked for weight and a sample of its contents be tested for purity. At today's prices, one cannot be too careful.

Electronic charging scales are compatible with all refrigerants and have a tare function that may be zeroed (0.00 oz.). Some meters have a digital display in either pounds (lbs.) or kilograms (kg). Battery-powered charging meters generally include an AC adapter and have a "sleep mode" to ensure extended battery life.

FIGURE 6-23 Typical electronic charging meter (scale).

The following procedure assumes that the system has been properly prepared for charging procedures. If not, consult the appropriate heading for the proper procedure *before* continuing.

Connecting the Refrigerant Container

1. Make sure that all service valves are in the OFF or closed position. Check the service hose shut-off valves, manifold hand valves, compressor service valves, if equipped, and refrigerant source valve.
2. Connect the center manifold service hose to the supply refrigerant cylinder.
3. Open the refrigerant cylinder hand valve.
4. The system is now under a vacuum from the manifold to the service hose shut-off valve, and under a refrigerant charge from the cylinder to the hose shut-off valve.

Charging the System

1. Open the service hose shut-off valve.
2. Open the low- and high-side manifold hose shut-off valves.
3. Briefly open the high-side manifold hand valve. Observe the low-side gauge.
4. System blockage is indicated if the low-side gauge needle does not move from the vacuum range into the pressure range. If the system is not blocked, proceed with step 5. If the system is blocked, correct the blockage and reevaluate the system before continuing with step 5.
5. Start the engine and adjust the speed to about 1250 rpm.
6. Adjust the air conditioning controls for MAX cooling with the blower on HI.
7. Place the refrigerant cylinder on an approved scale and note the **gross weight**.
8. Open the low-side manifold hand valve to allow refrigerant to enter the system.
9. When the system is fully charged, close the low-side manifold hand valve.
10. Close the refrigerant cylinder service valve.
11. Close the service hose shut-off valve.
12. Remove the service hose from the refrigerant cylinder.
13. Note the gross weight now shown on the scale. Refrigerant used, by weight, is the difference between what the cylinder weighed in step 7 and what it now weighs.

Be sure to use the correct refrigerant. R-12 and R-134a are not compatible.

CAUTION: Keep the refrigerant cylinder in an upright position at all times. Liquid refrigerant must not be allowed to enter the compressor. This can cause serious damage and possible injury.

Gross weight is the weight of a substance or matter that includes the weight of the container.

14. Conduct performance tests or other tests as required.
15. Return the engine to its normal idle speed.
16. Turn off all air-conditioning system controls.
17. Shut off the engine.
18. Close all valves. Close the low- and high-side service hose shut-off valves. Back seat the compressor service valves, if equipped.
19. Use the recovery system to remove any residual refrigerant from the manifold and hoses.
20. Remove the manifold and gauge set.
21. Replace all protective caps and covers.

Classroom Manual
Chapter 6, page 183

TESTING REFRIGERANT FOR NONCONDENSABLE GAS

Recycled refrigerant in portable storage containers should be periodically checked for noncondensable gas (air) contamination. This includes recycled storage containers that are on recovery/recycle/recharge units. To determine if the refrigerant is contaminated the temperature/pressure relationship of the refrigerant gas is analyzed. The best time to proceed with this test procedure is at the beginning of the workday before beginning any refrigerant system service.

The procedure that follows outlines the steps for testing for noncondensable gases.

Procedure

Prior to checking a recovery tank of recycled refrigerant for noncondensable gases, the tank must first sit at room temperature above 65°F (18.3°C) for at least 12 hours to stabilize out of direct sunlight.

1. Identify the type of recycled refrigerant based on tank color (sky blue = R-134a; white = R-12).
2. Verify that the manifold set control knobs and hose shut-off valves are closed.
3. Connect the manifold gauge set auxiliary hose (yellow) to the recovery tank low-side (vapor) outlet (blue).
4. Using a contact or infrared thermometer, record the tank temperature within 4 in. (10 cm) of the tank.
5. Open the low-pressure valve on the tank and the low-pressure valve on the manifold line and manifold; record the recovery tank pressure.
6. Go to the chart in Figure 6-24 for the type of refrigerant being tested; compare the tank temperature to the chart.
   a. Based on the chart, determine the acceptable tank pressure.
   b. Tank pressure should be at or below the listed pressure.
   c. If pressure is above the temperature/pressure chart, attach the refrigerant identifier and test for impurities. If no impurities are identified (i.e., alternative refrigerant or HC) proceed to the next step. If the refrigerant pressure is at or below the chart pressure, the refrigerant contains no noncondensable gases and may be put back into service.
   Note that higher pressures indicate the presence of noncondensable gases.
7. Slowly vent the vapor from the top of the recovery tank into the recovery/recycling unit until pressure falls below the limit shown in the charts. If the recycled refrigerant is R-134a and the pressure is high after purging, shake the container and wait several minutes, then retest.
8. If the pressure still exceeds the pressure limits in the chart, recycle the entire contents of the container.
9. Label and store the refrigerant.

Freon is a generic term used to refer to R-12.

Some recycling equipment have either an automatic or a manual air purge operation that can take place during the normal unit recycling or evacuation process.

STANDARD PRESSURE TEMPERATURE CHART FOR R-134A

Temperature Fahrenheit	Pressure PSIG	Pressure kPa	Temperature Fahrenheit	Pressure PSIG	Pressure kPa	Temperature Celsius	Pressure kPa	Pressure PSIG	Temperature Celsius	Pressure kPa	Pressure PSIG
70	76	524	86	102	703	21.1	524	76	30.0	703	102
71	77	531	87	103	710	21.7	531	77	30.5	710	103
72	79	545	88	105	724	22.2	545	79	31.1	724	105
73	80	551	89	107	738	22.8	551	80	31.7	738	107
74	82	565	90	109	752	23.3	565	82	32.2	752	109
75	83	572	91	111	765	23.9	572	83	32.8	765	111
76	85	586	92	113	779	24.4	586	85	33.3	779	113
77	86	593	93	115	793	25.0	593	86	33.9	793	115
78	88	607	94	117	807	25.6	607	88	34.4	807	117
79	90	621	95	118	814	26.1	621	90	35.0	814	118
80	91	627	96	120	827	26.7	627	91	35.6	827	120
81	93	641	97	122	841	27.2	641	93	36.1	841	122
82	95	655	98	125	862	27.8	655	95	36.7	862	125
83	96	662	99	127	876	28.3	662	96	37.2	876	127
84	98	676	100	129	889	28.9	676	98	37.8	889	129
85	100	690	101	131	903	29.4	690	100	38.3	903	131

STANDARD PRESSURE TEMPERATURE CHART FOR R-12

Temperature Fahrenheit	Pressure PSIG	Pressure kPa	Temperature Fahrenheit	Pressure PSIG	Pressure kPa	Temperature Celsius	Pressure kPa	Pressure PSIG	Temperature Celsius	Pressure kPa	Pressure PSIG
70	80	551	86	103	710	21.1	551	80	30.0	710	103
71	82	565	87	105	724	21.7	565	82	30.5	724	105
72	83	572	88	107	738	22.2	572	83	31.1	738	107
73	84	579	89	108	745	22.8	579	84	31.7	745	108
74	86	593	90	110	758	23.3	593	86	32.2	758	110
75	87	600	91	111	765	23.9	600	87	32.8	765	111
76	88	607	92	113	779	24.4	607	88	33.3	779	113
77	90	621	93	115	793	25.0	621	90	33.9	793	115
78	92	634	94	116	800	25.6	634	92	34.4	800	116
79	94	648	95	118	814	26.1	648	94	35.0	814	118
80	96	662	96	120	827	26.7	662	96	35.6	827	120
81	98	676	97	122	841	27.2	676	98	36.1	841	122
82	99	683	98	124	855	27.8	683	99	36.7	855	124
83	100	690	99	125	862	28.3	690	100	37.2	862	125
84	101	696	100	127	876	28.9	696	101	37.8	876	127
85	102	703	101	129	889	29.4	703	102	38.3	889	129

FIGURE 6-24 Compare the refrigerant static storage tank temperature to the chart above. Refrigerant gas pressure should be below the pressure listed for the given tank pressure. Higher pressures indicate contamination.

TERMS TO KNOW

Can tap valve

Charge

Contaminated refrigerant

Dry nitrogen

Evacuate

Gross weight

Pound cans

Standing vacuum test

Stratify

Triple evacuation

CASE STUDY

A customer complained of an inoperative air conditioner and requested that Freon be added. The technician advised the customer that if refrigerant were needed, there must be a leak in the system. The customer responded "I have to add Freon every couple of months—just put it in."

The technician attempted to explain the problems with just adding refrigerant, such as loss of oil, harm to the environment, and possible damage to the air-conditioning system components, such as the compressor. The customer still insisted that he only wanted Freon.

Politely and tactfully, the technician refused the service. "You have come to the wrong place," she told the customer. "This service facility employs only ASE-certified technicians who are dedicated to their profession. To perform a service in an improper manner violates the essence of ASE certification."

The somewhat surprised, but impressed, customer left the facility without further ado. He returned the next day and had the air conditioner properly repaired.

ASE-STYLE REVIEW QUESTIONS

1. *Technician A* says that a system is contaminated if it contains more than 2 percent of a foreign substance.
   Technician B says that air is considered a contaminant if it exceeds 2 percent of the system capacity. Who is correct?
   A. A only
   B. B only
   C. Both A and B
   D. Neither A nor B

2. *Technician A* says that tobacco smoke will not affect refrigerant leak detection.
   Technician B says that an electronic halogen leak detector is the best method of leak detection.
   Who is correct?
   A. A only
   B. B only
   C. Both A and B
   D. Neither A nor B

3. *Technician A* says that special electronic leak detectors are available that are used for halogen gases.
   Technician B says that there are electronic leak detectors available that will detect CFCs as well as HFCs.
   Who is correct?
   A. A only
   B. B only
   C. Both A and B
   D. Neither A nor B

4. All of the following statements about injecting a dye solution into the air-conditioning system are true, *except*:
   A. Dye is approved for automotive air-conditioning system use.
   B. Dye is available in either yellow or red.
   C. A few days after injection, the dye will be visible at the point of the leak.
   D. The dye affects the overall performance of the system.

5. *Technician A* says that one need not evacuate the system if the "sweep and purge" method is used.
   Technician B says that one need not evacuate the system if it has been "opened" for less than 5 minutes.
   Who is correct?
   A. A only
   B. B only
   C. Both A and B
   D. Neither A nor B

6. After a repair that required opening the refrigerant system the system should be evacuated (pumped down) for a minimum of how many minutes?
   A. 5 minutes.
   B. 10 minutes
   C. 30 minutes
   D. 60 minutes

7. *Technician A* says that the air-conditioning system can be charged from the low side while the system is running.

Technician B says that the air-conditioning system can be charged from the high side while the system is not running.

Who is correct?

A. A only   C. Both A and B

B. B only   D. Neither A nor B

8. *Technician A* says that the minimum pressure recommended for leak testing an R-12 system is 60 psig (414 kPa).

Technician B says the minimum recommended pressure for leak testing an R-134a system is 50 psig (414 kPa).

Who is correct?

A. A only   C. Both A and B

B. B only   D. Neither A nor B

9. *Technician A* says that fluorescent tracer dye is easily detected using an ultraviolet lamp.

Technician B says that the dye should be removed from the system after leak detection.

Who is correct?

A. A only   C. Both A and B

B. B only   D. Neither A nor B

10. *Technician A* says that a standing vacuum test may be held to check an air-conditioning system for leaks.

Technician B says that a standing vacuum test does not reveal how many leaks there are in the system.

Who is correct?

A. A only   C. Both A and B

B. B only   D. Neither A nor B

ASE CHALLENGE QUESTIONS

1. When an air-conditioning system is opened for repairs or service, there is a possibility that _____ will enter system.

A. Air   C. Both A and B

B. Moisture   D. Neither A nor B

2. During initial system evaluation an unknown refrigerant is identified. Which of the following should a technician do?

A. Treat it like R-134a and recover and recycle it.

B. Treat it like R-12 and recover and recycle it.

C. Purge it into the atmosphere

D. Treat it as a contaminated refrigerant.

3. *Technician A* says that the evacuation process will remove dirt and debris from the refrigerant system.

Technician B says that the evacuation process will remove moisture and air from the refrigerant system.

Who is correct?

A. A only   C. Both A and B

B. B only   D. Neither A nor B

4. *Technician A* says that when testing for a refrigerant leak, the electronic leak detector probe should be held against the suspected leak area.

Technician B says that the electronic leak detector requires routine calibration and probe tip maintenance.

Who is correct?

A. A only   C. Both A and B

B. B only   D. Neither A nor B

5. *Technician A* says that comparing the temperature-pressure relationship of recycled refrigerant is a reliable method for finding noncondensable gas contamination.

Technician B says that if the recovery tank pressure is higher than the temperature-pressure relationship chart indicates, the refrigerant is OK to use.

Who is correct?

A. A only   C. Both A and B

B. B only   D. Neither A nor B

Name _____ **Date** _____

ANALYZING REFRIGERANT GAS SAMPLE PURITY

Upon completion of this job sheet, you should be able to diagnose the purity of a vehicle's refrigerant system. The analyzer measures and displays the purity percentage of the refrigerant being tested, as well as the percentage of air in the sample being tested.

NATEF Correlation

HEATING AND AIR CONDITIONING: Refrigerant Recovery, Recycling, and Handling; *Identify (by label application or use of a refrigerant identifier) and recover A/C system refrigerant.* **(P-1)**

Tools and Materials

Late-model vehicle

Service manual or information system

Safety glasses or goggles

Hand tools, as required

Yokogawa GA500-Plus Refrigerant Gas Analyzer or similar device

Describe the vehicle being worked on.

Year _____ Make _____ Model _____

VIN _____ Engine type and size _____

Procedure

The following procedure is for the GA500-Plus Gas Analyzer. Procedures may vary for other models. Always follow the operating instructions that accompany the equipment being used.

1. Connect the purity test unit power cable to the vehicle's battery.
2. Attach the sample hose to the vehicle's refrigerant system service fitting.
   a. Select the R-134a (or the R-12) hose and push the quick-connect coupler onto the fitting on the bottom of the analyzer. Turn the fitting knob to the left to retract the valve activating pin.
   b. Connect the sample hose service port coupler to the low side (vapor port) of the air-conditioning system and turn the knob to the full right position to open the valve port.
3. Fully depress the purity test unit pump button four times to start the test cycles.

4. Evaluate the sample using LED indicators and the LCD display.
   a. Record refrigerant purity:

   _____ R-134a _____ R-12 _____ HC _____ R-22 _____ AIR

   b. Percent of air contamination _____

   NOTE: Because air is a noncondensable gas the analyzer ignores its presence in calculating gas concentrations. Consequently the total of all displays may be greater than 100 percent.

5. To repeat the test, pump the button four times to restart the test cycles.
6. Disconnect the sample hose attached to the vehicle's R-134a system; push the pump button to purge the refrigerant in the analyzer.
7. Based on the results of this test, what procedure will you use to recover the refrigerant? _____

   _____

Instructor's Response _____

_____

_____

Name _____ **Date** _____

REFRIGERANT LEAK DETECTING WITH ELECTRONIC LEAK DETECTOR

Upon completion of this job sheet, you should be able to use an electronic leak detector to leak test an air-conditioning system.

NATEF Correlation

HEATING AND AIR CONDITIONING: A/C System Diagnosis and Repair; *Leak test A/C system; determine necessary action.* **(P-1)**

HEATING AND AIR CONDITIONING *Refrigerant Recovery, Recycling, and Handling; Identify (by label application or use of a refrigerant identifier) and recover A/C system refrigerant.* **(P-1)**

Tools and Materials

Late-model vehicle

Service manual or information system

Safety glasses or goggles

Hand tools, as required

Manifold and gauge set

Dye injector kit

Electronic leak detector

Describe the vehicle being worked on.

Year _____ Make _____ Model _____

VIN _____ Engine type and size _____

Procedure

1. Locate the vehicle refrigerant system information label under the hood of the test vehicle. Determine the type of refrigerant that is used in this system, the recommended charge volume, and the type and amount of refrigerant oil.

   a. Type of refrigerant _____

   b. Refrigerant charge amount _____

   c. Lubricant type _____

   d. Lubricant amount _____

2. If label(s) are not present or are unreadable, explain how to identify the refrigerant. Identify the refrigerant. _____

_____

_____

3. Explain why this system is being checked for leaks. _____

4. Operate the air conditioning for approximately 5 minutes.

5. Shut down the system and engine.

6. Review all the instructions supplied with the electronic leak detector.

7. If applicable, connect the detector to a power source.

8. Inspect the system, starting with the compressor discharge line.

9. Move the detector probe over as much of the condenser as possible. Record the results. _____

10. Move the detector probe over and around each fitting and along the bottom of each hose and line. Record the results. _____

_____

11. Move the detector probe over the evaporator area, internal and external, as much as possible. If the case duct drain is accessible move the sensor probe to the drain outlet. Record the results. _____

_____

12. Record the results of any leak detected and list the name of the line or component. __

_____

13. Recommend and discuss any repairs that must be made. _____

_____

Instructor's Response _____

_____

_____

Name _____ **Date** _____

AIR-CONDITIONING SYSTEM UNLOADED PERFORMANCE TEST

Upon completion of this job sheet, you should be able to perform an unloaded air-conditioning system performance test and determine whether the system is operating as designed.

NATEF Correlation

HEATING AND AIR CONDITIONING: A/C System Diagnosis and Repair; *Performance test A/C system; diagnose A/C system malfunctions using principles of refrigeration.* **(P-1)**

HEATING AND AIR CONDITIONING A/C System Diagnosis and Repair; *Identify refrigerant type; select and connect proper gauge set; record pressure readings.* **(P-1)**

Tools and Materials

Late-model vehicle

Service manual or information system

Safety glasses or goggles

Hand tools, as required

Manifold and gauge set

Thermometer

Describe the vehicle being worked on.

Year _____ Make _____ Model _____

VIN _____ Engine type and size _____

Procedure

NOTE: The following procedures outlined are for an R-134a refrigerant system. If this procedure is being performed on a cycling clutch system, all observations should be made as close to the end of the compressor on cycle as possible.

Pretest Procedure

It is important to prepare and inspect the vehicle prior to beginning the test.

1. Move the vehicle to a shaded area not in direct sunlight and allow it to cool down. The performance test should not be performed until the vehicle has reached ambient air temperature.

2. Inspect the condenser fins and clean them with a soft brush and non-pressurized running water. A condenser that restricts airflow can give the appearance, based on pressure readings, that the system is fully charged when it is actually low on refrigerant.

3. Close all windows and doors.

4. Open the vehicle hood.

5. Place a fender cover on the vehicle to protect the finish.

6. Remove the protective caps from the high- and low-side service ports.

    NOTE: Remove the caps slowly to ensure that no refrigerant escapes past a defective service valve, and inspect the O-ring seal on the cap.

7. Ensure that all manifold valves are closed (clockwise) to prevent refrigerant venting from service ports.

8. Turn the low-side hose hand valve fully counterclockwise to retract the Schrader depressor into the service port.

9. Clip on the low-side hose; push firmly until it clicks into place.
    a. Turn the low-side hose hand valve fully clockwise to extend the Schrader depressor.

10. Clip on the high-side hose; push firmly until it clicks on.
    a. Turn the high-side hose hand valve fully clockwise to extend the Schrader depressor.

11. Place a fan in front of the vehicle to assist in cooling the air-conditioning condenser.

12. Turn the vent fan on high and activate the air-conditioning system in the MAX Recirculation mode and set the temperature control to the coldest setting.

13. Next run the engine at 1,000 rpm and allow the system to stabilize for 15 minutes. Measure the air temperature leaving the center vents. The unloaded test strips the air in the passenger compartment of heat and humidity as the passenger compartment air is recirculated through the evaporator, allowing for the lowest possible vent temperatures to be achieved. The performance chart (see Figure 6-3) for the unloaded performance test requires that the engine speed be at 1000 rpm ($\pm$50 rpm) to be accurate.

Test Procedure

1. Connect an R-134a recovery/recycling station to the air-conditioning system high-side and low-side pressure test fittings.
   Record the ambient air temperature approximately 12 in. in front of the vehicle on the R-134a Air Conditioning Work Sheet.

2. With the engine running at 1,000 rpm check both the high- and low-side system pressures. If the vehicle is equipped with a rear air-conditioning system, ensure that the rear blower is off and record the result on the R-134a Air Conditioning Work Sheet.

3. Place a thermometer in the center dashboard vent and record the outlet air temperature on the R-134a Air Conditioning Work Sheet.

4. If the vehicle is equipped with a rear air-conditioning system, turn the rear blower on high and place a thermometer in the rear outlet vent and record the outlet air temperature. With the engine running at 1000 rpm check both the high- and low-side system pressures and record the result on the R-134a Air Conditioning Work Sheet.

5. Increase the engine speed to 3000 rpm and measure the low-side system pressure as well as the dashboard center, left, and right corner vent outlet air temperatures and record the results on the R-134a Air Conditioning Work Sheet.

    NOTE: The pressure gauge readings at 1000 rpm are not comparable to the observations at 3000 rpm.

6. Compare the readings recorded on the R-134a Air Conditioning Work Sheet to the temperature pressure chart in Figure 6-3 on page 214.

R-134A AIR CONDITIONING WORK SHEET

Initial Air-Conditioning System Evaluation

NOTE: The following procedures outlined are for an R-134a refrigerant system. If this procedure is being performed on a cycling clutch system, all observations should be made as close to the end of the compressor on cycle as possible.

1. Record the following information obtained with the vehicle running at 1,000 rpm

Ambient air temperature 12 in. in front of the vehicle	
High-side pressure (rear air OFF, if equipped)	
High-side pressure (rear air ON, if equipped)	

Low-side pressure (rear air OFF, if equipped)	
Low-side pressure (rear air ON, if equipped)	
Center duct vent outlet air temperature	
Left corner duct vent outlet air temperature	
Right corner duct vent outlet air temperature	
Rear duct vent outlet air temperature	

2. Record the following information obtained with the vehicle running at 3,000 rpm:

Low-side pressure (rear air OFF, if equipped)	
Low-side pressure (rear air ON, if equipped)	
Center duct vent outlet air temperature	
Left corner duct vent outlet air temperature	
Right corner duct vent outlet air temperature	
Rear duct vent outlet air temperature	

After Air-Conditioning System Repair Comparison

NOTE: The following procedures outlined are for an R-134a refrigerant system. If this procedure is being performed on a cycling clutch system, all observations should be made as close to the end of the compressor on cycle as possible.

1. Record the following information obtained with the vehicle running at 1000 rpm

Ambient air temperature 12 in. in front of the vehicle	
High-side pressure (rear air OFF, if equipped)	
High-side pressure (rear air ON, if equipped)	

Low-side pressure (rear air OFF, if equipped)	
Low-side pressure (rear air ON, if equipped)	
Center duct vent outlet air temperature	
Left corner duct vent outlet air temperature	
Right corner duct vent outlet air temperature	
Rear duct vent outlet air temperature	

2. Record the following information obtained with the vehicle running at 3000 rpm:

Low-side pressure (rear air OFF, if equipped)	
Low-side pressure (rear air ON, if equipped)	
Center duct vent outlet air temperature	
Left corner duct vent outlet air temperature	
Right corner duct vent outlet air temperature	
Rear duct vent outlet air temperature	

Instructor's Response _____

_____

_____

Name _____ **Date** _____

MAXIMUM HEAT LOAD SYSTEM DIAGNOSTIC WORK SHEET

Upon completion of this job sheet, you should be able to test the air-conditioning system under maximum heat load while monitoring system pressures and temperatures and be able to analyze system efficiency and identify marginal or failed components in the air-conditioning system.

NATEF Correlation

HEATING AND AIR CONDITIONING: A/C System Diagnosis and Repair; *Performance test A/C system; diagnose A/C system malfunctions using principles of refrigeration.* **(P-1)**

HEATING AND AIR CONDITIONING: A/C System Diagnosis and Repair; *Identify refrigerant type; select and connect proper gauge set; record pressure readings.* **(P-1)**

Tools and Materials

Late-model vehicle

Service manual or information system

Safety glasses or goggles

Hand tools, as required

Refrigerant manifold and gauge set

Thermometer

Describe the vehicle being worked on.

Year _____ Make _____ Model _____

VIN _____ Engine type and size _____

Procedure

Follow the procedures outlined in the service manual and wear eye protection. Give a brief description of your procedure following each step:

1. Place the vehicle outside in direct sunlight.
2. Start the engine and run it at idle speed. Allow the engine to reach normal operating temperature.
3. Turn on the air conditioning and set the air-conditioning control panel to MAX AC, or select AC and Recirculation mode.
4. Open all doors and set the blower motor speed to HIGH.
5. Allow the air-conditioning system to stabilize for a minimum of 5 minutes.
6. Record the ambient air temperature approximately 1 ft. in front of the condenser in the table that follows.
7. Record the relative humidity in the table that follows.

8. Connect a refrigerant identifier and determine the type and purity of the sample.

   a. Refrigerant identified _____

   b. Purity of sample _____

   c. Percent of noncondensable gas (air) _____

9. Connect the manifold and gauge set to the high and low sides of the air-conditioning system.

Measurement	Specification	Before Repair	After Repair
Ambient air temperature			
Relative humidity			
AC center duct temperature			
Ambient air temperature to duct outlet temperature difference	Minimum 30°F*		
High-side pressure			
Low-side pressure			
Condenser inlet temperature			
Condenser outlet temperature			
Condenser temperature difference	20°F Min 50°F Max**		
Evaporator inlet temperature			
Evaporator outlet temperature			
Evaporator temperature difference	−5°F Min +5°F Max***		

*Ambient air to duct air temperature drop+Min 30°F
**Condenser inlet to outlet temperature drop = 20°F Min 50°F Max
***Evaporator inlet to outlet temperature drop = −5°F Min + 5°F Max

Instructor's Response _____

_____

_____

Name _____ **Date** _____

SYSTEM EVACUATION

Upon completion of this job sheet, you should be able to evacuate an air-conditioning system.

NATEF Correlation

HEATING AND AIR CONDITIONING: Refrigerant Recovery, Recycling, and Handling; *Evacuate and charge an A/C system.* **(P-1)**

Tools and Materials

Vehicle with air-conditioning system in need of evacuation

Gauge and manifold set

Vacuum pump

Describe the vehicle being worked on.

Year _____ Make _____ Model _____

VIN _____ Engine type and size _____

Procedure

After each of the following procedures, briefly explain how you performed the task:

1. Connect the gauge manifold hoses to the air-conditioning system service ports.

   _____

   _____

2. What type connectors are found:
   On the low-side service port? _____

   On the high-side service port? _____

3. Observe the gauges.
   The low-side gauge reads _____

   The high-side gauge reads _____

   NOTE: If pressure is observed in step 2, perform Job Sheet 24 before proceeding with this job sheet.

4. Connect the manifold gauge hose to the vacuum pump.

5. What type connector is found on the vacuum pump?

   _____

   _____

6. Turn on the vacuum pump, open the appropriate hand valves, and evacuate the air-conditioning system for the length of time suggested in the service manual, __ hr/min.

7. Close the hand valves (opened in step 6) and turn off the vacuum pump. Observe the gauges. What is the reading in psig or kPa?

	Now	After 5 Min.	After 10 Min.	After 15 Min.
Low Side	_____	_____	_____	_____
High Side	_____	_____	_____	_____

8. Explain the conclusion of the results of step 7.

_____

_____

9. Carefully remove the manifold and gauge set hoses, ensuring that no ambient air enters the system, or proceed with Job Sheet 34. Charge A/c System with refrigerant.

Instructor's Response _____

_____

_____

Name _____ **Date** _____

CHARGE AIR-CONDITIONING SYSTEM WITH REFRIGERANT

Upon completion of this job sheet, you should be able to charge or recharge an air-conditioning system.

NATEF Correlation

HEATING AND AIR CONDITIONING: Refrigerant Recovery, Recycling, and Handling; *Evacuate and charge an A/C system.* **(P-1)**

Tools and Materials

Manifold and gauge set with hoses

Source of refrigerant

   Small "pound" cans

   Bulk source

   Recovery system

Describe the vehicle being worked on.

Year _____ Make _____ Model _____

VIN _____ Engine type and size _____

Procedure

After each of the following procedures, briefly explain how you performed the task.

1. Connect the gauge manifold hoses to the air-conditioning system service ports.
2. What type connectors are found:

   On the low-side service port? _____

   On the high-side service port? _____

3. Observe the gauges.

   The low-side gauge reads _____

   The high-side gauge reads _____

   NOTE: If pressure is observed in step 3, there is refrigerant in the system. Either perform refrigerant recovery before proceeding, or "top off" the refrigerant as instructed by your instructor following this job sheet.

4. Connect the manifold gauge hose to the refrigerant source.
5. What type connector is found on the refrigerant source?

   _____

   _____

6. Open the appropriate hand valves and charge the air-conditioning system following the equipment manufacturer's instructions or those outlined in the service manual, as applicable.

7. After charging, close the hand valves (opened in step 6). Observe the gauges. What is the reading in psig and kPa?

	psig	kPa
Low Side	_____	_____
High Side	_____	_____

8. Explain the resultant low- and high-side pressures noted in step 7. (Are they normal?)

_____

_____

9. Carefully remove the manifold and gauge set hoses ensuring that no refrigerant is allowed to escape.

Instructor's Response _____

_____

_____

Name _____ **Date** _____

AIR-CONDITIONING SYSTEM DIAGNOSIS

Upon completion of this job sheet, you should be able to make basic air-conditioning system diagnostic checks.

NATEF Correlation

HEATING AND AIR CONDITIONING: A/C System Diagnosis and Repair; *Diagnose unusual operating noises in the A/C system; determine necessary action.* **(P-1)**

HEATING AND AIR CONDITIONING: Heating, Ventilation, and Engine Cooling Systems Diagnosis and Repair; *Inspect engine cooling and heater system hoses and belts; perform necessary action.* **(P-3)**

HEATING AND AIR CONDITIONING: Operating Systems and Related Controls Diagnosis and Repair; *Inspect and test an A/C-heater control panel assembly; determine necessary action.* **(P-1)**

Tools and Materials

A vehicle with an air-conditioning system

Selected air-conditioning service tools

Describe the vehicle being worked on.

Year _____ Make _____ Model _____

VIN _____ Engine type and size _____

Procedure

The following observations are to be made with the engine OFF. Record your procedure and findings in the space provided.

1. Inspect the compressor drive belt for condition and tightness.

   Type belt _____ Number of belts _____

   Procedure _____

   Findings _____

2. Inspect the water pump direct drive fan (if applicable).

   Procedure _____

   Findings _____

3. Inspect all coolant carrying hoses, fittings, and shut-off valve.

   Procedure _____

   Findings _____

4. Inspect all refrigerant-carrying hoses and fittings.

Procedure _____

Findings _____

5. Inspect all vacuum hoses, fittings, and components.

Procedure _____

Findings _____

Make the following checks with the engine running and the air-conditioning system controls to MAX cool. Give your procedure and findings in the space provided.

6. Check the operation of the compressor clutch.

Procedure _____

Findings _____

7. Check the operation of the electric cooling fan, if applicable.

Procedure _____

Findings _____

8. Check blower motor operation; LO, LO-MED, HI-MED, and HI.

Procedure _____

Findings _____

9. Check for proper airflow from outlets.

Procedure _____

Findings _____

10. Check refrigerant flow in sight glass, if applicable.

Procedure _____

Findings _____

11. Carefully check suction and liquid line temperature.

Procedure _____

Findings _____

12. Check temperature of airflow from dash outlets.

Procedure _____

Findings	OUTLET	TEMP°F	TEMP°C
	Right	_____	_____
	Center	_____	_____
	Left	_____	_____

On conclusion of testing, turn the air-conditioning system OFF and stop the engine. Remove all tools, such as the thermometer used in step 12.

Instructor's Response _____

_____

_____

Name _____ Date _____

OPERATION AND MAINTENANCE OF REFRIGERANT RECOVERY/RECYCLING EQUIPMENT

Upon completion of this job sheet, you should be able to verify the correct operation and determine the required maintenance of refrigerant-handling equipment.

NATEF Correlation

HEATING AND AIR CONDITIONING: Refrigerant Recovery, Recycling, and Handling; *Verify correct operation and maintenance of refrigerant-handling equipment.* **(P-1)**

Tools and Materials

Owner's manual for Refrigerant Recovery/Recycling Unit

Safety glasses or goggles

Hand tools, as required

Refrigerant Recovery/Recycling Unit

Describe the vehicle being worked on.

Year _____ Make _____ Model _____

VIN _____ Engine type and size _____

Procedure

Follow procedures outlined in the service manual for the Refrigerant Recovery/Recycling Unit. Give a brief description of your procedure following each step.

1. Based on the service information for the unit, what is the recommended service interval for the filter(s) and desiccant? _____

   _____

   _____

   a. Is the unit equipped with a filter change indicator? ☐ Yes ☐ No
   b. Does the filter require maintenance? ☐ Yes ☐ No

2. Briefly explain the service steps involved in replacing the filter(s).

   _____

   _____

   _____

   _____

3. Does the vacuum pump require routine oil change maintenance, or is it an oilless pump?

_____

_____

_____

_____

4. What is the recommended interval for changing vacuum pump oil and how is an oil change performed?

_____

_____

_____

_____

5. Locate and record the recovery tank date code.
   a. What is the current date code on the tank? _____

   b. What is the date the tank requires recertification? _____

   c. What color is the tank? _____

Instructor's Response _____

_____

Name _____ **Date** _____

CHECK STORED REFRIGERANT FOR NONCONDENSABLE GASES AND LABEL CONTAINER

Upon completion of this job sheet, you should be able to test recycled refrigerant for non-condensable gases. Label and store refrigerant.

NATEF Correlation

HEATING AND AIR CONDITIONING: Refrigerant Recovery, Recycling, and Handling; *Label and store refrigerant.* **(P-1)**

HEATING AND AIR CONDITIONING: Refrigerant Recovery, Recycling, and Handling; *Test recycled refrigerant for noncondensable gases.* **(P-1)**

Tools and Materials

Refrigerant recovery tank containing recycled refrigerant

Safety glasses or goggles

Hand tools, as required

Manifold gauge set or Refrigerant Recovery/Recycling Unit

Describe the vehicle being worked on.

Year _____ Make _____ Model _____

VIN _____ Engine type and size _____

Procedure

Prior to checking a recovery tank of recycled refrigerant for noncondensable gases, the tank must first sit at room temperature above 65°F (18.3°C) for at least 12 hours to stabilize out of direct sunlight.

Task Completed

1. Identify the type of recycled refrigerant.

   _____

   _____

2. Verify that the manifold set control knobs and hose shut-off valves are closed. ☐

3. Connect the manifold gauge set auxiliary hose (yellow) to the recovery tank low-side outlet (blue).

4. Using a contact or infrared thermometer, record the tank temperature within 4 in. (10 cm) of the tank.
   a. Tank temperature _____°F _____°C

☐ 5. Open the low-pressure valve on the tank and low-pressure valve on the manifold line and manifold. Record the recovery tank date code.
   a. Tank Pressure _____ psig _____ kPa

6. Go to the chart below for the type of refrigerant being tested; compare the tank temperature to the chart.
   a. What is the chart pressure listed? _____

   b. Is the tank pressure at or below listed pressure? ☐ Yes ☐ No

☐ c. If the pressure is above the limit, attach the refrigerant identifier and test for impurities. If no impurities are identified (i.e., alternative refrigerant or HC), proceed to the next step. If the refrigerant pressure is at or below the chart pressure, the refrigerant contains no noncondensable gases and may be put back into service.

   NOTE: Higher pressures indicate the presence of noncondensable gases.

☐ 7. Slowly vent the vapor from the top of the recovery tank into the recovery/recycling unit until pressure falls below the limit shown in charts. If the recycled refrigerant is R-134a and the pressure is high after purging, shake the container and wait several minutes, then retest.

☐ 8. If the pressure still exceeds the pressure limits in the chart, recycle the entire contents of the container.

☐ 9. Label and store refrigerant.

STANDARD PRESSURE TEMPERATURE CHART FOR R-134A

Temperature Fahrenheit	Pressure PSIG	Pressure kPa	Temperature Fahrenheit	Pressure PSIG	Pressure kPa	Temperature Celsius	Pressure kPa	Pressure PSIG	Temperature Celsius	Pressure kPa	Pressure PSIG
70	76	524	86	102	703	21.1	524	76	30.0	703	102
71	77	531	87	103	710	21.7	531	77	30.5	710	103
72	79	545	88	105	724	22.2	545	79	31.1	724	105
73	80	551	89	107	738	22.8	551	80	31.7	738	107
74	82	565	90	109	752	23.3	565	82	32.2	752	109
75	83	572	91	111	765	23.9	572	83	32.8	765	111
76	85	586	92	113	779	24.4	586	85	33.3	779	113
77	86	593	93	115	793	25.0	593	86	33.9	793	115
78	88	607	94	117	807	25.6	607	88	34.4	807	117
79	90	621	95	118	814	26.1	621	90	35.0	814	118
80	91	627	96	120	827	26.7	627	91	35.6	827	120
81	93	641	97	122	841	27.2	641	93	36.1	841	122
82	95	655	98	125	862	27.8	655	95	36.7	862	125
83	96	662	99	127	876	28.3	662	96	37.2	876	127
84	98	676	100	129	889	28.9	676	98	37.8	889	129
85	100	690	101	131	903	29.4	690	100	38.3	903	131

STANDARD PRESSURE TEMPERATURE CHART FOR R-12

Temperature Fahrenheit	Pressure PSIG	Pressure kPa	Temperature Fahrenheit	Pressure PSIG	Pressure kPa	Temperature Celsius	Pressure kPa	Pressure PSIG	Temperature Celsius	Pressure kPa	Pressure PSIG
70	80	551	86	103	710	21.1	551	80	30.0	710	103
71	82	565	87	105	724	21.7	565	82	30.5	724	105
72	83	572	88	107	738	22.2	572	83	31.1	738	107
73	84	579	89	108	745	22.8	579	84	31.7	745	108
74	86	593	90	110	758	23.3	593	86	32.2	758	110
75	87	600	91	111	765	23.9	600	87	32.8	765	111
76	88	607	92	113	779	24.4	607	88	33.3	779	113
77	90	621	93	115	793	25.0	621	90	33.9	793	115
78	92	634	94	116	800	25.6	634	92	34.4	800	116
79	94	648	95	118	814	26.1	648	94	35.0	814	118
80	96	662	96	120	827	26.7	662	96	35.6	827	120
81	98	676	97	122	841	27.2	676	98	36.1	841	122
82	99	683	98	124	855	27.8	683	99	36.7	855	124
83	100	690	99	125	862	28.3	690	100	37.2	862	125
84	101	696	100	127	876	28.9	696	101	37.8	876	127
85	102	703	101	129	889	29.4	703	102	38.3	889	129

Instructor's Response _____

_____

_____

Chapter 7

DIAGNOSIS OF THE REFRIGERATION SYSTEM

BASIC TOOLS
Basic mechanic's tool set
Manifold and gauge set
Flare nut wrench set
Springlock coupling tool set
Vacuum pump
Refrigerant recovery system
Test lamp
Jumper wire
Thermometers (2)
Large fan (if required)

UPON COMPLETION AND REVIEW OF THIS CHAPTER, YOU SHOULD BE ABLE TO:

- Determine if the air conditioning malfunction is due to an electrical or mechanical failure.

- Determine the "state-of-charge" of refrigerant in the air-conditioning system.

- Determine if the air conditioner is a cycling clutch or noncycling clutch system.

- Perform functional testing of the electrical and mechanical systems.

- Understand general troubleshooting procedures and practices.

AIR CONDITIONING DIAGNOSIS

Servicing the automotive air-conditioning system requires a good working knowledge of the purpose and function of the individual components that make up the total system. This includes the action and reaction of both the mechanical and electrical systems and subsystems.

Air-conditioning systems vary from vehicle to vehicle by year and model; therefore, no standard diagnostic procedures are possible. All automotive air-conditioning system diagnostics, however, share a few common prerequisites. These prerequisites include:

- Determine if the problem is electrical or mechanical, or both.
- Determine that the system is properly charged with refrigerant, that it is not under-charged or overcharged.
- Determine the type system: cycling clutch or noncycling clutch type.
- Perform a **functional test** of the air distribution system to determine proper operation. This should be accomplished before proceeding with the diagnosis procedures. The components of the air-distribution system include the blower motor, switches, vacuum lines, air **ducts**, and mode doors.
- Perform an unloaded performance test of the system to verify correct low- and high-side pressures based on the temperature-pressure chart discussed in Chapter 6.

NOTE: Consider the ambient temperature and relative humidity conditions.

SYSTEM INSPECTION

If the air-conditioning system malfunction is due to abnormally high or low system pressures, inspect the following (Photo Sequence 10):

1. Visually check the condenser. Make certain that the airflow is not blocked by dirt, debris, or other foreign matter.

> When in doubt, refer to the manufacturer's specifications.

> A **functional test** is another term used for a system performance test that compares readings of the temperature and pressure under controlled conditions to determine if an air-conditioning system is operating at full efficiency.

> **Ducts** are tubes or passages used to provide a means to transfer air from one point to another.

TYPICAL PROCEDURES FOR INSPECTING AN AIR-CONDITIONING SYSTEM

P10-1 Visually check for signs of a refrigerant leak as may be noted by an oil stain.

P10-2 Inspect the hoses for cuts or other obvious damage.

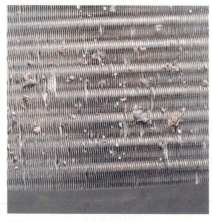

P10-3 Inspect the condenser for debris or other foreign matter.

P10-4 Carefully feel the hoses to determine if there is a temperature differential.

Caution: Some hoses may become extremely hot.

P10-5 A restriction will often result in frosting at the point of restriction on the high side.

P10-6 Ensure that the fan(s) are operating properly.

FIGURE 7-1 Check for refrigerant leaks.

SERVICE TIP: This determination may be made by performing a procedure known as an "insufficient cooling quick check."

SPECIAL TOOLS
Electronic leak detector
Belt tension gauge

A visual inspection often reveals a problem.

A **tension gauge** is a tool for measuring the tension of a belt based on deflection rate.

2. Inspect between the condenser and radiator. Check for any foreign matter. Clean, if necessary.
3. Visually check for bends or kinks in the tubes and lines that may cause a restriction. Check the condenser, refrigerant hoses, and joining tubes.
4. Using a proper leak detector (Figure 7-1), check for refrigerant leaks. It is good practice to leak check the air-conditioning system any time it has a low charge or a leak is suspected. It should also be checked for leaks whenever service operations have been performed that require "opening" the system.
5. Check the air distribution duct system for leaks, restrictions, or binding mode doors. Insufficient airflow may also indicate a clogged or restricted evaporator core.
6. Visually check for proper clutch operation. A slipping clutch may be caused by low voltage due to a loose wire, a defective control device, or improper clutch air gap.
7. Use a belt **tension gauge** to check for proper drive belt tension. Consult the manufacturer's specifications for proper belt tension (Figure 7-2).

CAUTION: Follow specific instructions when using a leak detector.

FIGURE 7-2 Using a belt tension gauge to check for proper belt tension.

ELECTRICAL DIAGNOSIS AND TESTING

Remove rings, watches, and jewelry when working on electrical components.

The following is a typical systematic approach for applying the diagnostic procedures used to determine air-conditioning system electrical problems. Those who lack a general knowledge of automotive electricity and electronics may review this information in Chapter 2 of the Classroom Manual and Chapter 3 of the Shop Manual in the *Today's Technician* series, *Automotive Electricity and Electronics.* A basic understanding of Ohm's law is needed to understand how and why a circuit operates. Review Ohm's law and the theory of series, parallel, and series-parallel circuits.

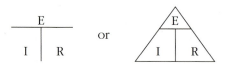

E = Voltage. It is measured in volts; most automotive circuits are measured in DC volts (direct current)

I = Current. It is measured in amperes

R = Resistance. It is measured in ohms

$E = I \times R$

$I = E/R$

$R = E/I$

There are eight basic conditions that one may encounter relating to the electrical system of the automotive air conditioner. These conditions are:

1. Everything works electrically, but there is poor or no cooling
2. Nothing works
3. Only the clutch works
4. Only the evaporator blower works
5. Only the cooling fan works
6. The clutch does not work
7. The evaporator blower does not work
8. The cooling fan does not work

Because of the complex electrical wiring of most factory installed air-conditioning systems (Figure 7-3 and Figure 7-4), it is generally necessary to consult electrical schematics of appropriate manufacturer's shop manuals for proper troubleshooting and testing procedures. The following, however, may be considered typical for basic testing of most systems.

If everything works but there is poor or no cooling, the problem is likely to be in the vacuum, refrigeration, or duct system. One might also suspect an engine cooling system problem. Refer to Chapter 11 of this manual as well as of the Classroom Manual for suggested troubleshooting techniques and procedures.

If nothing works, the complaint is obvious. The system is not cooling, and there is no air moving over the evaporator. A visual inspection will reveal that neither the compressor nor the evaporator fan motor is working, although the cooling fan may or may not be working. It will also be obvious that the blower motor is not running. The logical conclusion, then, is that the main fuse, circuit breaker, or fusible link is defective. The problem could also be in the master control or wiring.

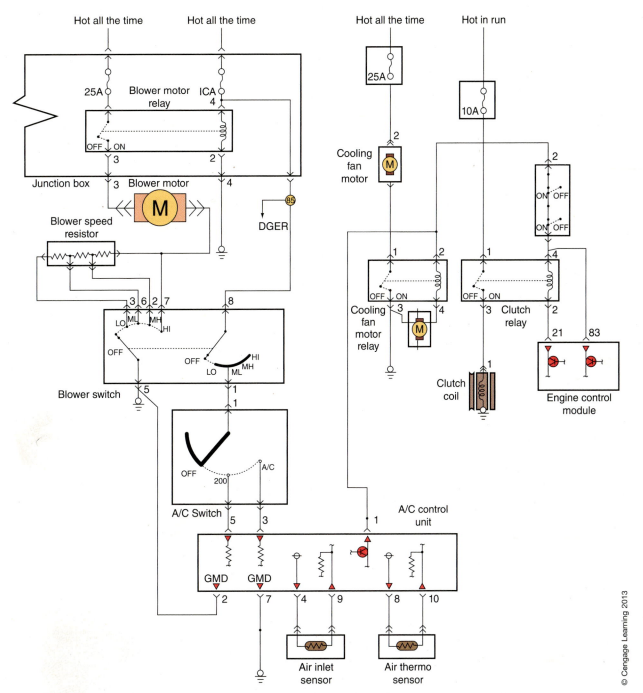

FIGURE 7-3 A typical import automotive air-conditioning system schematic.

Troubleshooting the Blower Motor Circuit

Before troubleshooting, check the battery charge (Figure 7-5). The voltage should be 12.4 volts or more. If not, recharge or replace the battery before proceeding. If the engine starts and runs, however, this test is not necessary. Proceed as follows for systematic testing of the blower motor electrical circuit. Select DC volts on the digital multimeter for the following steps.

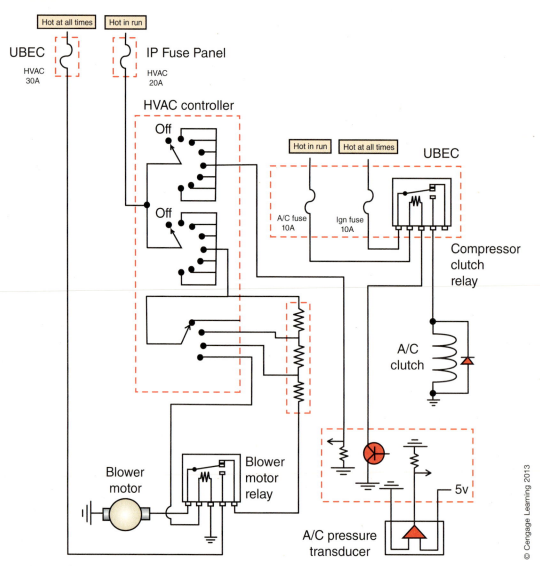

FIGURE 7-4 A typical domestic automatic air-conditioning system schematic.

Within the figure:

Hot at all times | Hot in run

UBEC | IP Fuse Panel

HVAC 30A | HVAC 20A

HVAC controller

Off

Off

Hot in run | Hot at all times | UBEC

A/C fuse 10A | Ign fuse 10A

Compressor clutch relay

A/C clutch

Blower motor | Blower motor relay

A/C pressure transducer

5v

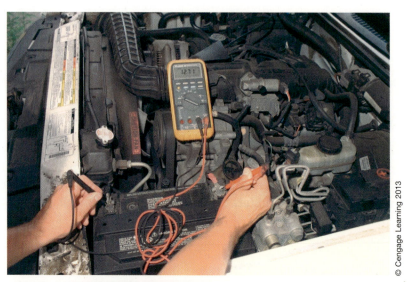

FIGURE 7-5 Testing battery voltage.

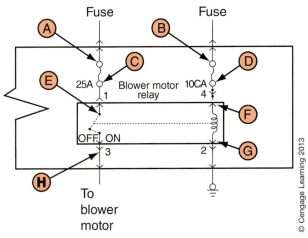

FIGURE 7-6 Test points for fuses and blower motor relay.

Testing the Fuses and Blower Motor Relay (Figure 7-6)

1. Connect black test lead of a digital multimeter (DMM) to **ground** and the other test lead to the hot side of the fuse or circuit breaker, point "A." Turn the ignition switch to the ON position, the air-conditioning switch to A/C, and the blower motor switch to high speed. Note the DMM reading. Next, disconnect the test lead from point "A" and connect it to point "B." Note the reading and disconnect the DMM.
   a. If there are 0 volts at either point "A" or "B," a broken, disconnected, or defective wire or a defective fuse is indicated.
   b. If there are less than 12 volts at either point, look for a loose or corroded (high-resistance) connection before the fuse that is causing the low voltage.
   c. If there are 12 volts or more at both points, proceed with step 2.
2. Connect the DMM from ground to the other side of the fuse, point "C," and note the reading.
   a. The reading should be the same as at point "A." If there are 0 volts, the fuse is defective and must be replaced.

 WARNING: Before replacing the fuse, turn off the air conditioner and ignition switch.

   b. Disconnect the DMM lead from point "C" and attach it to point "D." Note the reading, then disconnect the DMM positive (red) lead.
   c. The reading should be the same as at point "B." If there are 0 volts, the fuse is defective and must be replaced. (See Warning above.)
   d. If 12 volts or more are noted at points "C" and "D," proceed with step 3.
3. Connect the DMM lead to point "E." The voltage reading should be the same as at point "A." Disconnect the DMM.
   a. If there are 0 volts, look for a broken or disconnected wire.
   b. If less than 12 volts, look for a loose or corroded terminal.
   c. If 12 volts or more are noted, proceed with step 4.
4. Connect the DMM to point "F" and note the voltage. Disconnect the DMM.
   a. If 0 volts are noted, the switch is open. Skip to step 5.

   NOTE: It should be closed.

   b. If 12 volts or more are noted, the relay has power to the coil. Proceed with further testing.

Ground is a term used to describe the negative (−) side of an electrical system.

Wires separated inside a connector are not easily detected by visual inspection.

© Cengage Learning 2013

5. Connect one DMM lead to point "F" and the other lead to point "G." This is a voltage drop test of the relay coil assembly; your reading should be close to source voltage (12 volts).
   a. If there are 0 volts, the relay coil is internally open or the ground circuit is open. Inspect the ground circuit. If it is okay, the blower relay needs to be replaced.
   b. If 12 volts or more are noted, proceed to step 6.
   c. If less than 12 volts are noted, there is excessive resistance in the relay coil circuit. Perform voltage drops on the wiring, switches, and connections (each should drop less than 0.2 volt).
6. Connect the DMM red lead to point "H" and black lead to ground.
   a. If 12 volts or more are noted, the blower motor should operate. If the blower motor does not operate, proceed to the diagnostic steps under "Testing Blower Motor" in the next section.
   b. If there are 0 volts, replace the relay.
   c. If less than 12 volts are detected, check for loose or corroded connections. If connections are okay, replace the relay.

Testing the Blower Motor (Figure 7-7). Turn the ignition switch to the ON position, the air conditioning switch to A/C, and the blower motor switch to high speed. Set the DMM to a DC volt scale.

1. Connect the DMM to point "A" of the blower motor.
   a. The voltage should be the same as measured at point "H" of the previous test, 12 volts or more.
   b. If 0 volts are noted, look for a disconnected wire or connection.
   c. If less than 12 volts, look for a loose or corroded wire or connector.
   d. If there are 12 volts or more, proceed with step 2.

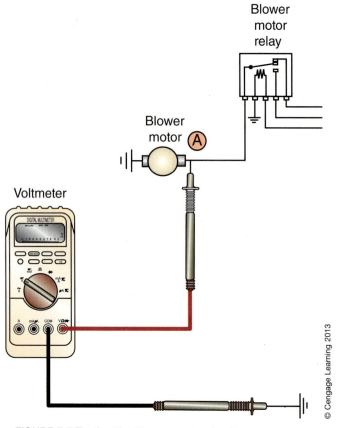

FIGURE 7-7 Testing the blower motor circuit.

© Cengage Learning 2013

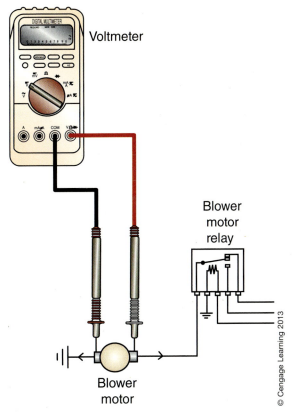

Voltmeter

Blower
motor
relay

Blower
motor

© Cengage Learning 2013

FIGURE 7-8 Blower motor voltage drop test.

Wiring and switches in a circuit should have a voltage drop of less than 0.2 volt, and the entire circuit's total voltage drop (aside from the load device in the circuit which should drop close to source voltage) should not add up to more than 0.5 volt in general. Remember, the sum of all voltage drops must add up to source voltage.

2. Connect the DMM in parallel as depicted in Figure 7-8.
   a. If the meter reads 12 volts or more and the blower motor will not operate, replace the blower motor. If the blower motor operates but turns slowly, proceed to step 3.
   b. If the voltage is 0 volts, check for an opening in the ground circuit wire. If the ground wire is okay, replace the blower motor.
   c. If the voltage is less than 12 volts, check for unwanted resistance in the connectors, wiring of the blower motor circuit, and repair fault. Use a voltage drop test of the suspect area to confirm the diagnosis (wiring should drop less than 0.1 volt).
3. If the blower motor operates slowly, check the amperage draw of the blower motor and compare the results to the manufacturer's specifications. Connect the DMM as outlined in Figure 7-9, and set the meter to the ampere (A) scale.
   a. Disconnect the blower motor harness connection from the motor, and connect a jumper lead from the ground (negative) terminal of the harness connector to the blower motor ground (negative) terminal.
   b. Connect a fused jumper lead (the fuse rating should be lower than the rating on the internal fuse of the meter and that of the circuit being tested) from the blower motor harness positive connector terminal to the meter positive (red) lead and connect the meter negative (black) lead to the positive terminal on the blower motor. The meter should now be wired in series with the blower motor.
   c. Compare the amperage draw to the manufacturer's specifications; if they are higher than specifications, replace the blower motor.

A defective ground wire accounts for a high percentage of electrical failures.

SERVICE TIP:
Digital multimeters have internal fuses to protect their circuits. Never connect a meter to a circuit in a series if the possible current level in the circuit could exceed the amperage rating of the internal meter fuse. If in doubt, use a fused jumper lead with a fuse rated at less than the internal meter fuse. This way, if the current draw is higher than expected, the jumper lead fuse will blow.

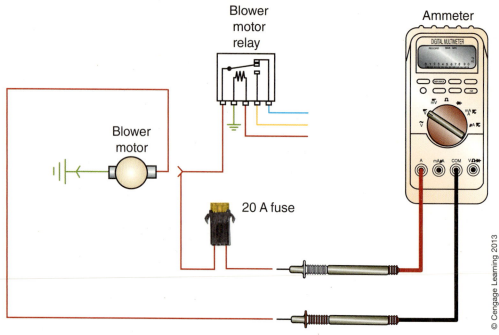

FIGURE 7-9 Blower motor current test with meter in series.

Testing the Blower Motor Speed Resistor (Figure 7-10). Though blower motor speed resistor resistance values differ between vehicle makes, they are typically as suggested in this procedure. Before conducting this test, disconnect all wires from the resistor.

NOTE: DO NOT connect the DMM to the disconnected wires.

1. Select the ohm scale of the DMM (on manual DMM selected the 200 Ω scale). Connect one lead of the DMM to terminal 2 of the blower motor resistor.
   a. Touch the other DMM test lead to terminal 4 of the blower motor resistor. The resistance should be within manufacturer specifications (i.e., 0.31–0.35 Ω).
   b. Touch the other DMM test lead to terminal 1 of the blower motor resistor. The resistance should be within manufacturer specifications (i.e., 0.88–1.0 Ω).
   c. Touch the other DMM test lead to terminal 3 of the blower motor resistor. The resistance should be within manufacturer specifications (i.e., 1.8–2.1 Ω).
2. If the resistance is not as specified by the manufacturer, the blower motor resistor should be replaced.

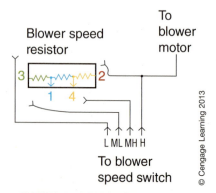

FIGURE 7-10 Test points for blower motor speed resistor.

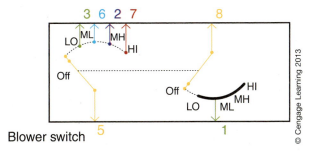

FIGURE 7-11 Test points for blower motor switch.

Testing the Blower Motor Switch (Figure 7-11). Methods of testing the blower motor switch, depending on vehicle make and model, are outlined in the manufacturer's service manual. Procedures given here assume reasonable access to the control. Remove the connectors from the switch.

Test 1:

1. Select the ohm scale of the DMM (on manual DMM select the 200 Ω scale). Connect one test lead to terminal 5 of the blower motor switch.

   a. Connect the other test lead to terminal 3 of the blower motor switch and, while observing the meter, turn the switch to LO.

   NOTE: The meter should have gone from infinity (∞) to a very low resistance, such as 0 Ω.

   b. Disconnect the test lead and connect it to terminal 6 of the blower motor switch and, while observing the meter, turn the switch to ML.

   NOTE: The meter should have gone from infinity (∞) to a very low resistance, such as 0 Ω.

   c. Disconnect the test lead and connect it to terminal 2 of the blower motor switch and, while observing the meter, turn the switch to MH.

   NOTE: The meter should have gone from infinity (∞) to a very low resistance, such as 0 Ω.

   d. Disconnect the test lead and connect it to terminal 7 of the blower motor switch and, while observing the meter, turn the switch to HI.

   NOTE: The meter should have gone from infinity (∞) to a very low resistance, such as 0 Ω.

Test 2:

1. Connect one test lead to terminal 1 of the blower motor switch.

   a. Connect the other test lead to terminal 8 of the blower motor switch.

   b. While observing the meter, turn the switch to LO.

   NOTE: The meter should go from infinity (∞) to a very low resistance, such as 0 Ω.

   c. While observing the meter, turn the switch to ML, MH, then to H.

   NOTE: The meter should remain on a very low resistance, such as 0 Ω.

If the blower motor speed control switch failed either of the above tests, it must be replaced.

Testing the A/C Switch (Figure 7-12). Methods of testing the A/C switch, depending on vehicle make and model, are outlined in the manufacturer's service manual. Procedures given here are typical and assume reasonable access to the switch.

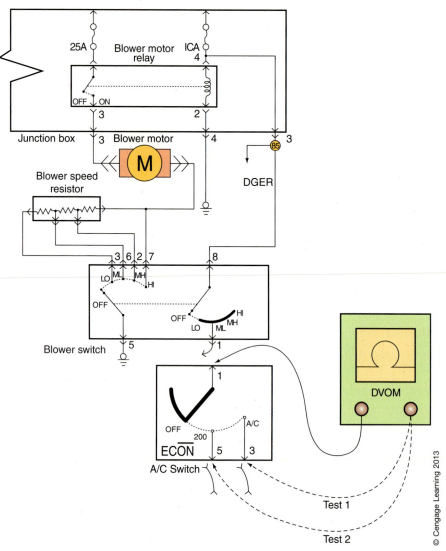

Be sure that the test lamp is not defective; that it will light when it is expected to do so.

FIGURE 7-12 Testing the A/C switch.

1. Gain access to the A/C switch and disconnect all wires.
2. Select the ohm scale of the DMM (on manual DMM select the 200 Ω scale).
3. Connect one test lead of the DMM to terminal 1 of the A/C switch.
   a. Connect the other test lead to terminal 5 of the A/C switch.
   b. While observing the meter, turn the switch to ECO.

   NOTE: The meter should go from infinity (∞) to a very low resistance, such as 0 Ω.

4. Disconnect the test lead from terminal 5 and connect it to terminal 3. While observing the meter, turn the switch to A/C.

   NOTE: The meter should go from infinity (∞) to a very low resistance, such as 0 Ω.

5. If the A/C switch failed either of the above tests, it must be replaced.

AIR CONDITIONING PRESSURE DIAGNOSTICS

If you are able to verify a customer complaint of poor passenger compartment cooling, compare the refrigerant system gauge readings to a manufacturer's published diagnostic information. In some cases a manufacturer may provide a diagnostic graph broken into the pressure zones for the refrigerant system high- and low-side pressures (Figure 7-13).

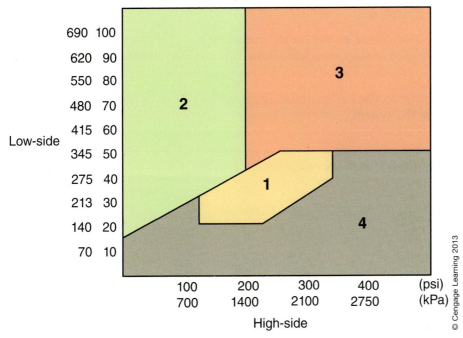

FIGURE 7-13 Refrigerant system high- and low-side pressure zone diagnostic graph.

In the following sections conditions 1 through 7, discussed in the Class Manual, may be condensed into four major system operation pressure zones.

Pressure Zone 1

If the high- and low-side refrigerant pressures fall into the Zone 1 area, the following should be checked. The high- and low-side pressures may be normal or slightly less than normal based on ambient air temperature. If the refrigerant system is not functioning as designed, the following items should be checked for proper operation:

- Refrigerant system slightly undercharged
- Refrigerant gas contamination
- Air duct delivery problem

Pressure Zone 2

If the high- and low-side refrigerant pressures fall into the Zone 2 area, the following should be checked. If the low-side pressure is higher than normal and the high-side pressure is lower than normal, the following items should be checked:

- Refrigerant system undercharged
- Defective or malfunctioning compressor

Pressure Zone 3

If the high- and low-side refrigerant pressures fall into the Zone 3 area, the following should be checked. If the low- and the high-side pressures are both higher than normal, the following items should be checked:

- Refrigerant system overcharged
- Thermostatic expansion valve defective
- Condenser cooling fan operation
- Restricted airflow across the condenser

Pressure Zone 4

If the high- and low-side refrigerant pressures fall into the Zone 4 area, the following should be checked. If the low-side pressure is lower than normal and the high-side pressure is higher than normal, the following items should be checked:

- A restriction in the air-conditioning system
- Debris in the system

DEFECTIVE COMPONENTS

The following is an overview of what the technician may expect if a component of the air-conditioning system is found to be defective.

Evaporator

A defective evaporator produces an insufficient supply of cool air. This symptom is often the result of a leak in the evaporator core. Other causes include:

- Dirt- or debris-plugged core (clean the core)
- Cracked or broken case (replace or caulk the case)
- Leaking seal or O-ring (replace the seal or O-ring)
- Restricted refrigerant passage ways through the core

Compressor

Classroom Manual
Chapter 7, page 233

Inspect the compressor for visual indications of a loss of oil.

A malfunctioning compressor will be indicated by one or more of the following symptoms:

- Noise, indicating premature failure
- Seizure, usually due to oil loss
- Leakage, due to defective seals, gaskets, or O-rings
- Low suction and low high-side pressure caused by an undercharge of refrigerant or a restriction in the low side of the system
- High suction and low discharge pressure usually caused by a defective compressor valve plate or gasket assembly

Some noise is to be expected from most compressors and may be considered normal during regular operation. Irregular rattles and noises, however, are an indication of early failure due to loss of oil or broken parts.

When the air-conditioning system control calls for cooling and the compressor is inoperative, verify that 12 volts are present at the clutch terminals. To check for compressor seizure:

1. Turn off the engine.
2. De-energize the compressor clutch.
3. Try to rotate the drive plate by hand (Figure 7-14).

NOTE: Compressors that have not been used for a long period of time may "stick." If this is the case, turning it four or five times in both directions should free it up. If it will not rotate or takes great effort, the compressor may have an internal defect. Also, if there is no resistance to turning, the shaft may be broken internally.

If the compressor clutch is slipping, it may be due to an internal compressor problem or it may be due to an incorrect clutch air gap. Before condemning the compressor, determine if the air gap is correct. Low voltage or a defective clutch coil may also cause clutch slipping problems. First, check to ensure that there are at least 12 volts available at the clutch coil. Next, disconnect the clutch coil and one side of the diode, if applicable, then

FIGURE 7-14 Rotate the drive plate (armature) by hand.

check its resistance. Refer to the manufacturer's specifications for specific values; however, the resistance should generally be between 3 and 4 ohms (3–4 Ω). Refer to Job Sheet 49 for procedures.

Low discharge pressure can be caused by poor internal compressor sealing, such as the valve plate or valve plate gasket set. It can also be caused by a restriction in the compressor or in the low side of the system. Yet another cause of low discharge pressure is an insufficient charge of refrigerant. All possibilities should be explored before diagnosing the problem as a defective compressor.

Condenser

There are three possible malfunctions of a condenser:

1. A leak due to rust and corrosion or by being struck by a sharp object or stone.
2. A restriction if a tube has been bent when struck with an outside object, such as a stone, with insufficient force to cause a leak but sufficient to kink or collapse a tube.
3. Restricted airflow through the condenser caused by dirt, debris, or foreign matter. When the airflow through the condenser is restricted or blocked, high discharge pressures will result. This high discharge pressure may or may not be noticeable on a high-side gauge connected to the vehicle service port depending on the location of the service port. If the gauge is connected before the restriction, high-side gauge pressure will be higher than normal. But if the gauge is connected after the restriction, the high-side gauge reading may be in the low to normal range. Always think about where the gauges are physically connected to the system and what these readings are telling you. Improper gauge reading interpretation can lead to a lengthy and often incorrect diagnosis.

NOTE: Though not easily detected, the outlet tube of the condenser may be slightly cooler than the inlet tube. Carefully feel the temperature by hand. It should go from a hot inlet at the top to a warm outlet at the bottom. A proper heat exchange should result in an even gradient across the surface of the condenser. Some technicians prefer the use of a laser-sighted, digital readout (DRO), infrared (IR) thermometer (Figure 7-15) for taking temperature measurements. This type of thermometer provides an immediate and accurate indication of the surface temperature of any object, in °F or °C, simply by pointing the

Classroom Manual
Chapter 7, page 232

Look for a "frost patch," indicating a restriction.

SERVICE TIP: A restricted condenser may also result in excessive compressor discharge pressure. A partial restriction can cause a temperature change and even frost or ice to form immediately after the restriction. In this case, the restriction is serving as a metering device.

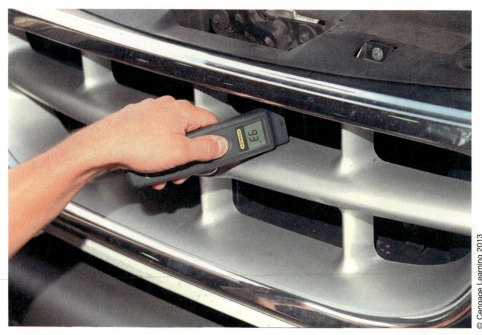

FIGURE 7-15 Using a laser-sighted, digital thermometer for measuring temperature gradient.

thermometer at the object. See Job Sheet 45 at the end of this chapter entitled "Heat Transfer through the Condenser."

Orifice Tube

Orifice tube failures are often indicated by low suction and discharge pressures. This condition results in insufficient cool air from the evaporator.

The common cause of orifice tube failure is a restriction. A less common cause is a clogged inlet screen due to contamination, corrosion particles, or refrigerant desiccant loose in the refrigerant system due to a defective accumulator. Regardless of the cause, the recommended repair is to replace the orifice tube (Figure 7-16) and the defective parts that caused the problem.

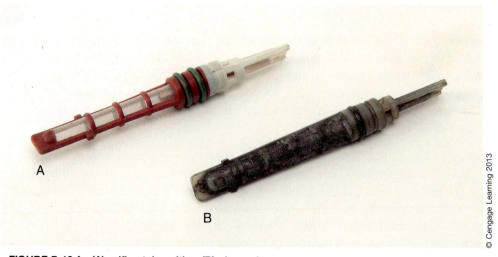

FIGURE 7-16 An (A) orifice tube with a (B) clogged screen.

Moisture contamination can also have similar symptoms and poor performance complaints such as a restricted orifice tube. If a refrigerant system is contaminated with moisture, the pressure in the system will swing between a vacuum to normal on the low-side gauge and between low to normal on the high-side gauge. The air-conditioning system may operate normally at first, but as the system runs, a no cooling condition may exist. This condition is often described as intermittent air conditioning operation by the driver. The moisture in the system freezes in the fixed orifice tube (FOT) or thermostatic expansion valve (TXV) and causes a temporary blockage. If moisture is suspected the accumulator assembly must be replaced and the system must be evacuated.

Thermostatic Expansion Valve

Most TXV failures are indicated by the same symptoms as orifice tube failures. Many TXV failures, however, are due to a malfunctioning of the power element. This failure usually results in a closed valve that will not allow refrigerant to enter the evaporator. Regardless of the cause of failure, except for a clogged inlet screen, replacement of the valve is the recommended repair.

The inlet screen of the TXV can also become plugged due to contamination, corrosion particles, or refrigerant desiccant loose in the system due to a defective receiver-drier (Figure 7-17).

Refrigerant Lines

Refrigerant line restrictions are generally indicated by one or more of the following symptoms:

- Suction line: A restriction of the suction line causes low suction pressure, low discharge pressure, and little or no cooling. The evaporator is starved of refrigerant.
- Discharge line: A restricted discharge line generally causes the pressure relief valve to open to release excess pressure to the atmosphere. Pressure relief valves are generally self-reseating whenever the pressure drops to a predetermined safe level.
- Liquid line: A liquid line restriction has the same general symptoms as a suction line restriction—low suction pressure, low discharge pressure, and little or no cooling from the evaporator.

Classroom Manual
Chapter 7, page 230

The suction line is usually the largest hose or tube.

The liquid line is usually the smallest hose or tube.

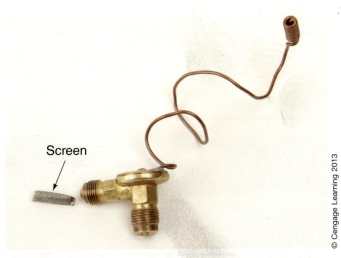

Screen

© Cengage Learning 2013

FIGURE 7-17 Inlet screen of a thermostatic expansion valve (TXV).

CAUSES OF FAILURE

Following are some of the common causes of failure found in an automotive air-conditioning system:

- Leaks; undercharge of refrigerant
- High pressure; overcharge of refrigerant in the system, air in the system, excess oil in the system
- Poor connections
- Restrictions
- Contaminants
- Moisture
- Defective component

FUNCTIONAL TESTING

A functional test may be performed to determine the operating conditions of the air-conditioning control head (Figure 7-18) as well as the air distribution system. The functional test consists of checking the operation of the blower, heater, and air-conditioning control assembly mode lever. This test also includes comparing mode lever and switch positions in relation to air delivery.

DIAGNOSING H-BLOCK THERMOSTATIC EXPANSION VALVE SYSTEM

The ideal ambient temperature to be testing expansion valves is 70–85°F (21–27°C). The following test procedure will quickly detect expansion valve malfunctions such as stuck open, stuck closed, or loss of gas charge in the power dome.

1. Connect the manifold and gauge set or charging station to the vehicle refrigerant system service ports and verify the refrigerant charge level.
2. Verify that all vehicle windows and doors are closed.
3. Start the engine and allow the vehicle to idle.
4. Activate the air-conditioning system, set the temperature control to the highest heat setting, turn the blower motor on high, and select Recirculation mode.
5. Allow the vehicle to run until the passenger compartment warms up. This is to put maximum heat load on the refrigerant system and create the need for maximum refrigerant flow through the evaporator (expansion valve will open).
6. If the refrigerant system has a sufficient charge, the high-pressure gauge should be 120–240 psig (827–1655 kPa) and the low-side gauge should read 30–50 psig (207–345 kPa). If pressure levels are correct, go to step 7. If pressures are out of range, replace the expansion valve.

FIGURE 7-18 A typical air-conditioning system control head.

7. If the low-side gauge reading is within the specified range, cool the TXV for 30 seconds by flowing liquid CO_2 (or another suitable super-cold substance) over the exterior of the TXV housing.

 WARNING: When working with liquid CO_2, wear eye protection, insulated protective gloves, and long sleeves to protect the skin and eyes from serious injury.

NOTE: Liquid CO_2 is the preferred method of super cooling the expansion valve and is available from welding supply shops. Ensure that you specify liquid CO_2; otherwise you may receive low-pressure CO_2 gas for beverage dispensing.

8. The low-side gauge reading should drop to 10 psig (69 kPa). If pressure does not drop, replace the expansion valve.

9. Allow the system to stabilize and the TXV to thaw. The low-side pressure should return to 30–50 psig (207–345 kPa). If pressure does not return to these levels, replace the expansion valve.

NOTE: If the expansion valve appears to be intermittently sticking, it may be caused by moisture contamination. If moisture contamination is suspected, the receiver-drier needs to be replaced, not the TXV, and the system needs to be thoroughly evacuated.

10. Perform an unloaded system performance test.

Diagnosing Thermostatic Expansion Valve Systems

The following procedure may be followed to diagnose the TXV system performance:

1. Connect the manifold gauge set to the system.
2. Start the engine and adjust the engine speed to fast idle, 1000–1200 rpm.
3. Place a large fan in front of the condenser to substitute for normal ram airflow (Figure 7-19).
4. Operate the air conditioner to "stabilize" the system:
   a. Adjust all controls for MAX cooling.
   b. Operate the system for 10–15 minutes.
5. Observe and note the gauge readings.

FIGURE 7-19 Place a fan in front of the vehicle to provide ram air.

Classroom Manual
Chapter 7, page 230

Interpreting gauge readings will become "second nature" to the technician.

SPECIAL TOOL
Large floor fan

SERVICE TIP: Gauge readings vary and their application is often a matter of "professional opinion." Gauge readings should, then, only be used as a guide in diagnostic procedures. Proceed with step 5 for abnormally low low-side gauge readings. Go to step 9 for abnormally high low-side gauge readings.

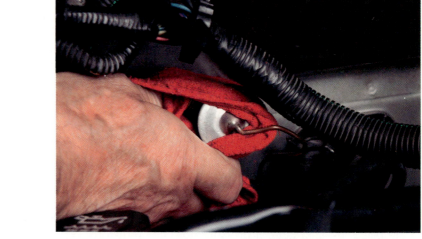
FIGURE 7-20 Place a warm rag around the TXV body.

SERVICE TIP:
A CO$_2$ pressure tank equipped with a low-pressure liquid regulator and a venting hose may be used to chill the remote bulb or the internally regulated H-Block TXV by bleeding low-pressure liquid CO$_2$ over the valve or bulb.

SERVICE TIP:
Replace the TXV if it fails the bulb warming/cooling test. If the problem in step 4 was low pressure, check the inlet screen for foreign matter before replacing the valve.

Rock salt (sodium chloride, NaCl) in ice water (H$_2$O) will produce a freezing (32°F [0°C]) liquid temperature for testing purposes.

6. If the low-side gauge reading is abnormally low, place a warm rag (125°F [52°C]) around the valve body (Figure 7-20).
7. Observe the low-side gauge.
   a. If the pressure rises, there is moisture in the system. Correct as required.
   b. If the pressure does not rise, proceed with step 8.
8. Remove the remote bulb and place it in a warm (125°F [52°C]) rag (Figure 7-21).
9. Observe the low-side gauge.
   a. If the pressure rises, the remote bulb was probably improperly placed. Reposition the bulb, tighten it, and retest the system.
   b. If the pressure does not rise, proceed with step 10.
10. If the low-side gauge reading, step 4, was abnormally high, remove the remote bulb from the evaporator outlet tube and place it in an ice water (H$_2$O) bath.
11. If the low-side presure drops to normal or near normal, the problem may be:
   a. Lack of insulation of the remote bulb. Reinstall, insulate, and retest.
   b. Improperly placed remote bulb. Reposition the remote bulb, insulate, and retest.
12. Conclude the test.
   a. Turn off all air-conditioning controls.
   b. Reduce the engine speed and turn off the engine.
   c. Remove the manifold and gauge set.

FIGURE 7-21 Place the remote bulb in a warm rag.

REFRIGERANT SYSTEM CHARGE LEVEL DETERMINATION TEMPERATURE METHOD

There are several methods available for determining the proper charge level of the vehicle refrigerant system. The following test procedure for an FOT air-conditioning system determines the proper refrigerant charge level based on system pressures and the temperature of the high-pressure liquid line. This test procedure is for FOT systems only and is not designed to be used with expansion valve systems. The technician will compare system pressures and monitored line temperature to the FOT charge determination chart (Figure 7-22). Both the procedure and the chart are based on ambient air temperature range of 70–85°F (21–29°C) in the service facility.

Procedure

1. Attach the manifold and gauge set or recovery/recycling/charge station to the vehicle refrigerant service ports. Ambient air temperature must be in the range of 70–85°F (21–29°C).
2. Attach a clamp-on thermocouple onto the vehicle high-pressure liquid line just before the FOT inlet point. The closer the temperature is recorded to the FOT inlet, the more accurate the temperature reading will be.
3. Start the engine and run it at idle speed. Turn on the air conditioning and set the A/C control panel to Fresh-Air mode and the temperature level to the coldest setting.
4. Open all doors and set the blower motor speed to HI.
5. Allow the air-conditioning system to stabilize for a minimum of 5 minutes.
6. Increase the high-pressure liquid line pressure to 260 psig (1793 kPa) by restricting airflow across the air-conditioning system condenser.
   a. Place a piece of cardboard in front of the air-conditioning condenser. Cover only enough of the condenser to raise and maintain the liquid line pressure at the level specified above.
7. Record the high-pressure gauge reading.

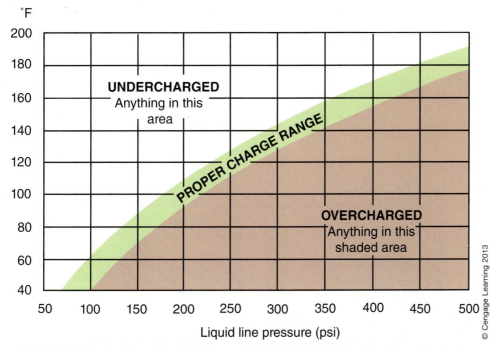

FIGURE 7-22 Fixed orifice tube (FOT) charge determination chart is based on ambient air temperature range of 70°F to 85°F (21°C to 29°C).

8. Record the high-pressure liquid line inlet temperature just before the FOT.
9. Referring to the Charge Determination Chart, record the ideal high-pressure range for the line temperature recorded in step 8.
10. Compare the pressure recorded in step 7 to the idle pressure range determined in step 9.
   a. If the refrigerant system is operating in the proper charge range, the refrigerant system is functioning correctly.
   b. If the refrigerant system is operating in the undercharge range, add 2 oz. of refrigerant (0.057 kg) R-134a to the refrigerant system and repeat steps 5–10.
   c. If the refrigerant system is operating in the overcharge range, reclaim 2 oz. of refrigerant (0.057 kg) R-134a from the refrigerant system using the recovery/recycling/charge station and repeat steps 5–10.

INSUFFICIENT COOLING: CYCLING CLUTCH ORIFICE TUBE (CCOT)

The quick test procedure in CCOT systems helps to determine if the air-conditioning system is properly charged with refrigerant. This test can only be performed if the ambient temperature is above 70°F (21°C).

The quick test can simplify system diagnosis by verifying the problem of insufficient refrigerant. This quick test also eliminates a low refrigerant charge as a source of the problem. The quick test is performed as follows:

1. Start the engine. Allow it to warm at normal **idle speed**.
2. Open the hood and doors.
3. Select the NORM mode.
4. Move the lever to the full COLD position.
5. Select HI blower speed.
6. Feel the temperature of the evaporator inlet after the orifice tube (Figure 7-23).
7. Feel the temperature of the accumulator surface when the compressor is engaged.
   a. Both surfaces (steps 6 and 7) should be at the same temperature. If they are not the same, check for other problems.
   b. If the inlet of the evaporator is cooler than the suction line accumulator surface or if the inlet has frost accumulation, a low refrigerant charge is indicated.

Classroom Manual

Chapter 7, page 217

Idle speed is the speed (rpm) at which the engine runs while at rest (idle).

SERVICE TIP:
If the compressor is cycling, wait until the clutch is engaged.

FIGURE 7-23 Feel the evaporator inlet after the orifice tube.

© Cengage Learning 2013

8. If a low refrigerant charge is indicated, add 4 oz. (113 g) of refrigerant and repeat steps 6 and 7.
9. Add 4 oz. (113 g) at a time until both surfaces feel the same temperature.

NOTE: It is normal for an accumulator to sweat if the system is properly charged with refrigerant. It means that the evaporated refrigerant is absorbing heat from the ambient air surrounding the accumulator and suction line. This heat, added to the heat adsorbed in the evaporator, is called superheat and does not cause a change in pressure. Overall sweating, then, does not indicate that there is a restriction in the accumulator.

DIAGNOSING ORIFICE TUBE SYSTEMS

If the indication is "no cooling" or "insufficient cooling" from the air-conditioning system, inspect the air-conditioning and cooling systems for defects as follows:

Preliminary Checks (All Models)

Inspect the system components before connecting the manifold and gauge set. Procedures for diagnosing the General Motors Cycling Clutch Orifice Tube (CCOT) and Ford Fixed Orifice Tube (FFOT) systems follow. Diagnosing TXV systems is covered later in this chapter.

General Motors CCOT Diagnosis

1. Check to verify that the temperature door strikes both stops when the lever is moved rapidly from hot to cold and cold to hot.
2. Check for a loose, missing, or damaged compressor drive belt.
3. Check for loose or disconnected wiring or connectors.

   NOTE: Be sure fuses or circuit breakers are not defective.

4. Check to see if the cooling fan (Figure 7-24) is running continuously in all air-conditioning modes.

   NOTE: Make repairs as necessary, and recheck cooling.

5. If items checked in steps 1 through 4 are satisfactory and the system still does not cool:
   a. Set the temperature lever to full (MAX) cold.
   b. Move the selector lever to NORMAL A/C.
   c. Set the blower switch to HI.

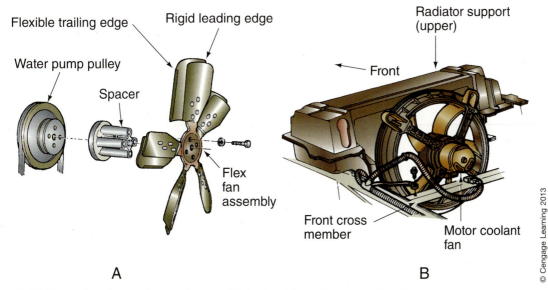

FIGURE 7-24 A typical engine cooling fan: (A) engine driven; (B) electrically driven.

© Cengage Learning 2013

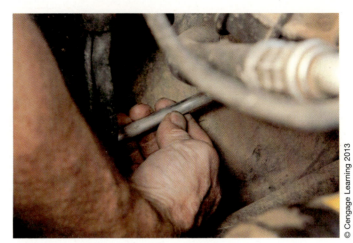

FIGURE 7-25 Feel the liquid line before the orifice tube.

Classroom Manual

Chapter 7, page 224

    d. Open the doors and hood.

    e. Warm the engine at 1500 rpm.

6. Perform a visual check for compressor clutch operation.

    a. If the clutch does not engage, proceed with the "Compressor Clutch Test."

    b. If the clutch engages or cycles, feel the liquid line before the orifice tube (Figure 7-25).

    c. If the tube is warm, proceed with step 7.

    d. If the tube is cold, check the high-side tubing for a restriction.

NOTE: A restriction will be marked with a drop in temperature or frost spot. If the tubing is restricted, repair, evacuate, and recharge the system.

7. If the system is equipped with variable displacement compressor, proceed with "Diagnosing Variable Displacement Compressor Orifice Tube Systems." For all other models, feel the evaporator inlet and outlet tubes.

    a. If the inlet is colder than the outlet, the system may be undercharged. Check for and repair any leaks; evacuate, charge, and retest the system. If no leaks are found, proceed with "System Charge Test."

    b. If the outlet tube is colder than the inlet tube or both tubes are the same temperature, proceed with "Pressure Switch Test."

8. Normal system performance ranges for a cycling clutch system are 15–35 psig (103.42–241.32 kPa) on the low side, 145–200 psig (999.74–1378.95 kPa) on the high side, and the center duct vent output temperature should be in the range of 35–55°F (1.7–12.8°C).

SYSTEM CHARGE TEST

This test may be performed if cooling is not adequate and the static pressure of both the high and low side are below 50 psig (344.74 kPa) with the system off. While performing the system charge test, watch the high-side gauge for any indication of overcharging, such as excessively high pressure. Discontinue the test if the system pressure exceeds that expected for any given ambient temperature condition. Compare system pressure to the manufacturer's specifications. The Delta-T method described in Chapter 7 of the Class Manual may also be useful for FOT system diagnosis. (See Job Sheet 40 on "Refrigerant System Charge Level Temperature Method [Delta-T].")

1. Add a "pound" of refrigerant. Check the clutch cycle rate.

    a. If the clutch cycles more than eight times per minute (less than every 7 seconds), discharge the system and check for a plugged orifice tube or some other restriction.

    b. If the clutch cycles less than eight times per minute (more than every 8 seconds), proceed with step 2.

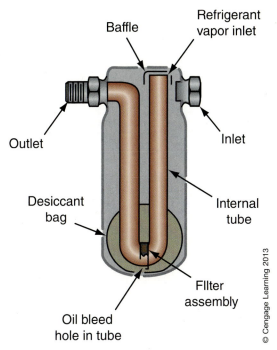

Baffle

Refrigerant vapor inlet

Outlet

Inlet

Desiccant bag

Internal tube

Filter assembly

Oil bleed hole in tube

© Cengage Learning 2013

FIGURE 7-26 A typical accumulator showing inlet and outlet tubes.

2. Feel the accumulator inlet and outlet tubes (Figure 7-26).
   a. If the inlet tube is warmer or the same temperature as the outlet tube, add 3–4 oz. (85–113 g) more refrigerant.
   b. If the inlet tube is colder than the outlet tube, add 3–4 oz. (85–113 g) more refrigerant.
3. Feel the inlet and outlet tubes of the accumulator.
   a. If the inlet tube is, again, warmer or if it is the same temperature as the outlet tube, add 3–4 oz. (85–113 g) more refrigerant.
   b. If the inlet tube is still colder than the outlet tube, add 3–4 oz. (85–113 g) more refrigerant.
4. Again, feel the accumulator inlet and outlet tubes.
   a. If the inlet tube is still warmer or at the same temperature as the outlet tube, add 3–4 oz. (85–113 g) more refrigerant.
   b. If the inlet tube is still colder than the outlet tube, recover the refrigerant and check for a clogged or restricted orifice tube.

 WARNING: While performing this test, watch for any indication of over-charging, such as excessively high discharge pressure. If high pressures occur, discontinue the test.

Pressure Switch Test

This test may be performed whenever the inlet and outlet tube temperatures are acceptable, but cooling is not sufficient.

1. Using a manifold and gauge set (Figure 7-27), check to determine if the clutch cycles on between 41 and 51 psig (283 and 352 kPa), and cycles off between 20 and 28 psig (138 and 193 kPa). (Refer to Figure 7-33.)
   a. If cycling is correct, proceed with "Performance Test."
   b. If the clutch cycles at pressures too low or too high, replace the pressure cycling switch.

CAUTION:
When adding refrigerant, *never* exceed the recommended capacity of the system.

Classroom Manual
Chapter 7, page 233

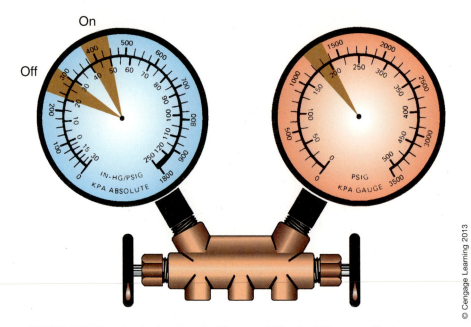

FIGURE 7-27 The clutch should cycle (A) on and (B) off within a specified range.

FIGURE 7-28 Disconnect the blower motor.

2. If the clutch runs continuously, disconnect the evaporator blower motor wire (Figure 7-28). The clutch should cycle off between 20 and 28 psig (138 and 193 kPa).
   a. If the pressure drops to below 20 psig (138 kPa), replace the pressure cycling switch.
   b. If the clutch cycles off, proceed to "Performance Test."

Performance Test

To conduct the performance test:

1. Set the temperature lever to full cold, selector to MAX A/C, and place the blower motor switch in the HI position.
2. Start the engine. Allow it to run at 2,000 rpm.
3. Close the doors and windows.

FIGURE 7-29 Measure discharge air temperature at the center register.

4. Place an auxiliary fan in front of the condenser.
5. Allow the system to stabilize, 5-10 minutes.
6. Place a thermometer in the register nearest the evaporator and check the temperature (Figure 7-29). The temperature should be 35–45°F (1.7–7.2°C) with an ambient temperature of 80°F (27°C).
7. If the outlet temperature is high, check the compressor **cycling time**.
   a. Check the cycling clutch switch operation.

   NOTE: Many of today's air-conditioning systems use an ambient air temperature sensor that will not allow the compressor to engage if outside temperatures are below a predetermined level, generally 50°F (10°C).

   b. If the clutch is energized continuously, discharge the system and check for a missing orifice tube, plugged inlet screen, or other restriction in the suction line.
   c. If the clutch cycles on and off or remains off for an extended period, discharge the system and check for a plugged orifice tube. Replace the tube, and evacuate, charge, and retest the system. Refer to Figure 7-33 and the manufacturer's cycling clutch rate for the vehicle.

DIAGNOSING VARIABLE DISPLACEMENT COMPRESSOR ORIFICE TUBE SYSTEMS

1. Connect the gauge set to the system.
2. Use a jumper wire to bypass the cooling fan switch.
3. Start and run the engine at about 1000 rpm.
4. Set the selector lever to NORM A/C and the blower motor switch to the HI position.
5. Measure the discharge air temperature at the center register as shown in Figure 7-29.
6. If the temperature is less than 60 °F (16 °C), proceed with step 8; if it is more than 60°F (16°C), check the pressure at the accumulator (low side).
   a. If this pressure is 35–50 psig (241–345 kPa), proceed with step 8 (Figure 7-30).
   b. If the pressure is greater than 50 psig (345 kPa), proceed to step 11.
7. If the accumulator pressure in step 6 was less than 35 psig (241 kPa), add a "pound" can of refrigerant.
   a. If the pressure is now more than 35 psig (241 kPa), leak test the system.

Cycling time is a term used to describe the total time from when the clutch engages until it disengages and engages again.

An overcharge of refrigerant is generally accompanied by excessive high-side pressure.

SERVICE TIP: If none of these conditions exist, the system was overcharged. Recover, evacuate, and recharge following approved procedures.

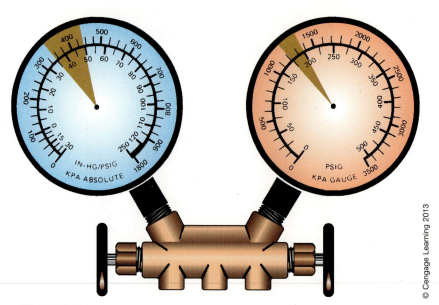

FIGURE 7-30 Pressure between 35 and 50 psig (241 and 345 kPa).

    b. If the pressure is still low, discharge the system and examine the orifice tube for a restriction. If it is plugged or otherwise defective, replace the orifice tube, and evacuate, charge, and retest the system.

8. Set the selector lever to the DEF mode. Disconnect the engine cooling fan and allow the compressor to cycle on the high-pressure cut-out switch.

    a. If a compressor knocking noise is noted on clutch engagement, the system oil charge is high. If this is the case, discharge the system, flush all components, and charge with the appropriate amount of refrigerant and oil.

    b. If no compressor noise was heard in step 8a, set the selector lever to MAX cooling. Adjust the blower control to its LO setting.

9. Idle the engine for 5 minutes at 1000 rpm. If the pressure at the accumulator is now 29–35 psig (200–241 kPa), the system is operating properly (Figure 7-31).

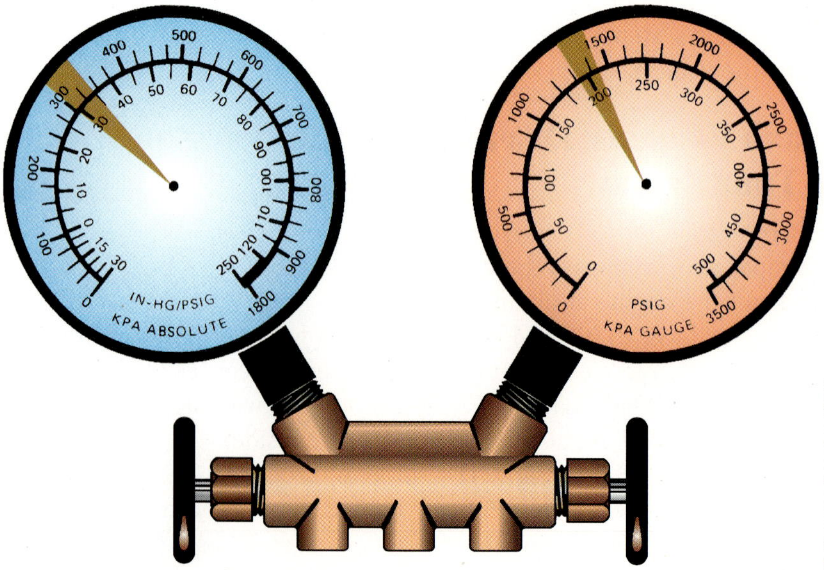

> Slight compressor noises are to be expected.

FIGURE 7-31 Pressure at the accumulator of 29–35 psig (200–241 kPa).

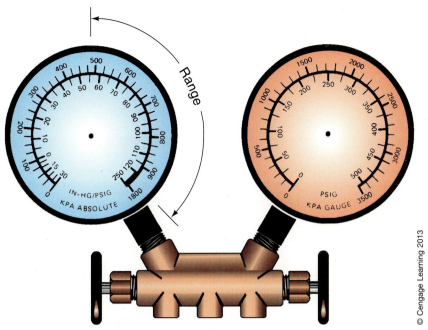

FIGURE 7-32 Pressure range of 50–160 psig (345–1,103 kPa).

10. If the pressure at the accumulator is below 28 psig (193 kPa), discharge the system. Replace the compressor control valve, and evacuate, charge, and retest the system.
    a. If the pressure is above 36 psig (248 kPa), discharge the system. Replace the compressor control valve, and evacuate, charge, and retest the system.
    b. If step 10a did not prove effective—the pressure is still above 36 psig (248 kPa)—replace the compressor.
11. In step 6, if the pressure at the accumulator (Figure 7-32) was above 50 psig (345 kPa) and below 160 psig (1103 kPa), discharge the system and check for a missing orifice tube.
    a. If the orifice tube is not missing, replace the compressor control valve. Evacuate, charge, and retest the system.
    b. If the condition is not corrected with a new control valve, replace the compressor.
12. If the pressure in step 6 was higher than 160 psig (1103 kPa), the system is overcharged. Discharge, evacuate, charge, and retest the system.

DIAGNOSING FORD'S FOT SYSTEM

Proper diagnosis of Ford's FOT system may be accomplished by observing the clutch cycle rate, time on plus time off, and system pressures. Low and high pressure will vary somewhat between the low and high points as the clutch cycles on and off.

Prepared charts (Figure 7-33) are used to compare the system pressures and clutch cycle rate to determine if the system pressures and clutch cycle rate are as specified.

Most Ford lines with an electric cooling fan use an electronic module to control the fan and clutch circuits. For electrical troubleshooting and repairs, refer to the manufacturer's appropriate wiring diagrams for specific model year vehicles. The following diagnosis assumes that the cooling fan and clutch electrical circuits are functioning properly.

The accuracy of clutch cycle timing for the FFOT system depends on the following conditions:

1. The in-car temperature must be stabilized at 70–80°F (21–27°C).
2. MAX air conditioning with RECIR (recirculating) air must be selected.

CAUTION:
Replace any lubricant that may have been removed from the air-conditioning system during the refrigerant recovery process plus any that may have been lost due to a leak. One of the major causes of compressor failure is lack of lubricant.

FFOT is an acronym for Ford Fixed Orifice Tube.

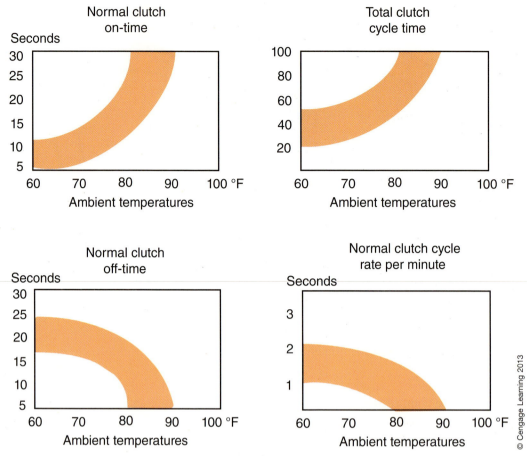

FIGURE 7-33 Typical diagnostic and testing charts for a cycling clutch system.

3. MAX blower speed must be selected.

4. The engine should be running at 1500 rpm for a minimum of 10 minutes.

The lowest pressure noted on the low-side gauge, as observed as the clutch is disengaged, is the low-pressure setting of the clutch cycling pressure switch. Conversely, the pressure recorded when the clutch first engages is the high-pressure setting for the clutch cycling pressure switch.

Compressor clutch cycling will not normally occur if the ambient temperature is above 100°F (38°C), and in some instances above 90°F (21°C), depending on conditions such as relative humidity and engine speed. Also, clutch cycling does not usually occur when the engine is operating at curb idle speed.

If the system contains no refrigerant or is extremely low on refrigerant, the clutch may not engage. If, on the other hand, a clutch cycles frequently, it is an indication that the system is undercharged or the orifice tube is restricted.

FFOT System Diagnosis

If poor or insufficient cooling is noted and the system does not have an electrodrive engine cooling fan, proceed to step 2. If it is equipped with an electrodrive engine cooling fan as shown in Figure 7-24, check to determine if the clutch energizes. If the clutch does not energize, check the electrical clutch circuit.

1. If the clutch energizes, determine if the cooling fan operates when the clutch is engaged.
   a. If it does not, check the cooling fan electrical circuit.
   b. If the fan operates properly, proceed to step 2.

FIGURE 7-34 A vacuum leak can be a source of trouble.

© Cengage Learning 2013

2. Check for a loose, missing, or damaged compressor drive belt.
3. Inspect for loose, disconnected, or damaged clutch, clutch cycling wires, or connectors.
4. Check the resistor connections, if equipped.
5. Check the connections of all vacuum hoses (Figure 7-34).
6. Check for blown fuses and proper blower motor operation.
7. Be sure all vacuum motors and temperature doors provide full travel.
8. Inspect all control electrical and vacuum connections.

   NOTE: Repair all items as necessary and recheck the system.

9. If cooling is still inadequate, refer to the pressure-cycle time charts.
   a. Hook up a manifold gauge set.
   b. Set the selector lever to MAX A/C and the blower switch to HI.
   c. Set the temperature lever to full cold.
   d. Close all doors and windows.
10. Insert a thermometer in the center grille outlet.
    a. Allow the engine to run for 10–15 minutes at approximately 1500 rpm with the compressor clutch engaged.
    b. Check and note the discharge temperature.
    c. Check and record the outside ambient temperature.
11. With a watch, time the compressor on and off time. Compare the findings with the appropriate chart.
    a. If the clutch does not cycle rapidly, proceed with step 15.
    b. If the clutch cycles rapidly, bypass the clutch cycling switch with a jumper wire. The compressor should now operate continuously.
12. Feel the evaporator inlet and outlet tubes.
    a. If the inlet tube is warm or if the outlet tube is colder after the orifice tube, leak test the system. Repair leaks, and evacuate, charge, and retest the system.
    b. If no leaks are found, add approximately 4 oz. (113 g) of refrigerant.
13. Again, feel the inlet and outlet tubes.
    a. If the inlet tube is colder, add 4 oz. (113 g) of refrigerant.
    b. Once more, check the inlet and outlet tubes. Continue to add refrigerant in 4 oz. (113 g) increments until the tubes feel equal in temperature and are about 28–40°F. (–2–4°C).
    c. If, in step 12, the inlet tube was equal to the outlet tube (approximately 28–40°F [–2–4°C], add 8–12 oz. (226–339 g) of refrigerant (Figure 7-35).
    d. Check the outlet discharge temperature for a minimum of 50°F (10°C).
14. If, in step 11, the outlet tube temperature was equal to the inlet tube temperature, 28–40°F (–2–4°C), replace the clutch cycling switch and retest the system.

Ambient: surrounding air. Ambient temperature refers to surrounding air temperature.

Ambient air contains moisture in the form of "humidity."

FIGURE 7-35 Typical "pound" cans actually contain 12 oz. (339 g) of refrigerant.

15. Feel the evaporator inlet and outlet tubes.
    a. If the inlet tube is warm or if the outlet tube is colder after the orifice tube, perform steps 12 and 13 to restore the system.
    b. If the inlet and outlet tubes are at the same temperature, 28–40°F (−2–4°C), or if the outlet tube after the orifice tube is slightly colder than the inlet tube, check for normal system pressure requirements.
16. If the compressor cycles within limits, the system is functioning properly.
    a. If the compressor cycles on high or low pressures, on above 52 psig (359 kPa) and off below 21 psig (145 kPa), replace the clutch cycling switch and retest the system.
    b. If the compressor runs continuously, disconnect the blower motor wire. Check for the compressor cycling OFF at 21–26 psig (145–179 kPa) suction pressure (Figure 7-36).
    c. If so, reconnect the blower motor wire. The system is functioning properly.
    d. If the suction pressure fell below 21 psig (145 kPa) when the blower motor wire was disconnected, replace the clutch cycling switch and retest the system.

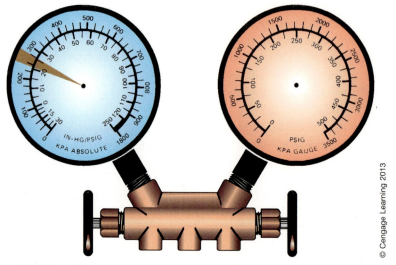

FIGURE 7-36 Compressor cycles off at 21–26 psig (145–179 kPa).

POOR COMPRESSOR PERFORMANCE

Refer to the FFOT chart (see Figure 7-33). Some of the other problems relating to poor compressor performance include:

- Clutch slippage
- Loose drive belt
- Clutch coil open
- Dirty control switch contacts
- High resistance in clutch wiring
- Blown fuse or open circuit breaker

Additional problems associated with compressors include:

- Cycling switch
- Clutch seized
- Accumulator refrigerant oil bleed hole plugged
- Refrigerant leaks

Classroom Manual
Chapter 7, page 239

CUSTOMER CARE: Customers often want a "quick fix" at the lowest possible price. Do not sacrifice your integrity as a technician and succumb to customer demands. One example of this is the use of refrigerant stop-leak products to seal refrigerant leaks that are difficult to find or expensive to repair. These products can potentially damage service equipment and system components. In the end, neither the technician nor the customer will be satisfied.

CASE STUDY

The customer complains that the air conditioning blows warm air. Recent service history shows that the air-conditioning compressor, receiver dryer, and condenser had been replaced.

The technician attached a manifold and gauge set to the air-conditioning system and found low-side pressure to be 40 psi and high side pressure to be 450 psi after the air conditioning was operated for a few minutes. First the technician recovered and recharged the system with the amount of refrigerant specified by the manufacturer to verify that the system had not been overcharged during the previous services. But the pressures did not change.

The technician inspected for debris in front of the condenser and between the condenser and the radiator. Next the technician placed a shop fan in front of the vehicle and sprayed the condenser with water. The high-side pressure immediately dropped to 260 psi. The technician checked to see if the cooling fan was being commanded on and determined that the cooling fan was not coming on when a high pressure condition existed in the air conditioning high-side. Upon further diagnosis the fan control module was found to be faulty and was replaced.

Once the repair was made the system pressures were 26 psi on the low-side and 220 psi on the high-side and the air-conditioning system functioned properly.

This case is an example of how something as simple as checking the operation of the electric cooling fan can be overlooked. If all the causes of a system malfunction condition are not checked it may lead to the misdiagnosis, as was indicated by the previous repairs. This was a summary of an actual case posted on http://www.iATN.net.

TERMS TO KNOW

Cycling time
Ducts
Functional test
Ground
Idle speed
Tension gauge

ASE-STYLE REVIEW QUESTIONS

1. Refer to Figure 7-37 below.

   Technician A says that the compressor clutch diode in the clutch electrical circuit is in series with the compressor clutch coil.

   Technician B says that a blown HVAC 20 A fuse in the IP fuse block will prevent blower motor operation in all speeds *except* high.

   Who is correct?

   A. A only
   B. B only
   C. Both A and B
   D. Neither A nor B

2. A test lamp has been connected from the blower motor housing to the "hot" side of the fuse or circuit breaker:

   Technician A says that if the lamp does not light, a defective ground wire is indicated.

   Technician B says that if the lamp lights, a defective blower motor is indicated.

   Who is correct?

   A. A only
   B. B only
   C. Both A and B
   D. Neither A nor B

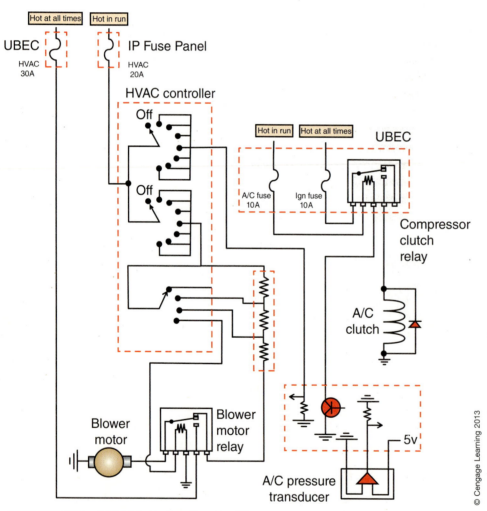

© Cengage Learning 2013

FIGURE 7-37 For ASE-Style Review Question #1.

3. *Technician A* says that low suction and discharge pressure may indicate a low refrigerant charge.

Technician B says that low suction and discharge pressure may indicate a restriction in the system.
Who is correct?
   A. A only
   B. B only
   C. Both A and B
   D. Neither A nor B

4. *Technician A* says that a liquid line restriction has the same general symptoms as a suction line restriction.

Technician B agrees, adding, "An undercharge of refrigerant also has the same symptoms."
Who is correct?
   A. A only
   B. B only
   C. Both A and B
   D. Neither A nor B

5. *Technician A* says that air is a noncondensable gas.

Technician B says that air collects in the evaporator during the OFF cycle of the air-conditioning system.
Who is correct?
   A. A only
   B. B only
   C. Both A and B
   D. Neither A nor B

6. *Technician A* says that moisture in the system can cause poor or no cooling.

Technician B says that moisture in the system can cause harmful acids.
Who is correct?
   A. A only
   B. B only
   C. Both A and B
   D. Neither A nor B

7. *Technician A* says that a restriction in the suction line will cause excessive high-side pressure.

Technician B says that a restriction will result in a pressure change, usually marked by frost or ice.
Who is correct?
   A. A only
   B. B only
   C. Both A and B
   D. Neither A nor B

8. *Technician A* says that if the orifice tube is missing, the low-side pressure will be low.

Technician B says that if the orifice tube is plugged with debris, low-side pressure will be high.
Who is correct?
   A. A only
   B. B only
   C. Both A and B
   D. Neither A nor B

9. A manifold and gauge set is attached to an operating air conditioner. The low-side gauge reading is too high and the high-side reading is too low. What could be the problem?
   A. Refrigerant overcharge
   B. Insufficient refrigerant
   C. Expansion valve stuck open
   D. Defective compressor

10. *Technician A* says that the minimum flow test on a TXV is made by wrapping the remote bulb in a warm rag.

Technician B says that a maximum flow test may be made on a TXV by immersing the remote bulb in ice water (H_2O).

Who is correct?
   A. A only
   B. B only
   C. Both A and B
   D. Neither A nor B

ASE CHALLENGE QUESTIONS

1. Which of the following could cause very low pressures on the low side of a system equipped with an expansion valve?
   A. Expansion valve stuck closed
   B. Expansion valve stuck open
   C. Compressor valves not sealing
   D. Compressor clutch inoperative

2. All of the following may cause a low-pressure gauge to go into a vacuum, *except:*
   A. Clogged screen in the metering device
   B. Clogged screen in the accumulator
   C. Clogged screen in the receiver-drier
   D. Thermostatic expansion valve stuck closed

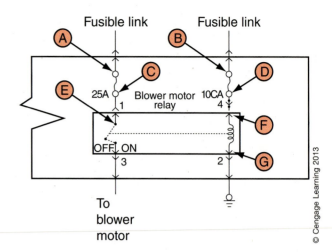

© Cengage Learning 2013

4. The blower motor does not operate. The voltage at point F on the illustration above is 0 volts. The most likely cause of this problem is a defective:
   A. 10-ampere fuse
   B. 25-ampere fuse
   C. Blower motor relay coil
   D. Blower motor relay contacts

5. An R-134a expansion valve system that is properly charged has the following gauge readings, high-side 165 psig and low-side 40 psig and the ambient air temperature is 85°F. Which of the following is the most likely cause?
   A. A restricted receiver drier
   B. A restricted condenser
   C. An out of adjustment temperature blend door
   D. A stuck open expansion valve

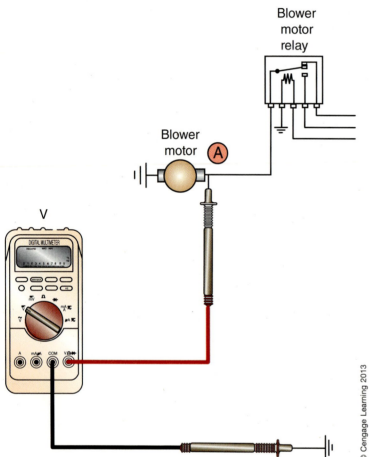

© Cengage Learning 2013

3. The blower motor does not operate although there are 12 volts available at the motor terminal in the illustration above. The most likely problem is:
   A. "Blown" fuse or open circuit breaker
   B. Defective relay ground
   C. Defective blower motor relay
   D. Loose or defective motor ground wire

Name _____ Date _____

INSPECT THE V-BELT DRIVE

Upon completion of this job sheet, you should be able to visually check the compressor drive V-belt and check its tension.

NATEF Correlation

HEATING AND AIR CONDITIONING: Refrigeration System Component Diagnosis and Repair; Compressor and Clutch; *Inspect A/C compressor drive belts; replace and adjust as needed.* **(P-1)**

Tools and Materials

A vehicle with V-belt drive

Service manual

Belt tension gauge

Describe the vehicle being worked on.

Year _____ Make _____ Model _____

VIN _____ Engine type and size _____

Procedure

1. List the different belts, and their purpose. _____
   _____

2. Visually inspect the belts and describe the condition of each. _____
   _____

3. Check the tension of the A/C belt. According to specifications, the tension should be _____. The tension is _____.

4. Based on the inspection above, what is your recommendation? _____
   _____

5. What procedure would you follow to adjust the belt tension? _____
   _____

Instructor's Response _____
_____
_____

Name _____ **Date** _____

INSPECT THE SERPENTINE DRIVE BELT

Upon completion of this job sheet, you should be able to visually check the compressor serpentine drive belt and check its tension.

NATEF Correlation

HEATING AND AIR CONDITIONING: Refrigeration System Component Diagnosis and Repair; Compressor and Clutch; *Inspect A/C compressor drive belts; replace and adjust as needed.* **(P-2)**

Tools and Materials

A vehicle with serpentine drive belt

Service manual

Belt tension gauge

Describe the vehicle being worked on.

Year _____ Make _____ Model _____

VIN _____ Engine type and size _____

Procedure

1. If more than one belt, list the different belts and their purpose(s). _____

   _____

2. Visually inspect the belt(s) and describe its (their) condition. _____

   _____

3. Check the tension of the belt(s). According to specifications, the tension should be
   _____. The tension is _____.

4. Based on the above inspection, what is your recommendation? _____

   _____

5. What procedure would you follow to adjust the belt tension?

   _____

Instructor's Response _____

_____

_____

Name _____ **Date** _____

REFRIGERANT SYSTEM CHARGE LEVEL TEMPERATURE METHOD (DELTA-T)

Upon completion of this job sheet, you should be able to test the air-conditioning system for proper refrigerant charge level while monitoring system pressures and temperatures. The following procedure is for a fixed orifice tube (FOT) system only and is not designed to be used with an expansion valve system. FOT charge determination chart is based on ambient air temperature range of 70–85°F (21–29°C).

NATEF Correlation

HEATING AND AIR CONDITIONING: A/C System Diagnosis and Repair; *Performance test A/C system; diagnose A/C system malfunctions using principles of refrigeration.* **(P-1)**

HEATING AND AIR CONDITIONING: A/C System Diagnosis and Repair; *Identify refrigerant type; select and connect proper gauge set; record pressure readings.* **(P-1)**

Tools and Materials

Late-model vehicle with an FOT refrigerant system

Service manual or information system

Safety glasses or goggles

Hand tools, as required

Refrigerant manifold and gauge set

Clamp-on thermocouple and Charge Determination Chart

Digital multimeter

Thermometer

Describe the vehicle being worked on.

Year _____ Make _____ Model _____

VIN _____ Engine type and size _____

Procedure

1. Attach manifold and gauge set or recovery/recycling/charge station to vehicle refrigerant service ports. Ambient air temperature must be in the range of 70–85°F (21–29°C).
2. Attach a clamp-on thermocouple onto the vehicle high-pressure liquid line just before the FOT inlet point. The closer the temperature is recorded to the FOT inlet the more accurate your temperature reading will be.
3. Start engine and run at idle speed. Turn on air conditioning and set AC control panel to Fresh-Air mode and temperature level to the coldest setting.
4. Open all doors and set blower motor speed to HI.

5. Allow air-conditioning system to stabilize for a minimum of 5 minutes.
6. Increase high pressure liquid line pressure to 260 psig (1793 kPa) by restricting airflow across air-conditioning system condenser.
   a. Place a piece of cardboard in front of the air-conditioning condenser. Cover only enough of the condenser to raise and maintain the liquid line pressure at the level specified above.
7. Record the high-pressure gauge reading. _____

   _____

8. Record the high-pressure liquid line inlet temperature just before the FOT. _____

   _____

9. Referring to the Charge Determination Chart, record the ideal high-pressure range for the line temperature recorded in step 8. _____

10. Compare the pressure recorded in step 7 to the idle pressure range determined in step 9.
   a. If the refrigerant system is operating in the proper charge range, the refrigerant system is functioning correctly.
   b. If the refrigerant system is operating in the undercharge range, add 2 oz. of refrigerant (0.057 kg) R-134a to the refrigerant system and repeat steps 5–10.
   c. If the refrigerant system is operating in the overcharge range, reclaim 2 oz. of refrigerant (0.057 kg) R-134a from the refrigerant system using the recovery/recycling/charge station and repeat steps 5–10.

Measurement	Specification	Before Repair	After Repair
A/C center duct temperature			
High-side pressure			
Low-side pressure			
High-pressure liquid line inlet Temperature just before FOT			
Low-pressure liquid line outlet Temperature just after FOT			
FOT inlet temperature minus FOT			
Outlet temperature			
Refer to the Charge Determination Chart. Record the ideal high-pressure range.			

Charge Determination Chart

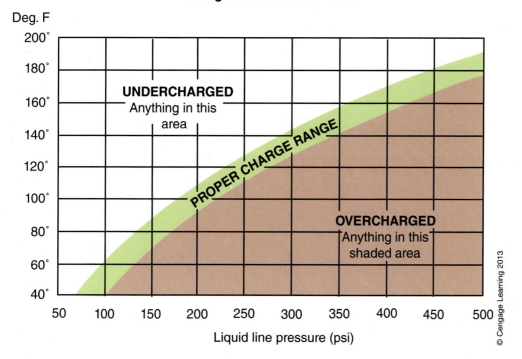

Instructor's Response _____

_____

_____

Name _____ Date _____

INSPECT THE CONDENSER

Upon completion of this job sheet, you should be able to visually check the air-conditioning system condenser for proper airflow.

NATEF Correlation

HEATING AND AIR CONDITIONING: Refrigeration System Component Diagnosis and Repair; Evaporator, Condenser, and Related Components; *Inspect A/C condenser for airflow restrictions; perform necessary action.* **(P-2)**

Tools and Materials

An air-conditioned vehicle

Contact thermometer or infrared thermometer

Describe the vehicle being worked on.

Year _____ Make _____ Model _____

VIN _____ Engine type and size _____

Procedure

1. Visually inspect the air-conditioning system and describe its overall condition.

   _____

2. In a well-ventilated area, start the engine, place the transmission in park, and turn on the air-conditioning system to MAX cooling. If a standard transmission, place in NEUTRAL and chock the wheels. Describe your procedure for accomplishing this step.

   _____

⚠️ **WARNING: Take care not to come into contact with moving parts, such as fan blades or belts, and heated metal, such as the exhaust manifold, when performing steps 3 through 5. In addition, air-conditioning lines may be very hot.**

3. Carefully place your hand on the condenser near the refrigerant inlet. Describe what you feel. Check and record the temperature. _____

   _____

4. Carefully place your hand on the condenser near the refrigerant outlet. Describe what you feel. Check and record the temperature. _____

   _____

5. When moving your hand across the condenser, refrigerant inlet to outlet, do you feel a change in temperature? If so, describe the change. Check and record the temperature at various locations. _____

_____

6. Based on the inspection above, what is your recommendation? _____

_____

7. What procedure would you follow to increase the airflow across the condenser?

_____

Instructor's Response _____

_____

_____

Name _____ **Date** _____

Heat Transfer through the Condenser

Upon completion of this job sheet, you should be able to test heat transfer through the air-conditioning condenser assembly and determine if there is an internal blockage.

NATEF Correlation

HEATING AND AIR CONDITIONING: A/C System Diagnosis and Repair; *Identify and interpret heating and air conditioning concerns; determine necessary action.* **(P-1)**

Tools and Materials

Late-model vehicle

Service manual or information system

Safety glasses or goggles

Hand tools, as required

Test vehicle

Infrared noncontact thermometer

DMM and temperature probe

Describe the vehicle being worked on.

Year _____ Make _____ Model _____

VIN _____ Engine type and size _____

Procedure

1. First, record all the temperatures at the condenser locations listed in the chart below before starting the vehicle.
2. Next, start the engine and run it at about 1000 rpm.
3. Turn the vent fan on high and activate the air-conditioning system in the MAX "Recirculation" mode; set temperature control to the coldest setting.
4. Record the temperatures at the various locations across the condenser and time intervals indicated.

	Temperature prior to Starting	Temperature after 1 Minute	Temperature after 5 Minutes	Temperature after 10 Minutes
Condenser upper left corner				
Condenser lower left corner				
Condenser center				
Condenser upper right corner				
Condenser lower left corner				
Condenser upper hose (compressor discharge line)				
Condenser lower hose (liquid line)				

5. What kind of heat transfer process does the condenser use? _____

_____

6. Which condenser hose becomes hot first? _____

Why? _____

_____

7. How does heat flow through the condenser? _____

_____

8. On the graph below show how heat flows through the condenser and indicate where the refrigerant inlet and outlet lines are located.

9. Were any blockages to the refrigerant flow through the condenser noted? _____

_____

_____

Instructor's Response _____

_____

_____

Name _____ **Date** _____

REMOVE AND REPLACE CONDENSER ASSEMBLY

Upon completion of this job sheet, you should be able to remove and reinstall the condenser, measure oil quantity, and determine the necessary action.

NATEF Correlation

HEATING AND AIR CONDITIONING: Heating, Ventilation, and Engine Cooling Systems Diagnosis and Repair; Evaporator, Condenser, and Related Components; *Remove and reinstall condenser; measure oil quantity; determine necessary action.* **(P-3)**

Tools and Materials

Late-model vehicle

Service manual or information system

Safety glasses or goggles

Hand tools, as required

Describe the vehicle being worked on.

Year _____ Make _____ Model _____

VIN _____ Engine type and size _____

Procedure

Access to the condenser assembly is gained by following the procedures outlined in the appropriate service manual. The following procedure is typical and assumes the procedure for access to the condenser is available. Give a brief description of your procedure following each step. Ensure that the engine is cold and wear eye protection.

1. Recover the refrigerant.
2. Remove the hood hold-down mechanism and any other cables or hardware that inhibit access to the condenser.
3. Remove the hot-gas line at the top of the condenser.
4. Remove and discard any O-rings.
5. Remove the liquid line at the bottom of the condenser.
6. Remove and discard any O-rings.
7. Remove and retain any attaching bolts or nuts holding the condenser in place.
8. Lift the condenser from the car.
9. Drain the oil from the evaporator into a calibrated cup. How much if any oil was removed?

_____

_____

10. For replacement, reverse the preceding procedure. How much oil does the manufacturer service information recommend adding to the system as part of the repair?

11. Evacuate the system, does it hold a vacuum for 5 minutes after the vacuum pump is turned off? What does this tell you?

12. Charge the system with refrigerant. How much refrigerant does the system require?

Instructor's Response _____

_____

_____

Chapter 8

COMPRESSORS AND CLUTCHES

BASIC TOOLS

Basic technician's tool set
Refrigerant recovery equipment
Manifold and gauge set with hoses
Safety glasses/goggles
Graduated container
Funnel
Fender covers

UPON COMPLETION AND REVIEW OF THIS CHAPTER, YOU SHOULD BE ABLE TO:

- Identify the various makes and models of compressors used in automotive air-conditioning service.

- Check and correct the oil level in various models of compressors.

- Leak test and replace shaft seals in various models of compressors.

- Leak test and correct shell and fitting leaks of various models of compressors.

- Troubleshoot and replace various types of compressors.

- Troubleshoot and make mechanical repairs to clutch coils and rotor assemblies.

The compressor (Figure 8-1) is thought of as the heart of the automotive air-conditioning system. Without the compressor, the system would not function. Actually, all five components of the system are essential if the system is to function properly. In addition to the connecting hoses and fittings, these five components are the compressor, condenser, metering device, evaporator, and accumulator or receiver-drier.

This chapter covers the compressor and clutch. The other components of the system are covered in detail in other chapters of this manual as well as the Classroom Manual. Refer to the index of either manual for further reference.

> Proper refrigerant is also essential for an air-conditioning system.

COMPRESSOR

There are more than 12 manufacturers of compressors with hundreds of different models and configurations available for use on the modern automobile. Some that are still in use date back to 1961. Others, which are seldom found, have been discontinued generally because of size and weight.

COMPRESSOR CLUTCH

There are various designs and manufacturers of compressor clutch assemblies today. This chapter highlights some of the models available and how to service the compressor clutch and coil assembly on those models. It is important, however, to cover in general terms how to inspect and diagnose a compressor clutch assembly, regardless of the model of compressor you are working on.

Classroom Manual
Chapter 8, page 258

TESTING COMPRESSOR CLUTCH ELECTRICAL CIRCUIT

Air-conditioning compressor clutch electrical circuits incorporate a clamping diode to save the electrical circuit from voltage spikes. It is imperative that this diode be checked any time a compressor or clutch assembly is replaced or if a fault was found on the electrical engagement circuit. A faulty diode will lead to the premature failure of the control module

Classroom Manual
Chapter 8, page 261

A590   **A6**   **BEHR PAD MOUNT**   **BMW**

C171   **SELTEC / TAMA/ DIESEL KIKI**   **FS-6**   **FS-10**

HITACHI   **HITACHI MJS170**   **HONDA**   **MERCEDES 10PA15 / 10PA17**

TYPICAL DA6, HR6, HR6-HE   **NIHON**   **NIPPONDENSO 2C90**   **PORSCHE**

TYPICAL R4   **6C17**

© Cengage Learning 2013

FIGURE 8-1 **Some of the many compressors that have been used over the years.**

if not detected (Figure 8-2). A digital storage oscilloscope or a digital multimeter with a MIN/MAX function may be used to detect excessive voltage spikes when the compressor is shut off. Generally, the voltage spike should not exceed 60 volts.

 WARNING: **If the vehicle is equipped with an air bag, refer to the appropriate service information under the supplemental passive restraint system to avoid accidental air bag deployment and the possible bodily harm that could result.**

Clutch Test

If the clutch is not operational, check the condition of the wiring and verify that the system contains the proper amount of refrigerant. Connect a jumper wire with an in-line fuse (Figure 8-3) between the positive (+) battery terminal and the clutch coil lead. If the clutch does not energize, proceed with step 1. If the clutch energizes, proceed with step 2.

1. Connect a jumper wire from the clutch coil to an engine ground (−). If the clutch does not energize, remove and repair it as necessary.

SERVICE TIP:
Prior to testing the compressor clutch operation you should verify that the compressor is not seized. One way to do this is with the engine off; grasp the front of the compressor clutch and see if you can turn it by hand, and do not forget to wear safety glass. Also, if the compressor has not been operated for some time a layer of rust may form between the clutch halves and may become dislodged when the compressor engages creating an eye injury hazard.

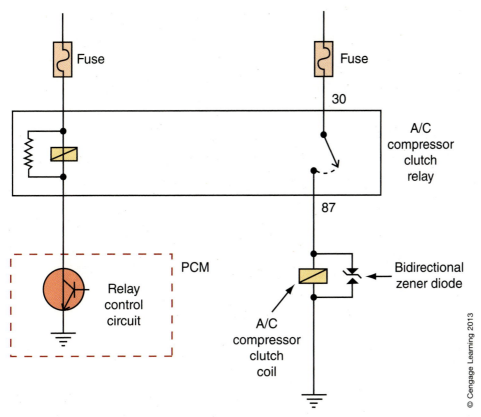

FIGURE 8-2 Wiring diagram for an air-conditioning clamping diode.

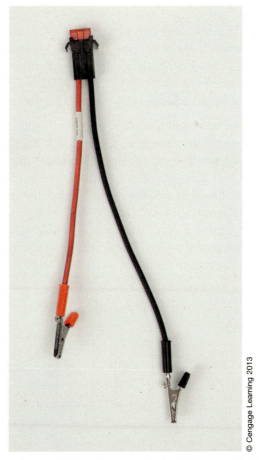

FIGURE 8-3 A typical jumper wire. The in-line fuse guards against an accidental short circuit.

FIGURE 8-4 Jump across the pressure switch connector.

2. If the clutch operates but cooling is not sufficient, allow the system to operate for a few minutes and check the low-side pressure at the accumulator access fitting.
   a. If the pressure is above 50 psig (345 kPa), proceed with step 3.
   b. If the pressure is below 50 psig (345 kPa), proceed with step 4.
3. Connect a jumper wire across the pressure switch connector (Figure 8-4).
   a. If the compressor operates, the switch is defective. Replace the switch and retest the system.
   b. If the compressor does not operate, check for an open or short circuit between the switch and clutch.
4. Connect a high-pressure gauge and check the high-side pressure.
   a. If the high-side pressure is above 50 psig (345 kPa), discharge the system and check for a plugged orifice tube or a restriction in the high side.
   b. If the high-side pressure is below 50 psig (345 kPa), the refrigerant charge is lost. Leak test, repair, evacuate, recharge, and retest the system.

Clutch Diode. A strong electromagnetic field is generated when electrical power is applied to the clutch. When this power is disconnected, the magnetic field collapses and creates high-voltage spikes. These spikes, harmful to the delicate electronic circuits, must be eliminated.

A diode, across the clutch coil, provides a path to ground for the electrical spikes as power is interrupted. This diode is usually taped inside the wiring harness across the 12-volt and ground leads (Figure 8-5).

Typical procedures for testing a diode are given in Photo Sequence 11. The diode may also be tested using a digital multimeter (DMM) following the procedures outlined in the instructions included with the meter.

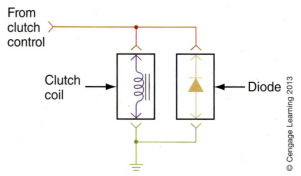

FIGURE 8-5 A diode in the clutch circuit prevents electrical spikes.

BENCH TESTING THE COMPRESSOR CLUTCH COIL AND DIODE

The following procedure may be followed by technicians using a digital ohmmeter. Many digital multimeters (DMM), however, have a "short cut" method for testing a diode. Follow instructions included with the DMM. If either the clutch coil or the diode fails the test, it is defective and must be replaced.

P11-1 Before bench testing an inoperative clutch coil, check for voltage present at the connector with the coil disconnected and controls set for COOL. **NOTE:** If 12 volts are available, proceed with the bench test, step 2. If 12 volts are not present, the problem is probably neither the clutch coil nor the diode.

P11-2 Carefully cut and remove the tape to expose the diode leads to the clutch coil.

P11-3 Isolate the diode by disconnecting one lead from the clutch coil.

P11-4 Connect an ohmmeter to the clutch coil leads or from the coil lead to ground, as applicable. Note the resistance. **NOTE:** 0 Ω indicates that the coil is shorted. Infinite (O.L.) Ω indicates that the coil is open.

P11-5 Connect an ohmmeter to the diode. Note the resistance. **NOTE:** If your meter is equipped with a diode test setting, use this setting instead of the ohm's setting for more accurate results. The diode setting reading is generally 0.7 volt in one direction and 0 volts or open in the other direction.

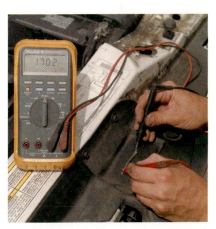

P11-6 Reverse the ohmmeter leads to the diode and, again, note the resistance. **NOTE:** There should be low ohms (due to internal resistance of the diode) when the ohmmeter is connected in one direction and infinite ohms resistance when connected in the other direction.

Compressor Clutch Amperage Draw and Resistance Test

1. Visually check the coil for loose connections or cracked insulation.
2. Inspect the clutch plate and hub assembly. Check for signs of looseness between the plate and hub.
3. Check the rotor and bearing assembly. With the belt removed, the clutch pulley should spin smoothly without roughness.
4. Check the bearing for signs of excessive noise, binding, or looseness. Replace the bearing, if necessary.
5. Check for oil on the friction surface. If oil is present, the compressor shaft seal is leaking and the compressor or seal must be replaced.
6. Check for a proper clutch air gap. Some compressor clutches use spacer shims to adjust this air gap.
7. Locate the air-conditioning compressor clutch wiring diagram for the vehicle you are servicing, as well as the amperage draw and resistance specifications for the compressor clutch coil.
8. Ensure the ignition switch is in the OFF position and remove the air-conditioning compressor clutch relay.
9. Using a digital multimeter (DMM) set to the DC ampere scale, connect the meter leads in series across the contact point circuit of the relay connector (these are the larger cavities). On an ISO relay connector this would be terminal #30 and terminal #87.
10. Start the engine and select air conditioning MAX mode. Record the amperage (I) draw and compare it to specifications, generally 2–3 amperes at 12 volts.
11. Turn the engine off and replace the relay.
12. Set your DMM to the DV volt scale and connect it in parallel across the compressor clutch connector.
13. Start the engine and select air conditioning MAX mode. Record the voltage drop and compare it to the battery voltage. The voltage drop and battery voltage should be within 0.5 volts of each other. Example: voltage drop 12.1 volts and battery voltage of 12.6 volts. If the voltage drop to battery voltage comparison is greater than specified, inspect for high circuit resistance.

It should be noted that a clutch failure may be due to other system failures such as a compressor that is binding, higher-than-normal high-side system pressure, excessive oil in the air-conditioning system, as well as a loose drive belt or faulty belt tensioner.

COMPRESSOR IDENTIFICATION

When servicing a compressor, it is important to be able to identify its type, manufacturer, and model number. This is especially true for identifying replacement parts, such as a compressor **shaft seal**. It is, perhaps, equally important if it is necessary to replace the compressor. Following is a brief overview, in alphabetical order, of most of the compressors that are currently available, new or rebuilt.

> **CUSTOMER CARE:** While under the hood servicing a vehicle, make a point of using a shop towel to wipe off information labels and major components and shrouds. If a customer opens the hood to inspect the vehicle after he picks it up, it is a sign to him that someone who cares has worked on it.

Diagnosis of Compressor Noise

When attempting to diagnose air conditioning related noises, it is important to know when the noise occurs and under what operating and temperature conditions. Noises related to the air-conditioning compressor are often misleading. Conditions that effect compressor noise are vehicle speed, engine speed, engine temperature, weather conditions, and engine load, to name a few.

Drive belt noise is speed sensitive and depending on belt tension can often be misdiagnosed as compressor noise. Drive belt tension can cause a noise when the compressor clutch is engaged, which may not be present when the compressor clutch is disengaged. Always check drive belt condition and tension before proceeding with compressor noise diagnosis.

It is important to select a quiet area and to duplicate the complaint conditions in order to precisely locate the source of the noise. Turn the air conditioning on and off several times and listen to the compressor. Use an engine stethoscope (Figure 8-6), an electronic ear, or a long screwdriver with the handle held to your ear to aid in localizing the area that the noise is emanating from.

Verify that the compressor clutch air gap (Figure 8-7) is set to specifications and that the pulley and clutch plate is properly aligned. In addition, make sure that the compressor clutch field coil is securely mounted to the compressor assembly.

It is possible to duplicate high ambient air temperature conditions by restricting airflow across the condenser and listening for unusual noises. By restricting airflow across the condenser with a piece of cardboard, high-side discharge pressure can be increased. Do not allow high-side pressure to increase above 400 psig (2760 kPa) or system damage may result. Sometimes a high-pressure relief valve will open prior to its designed pressure relief point, generating a very loud high-pitched noise. If the relief valve is the source of the noise, the refrigerant will have to be recovered from the system and the valve will have to be replaced. If the valve still does not seat properly after being replaced, the compressor assembly will have to be replaced.

SERVICE TIP: Some service procedures, such as replacing the air-conditioning compressor clutch coil assembly and bearing, can be performed on the vehicle if space permits.

SERVICE TIP: A **running design change** is a design change made during a current model/year production.

FIGURE 8-6 An engine stethoscope can aid in localizing the area the noise is emanating from.

© Cengage Learning 2013

FIGURE 8-7 Verify that the AC compressor clutch air gap is set to specifications and pulley is aligned correctly.

Inspect the refrigerant hose and line routing and points of interference, and verify that the mounting brackets are secure to eliminate line vibration as a cause of noise concern (Figure 8-8). At this time also inspect the lines for kinks and sharp bends that could cause a point of restriction and increased operational noise.

Internal compressor noises come from a variety of sources. Swash plate compressors can develop excessive internal clearance due to lack of lubrication, refrigerant overcharge, or from normal wear over time. It is possible to check for abnormal wear between the

FIGURE 8-8 Inspect refrigerant hose and line routing.

swash plate and the shoe discs without removing the compressor from the vehicle by following the following procedure:

- Attach a recovery/recycle/charge station to the vehicle and recover all the refrigerant from the system.
- Disconnect both the suction and the discharge hoses from the compressor assembly.
- Rotate the compressor clutch hub in a clockwise direction for several turns and note the resistance to rotation.
- Now rotate the compressor clutch hub in a counterclockwise direction for several turns and note the resistance to rotation. The compressor should immediately have the same resistance to rotation as it did in the clockwise direction.
- Any free play (backlash) noted on the initial reverse of direction indicates excessive free play clearance between the swash plate and shoe discs. The compressor will need to be replaced.

Compressor piston noise is often caused by excessive internal pressure on the piston head. The internal pressure on the piston head can be up to 30 percent higher than the high pressure recorded on the manifold and gauge set. An example of this would be if the high-side discharge pressure is 300 psig (2068 kPa), the internal head pressure could be 390 psig (2689 kPa). The typical R-134a compressor is designed to withstand up to an 8 to 1 ratio (8:1) or pressure differential. When the ratio goes above that point, a compressor may develop a piston knock and over time complete compressor failure. A crude method for calculating compressor compression ratios is to take the low-side gauge reading and add 15 psig (103 kPa) to the reading (i.e., 30 psig + 15 psig = 45). Next, measure and record the high-side discharge (not liquid line) pressure and add 15 psig (103 kPa) to the reading. If the service port is located on the refrigerant liquid line after the condenser, you will need to use a thermocouple temperature probe and record the line temperature. After recording the line temperature, use a refrigerant temperature to pressure conversion chart to obtain the pressure for the recorded temperature. Next, add 15 psig (103 kPa) to this determined pressure. The final step is to divide the calculated low-side pressure number by the calculated high-side pressure number.

In the preceding example, if the pressure recorded on the high-side discharge line is 400 psig (2758 kPa) and the low-side pressure is 30 psig (207 kPa), what would the compressor ratio be?

400 psig + 15 = 415   (2758 kpa + 103 = 2861)
30 psig + 15 = 45   (207 kpa + 103 = 310)
415/45 = a ratio of 9.2:1 (2861/310 = 9.2:1), which is an excessively high ratio for continuous compressor operation.

This is why maintaining proper system function and pressures as well as accurate charge levels is critical to overall system performance and life. Leaves and road debris clogging the condenser airflow can have a devastating impact on compressor life. In addition, even slight refrigerant overcharging errors on small-capacity systems or air (noncondensable gas) contaminated systems may dramatically affect system performance.

Cause of Compressor Failure

Service technicians need to determine the root cause of compressor failures or leaks, whenever possible. Compressor seals can develop leaks due to excessive system pressure and heat, system contamination, improper mounting torque, or compressor-to-mounting bracket alignment errors. The following questions need to be answered when determining the root cause of a failure:

- Did compressor damage result from previous service work, such as prying against the case or pulley (Figure 8-9), failure to properly torque down the case to the mounting bracket, or a warped case or mounting bracket?

FIGURE 8-9 AC compressor damage can result from prying against compressor case or pulley assembly.

Flange is another term used for mounting boss. It is a protective rim, collar, or edge on an object used to keep the object in place or secure it to another object.

A **mounting boss** is another term used for a mounting flange. It is a protective rim, collar, or edge on an object used to keep the object in place or secure it to another object.

The term *logo* refers to a company trademark, such as the Ford Motor Company "FORD" or General Motors' "GM."

- Was a high refrigerant system temperature caused by lack of refrigerant system lubrication, incorrect refrigerant oil, excessive refrigerant oil, poor heat transfer at the condenser (physical blockage or excessive oil volume), or refrigerant overcharge? In addition, a slipping compressor clutch or clutch bearing failure will also generate excessive heat buildup, resulting in compressor shaft seal leaks and possibly compressor failure.
- Is there refrigerant system contamination such as Type I system sealant additive, which causes seals to soften and swell; moisture, causing oxidation at the sealing areas; solid particulate contamination circulating with the refrigerant (desiccant, metal); blended refrigerants; or air contamination, causing high-operating pressures?

It is essential to follow manufacturer-recommended torque and mounting procedures for refrigerant system compressors. This is especially critical with today's alloy composite compressor housing, which can easily be distorted during mounting. When a compressor **flange** or **mounting boss** is installed in the mounting bracket, check for wobbling by gently shaking the assembly back and forth (Figure 8-10). The compressor should fit snugly against the bracket and not rock back and forth prior to being torqued into place. If a wobbling condition is detected, check for warpage; the acceptable range for bracket or compressor housing mounting point warpage is 0.0–0.030 in. A feeler gauge may be used to measure the amount of warpage. The compressor should be torqued into place in stages. If the torque specification for the mounting bolts is 40 lb. (13.5 N·m), first tighten all bolts in a circular pattern at 10 lb. (13.5 N·m), repeat the procedure at 20 lb. (27 N·m), and finally repeat the process again at the final torque setting of 40 lb. (54 N·m). This procedure should prevent case distortion and seal failure.

FIGURE 8-10 The AC compressor should fit snuggly against bracket and not rock back and forth prior to being torqued into place.

REFRIGERANT LUBRICANT

The lubricant in the air-conditioning compressor circulates through the system with the refrigerant when the system is in operation. Most manufacturers specify PAG oil for R-134a systems with a specific viscosity number, whereas a few manufacturers call for ester oil as the factory-recommended lubricant. Older R-12 air-conditioning systems called for mineral oil as the system lubricant. Refrigerant oil should be added to a system any time a major component (e.g., compressor, condenser, etc.) is replaced or after a large leak has been repaired. It is important to maintain the manufacturer-specified amount of lubricant in the system and not to overfill the system. Too little oil in the refrigerant system will lead to compressor failure. Too much refrigerant oil may cause poor cooling performance due to thermal exchange interference. Always refer to the specific vehicle manufacturer's service information and service bulletins when selecting refrigerant lubricant.

When servicing the hybrid electric vehicle refrigerant system do not assume that it takes the same refrigerant oil as the non-hybrid vehicle refrigerant system. When servicing a hybrid vehicle's refrigerant system it is imperative that the correct refrigerant oil be used. The hybrid electric vehicle uses insulated refrigerant oil designed to minimize the conductivity of electricity through the compressor case in the event of a circuit failure. Many Toyota hybrid electric vehicles call for ND-11 refrigerant oil. In addition, it is even more critical than ever to properly evacuate the refrigerant system after service to the required vacuum levels for proper moisture removal. Always refer to vehicle manufacturers' recommendations for correct refrigerant oil.

Always select the correct lubricant and amount for the system being serviced. The amount of refrigerant oil contained in a refrigerant system varies according to the design and size of the system. As a rough guideline the distribution of oil throughout the refrigerant system is 50% in the compressor, 10% in the condenser, 10% in the fluid container (receiver-drier/accumulator), 20% in the evaporator, and 10% in the suction hose (Figure 6-28). When a refrigerant component is replaced always refer to the service information for the recommended amount of oil that should be added to the system.

Refrigerant Oil Return Operation

A lubricant return operation should be performed prior to replacing any major refrigerant system component or compressor assembly. The intent of this operation is to allow the majority of the system oil to be returned to the air-conditioning compressor assembly.

CAUTION:
If excessive refrigerant oil loss has occurred, never perform the lubricant return operation because compressor damage may result.

In order to perform the refrigerant return operation, the air-conditioning system must be operating and there must be no evidence of a large amount of refrigerant oil loss.

Procedure

1. Start the engine and allow it to idle at 1,500 rpm.
2. Turn on the air-conditioning system and set it to MAX, or select the Recirculation mode.
3. Set the blower motor speed to HI.
4. Allow the engine to idle at 1,500 rpm for 10 minutes.
5. Turn off the engine.
6. Recover the refrigerant and record the amount of refrigerant oil removed from the system by the refrigerant recovery/recycle station. Add this amount of refrigerant oil back into the system along with the amount specified with the specific component being replaced. Special procedures for determining the amount of refrigerant oil to be added to the system when the compressor is replaced are detailed at the end of this section.
7. Replace the system component and evacuate and recharge the system according to manufacturer service procedures. Add the specified amount of refrigerant oil for the component being replaced plus the amount of refrigerant oil recorded during the recovery process in step 6.

After replacing the evaporator, condenser, or refrigerant storage container (receiver-drier/accumulator), add the manufacturer-specified amount of refrigerant oil.

An example of the amount of lubricant to be added is listed in the following table.

Part Replaced	Refrigerant Oil to Be Added to System
Condenser	1.2 fl. oz. (35 mL)
Refrigerant storage container (receiver-drier/accumulator)	0.3 fl. oz. (10 mL)
Evaporator	2.5 fl. oz. (75 mL)

When the air-conditioning system compressor is replaced, it is necessary to determine how much refrigerant oil was in the old compressor assembly and the amount removed during the refrigerant recovery process in order to determine how much clean fresh oil must be added to the system along with the new compressor. The total of these two amounts is added to the refrigerant system when the new compressor is installed (Figure 8-11). It is critical that just the right amount of refrigerant be added to the air-conditioning system. If too little oil is added the replacement compressor will fail, and if too much oil is added system performance will be affected. Always use fresh oil when adding lubricant to the system.

The following procedure covers removing refrigerant oil from an old compressor assembly.

1. Follow the steps in "Refrigerant Oil Return Operation" detailed earlier and recover/recycle the refrigerant from the air-conditioning system. Record the amount of refrigerant oil removed during this process.
2. Once the old air-conditioning compressor is removed, drain the old lubricant from the compressor into a graduated container and record the amount removed.
3. Drain the lubricant from the new compressor that is going to be installed into the vehicle into another clean container.
4. Add fresh, clean lubricant through the suction port of the new compressor. The amount to add is the amount recorded in step 1 plus the amount recorded in step 2.
5. Install the new compressor on the vehicle. Once the refrigerant lines have been connected, rotate the compressor clutch plate by hand several times to distribute the lubricant.

Some compressors have little or no oil reserve. Since oil is often lost with refrigerant, it is unwise to "top off" a system without determining the cause.

An "off-road" vehicle is one that is not licensed for street travel, such as a harvester or thrasher.

Compressors, like most other components with moving parts, require oil for lubrication to prevent damage.

Remove any other wires, such as superheat switch wire, from the compressor, if equipped.

Pin-type connectors are single or multiple electrical connectors that are round-or pin shaped and fit inside a matching connector. Modern connectors outside the passenger compartment use weather pack seals.

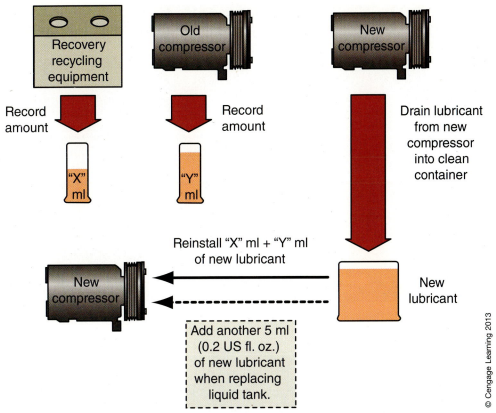

FIGURE 8-11 Before a replacement compressor is installed, you must first determine how much refrigerant oil was recovered during the recovery process, plus how much refrigerant was drained from the old compressor assembly. The total of these two amounts is added to the refrigerant system when the new compressor is installed.

REMOVING AND REPLACING THE COMPRESSOR

1. Follow all of the required procedures as outlined in "Refrigerant Oil Return Operation."
2. Remove the inlet and outlet hoses or service valves from the compressor.

   NOTE: The suction and discharge lines of many systems are connected to the compressor with a common manifold.

3. Remove the clutch lead wire.
4. Loosen and remove the belt(s).
5. Remove the mounting bolt(s) from the compressor bracket(s) and brace(s).
6. Remove the compressor from the vehicle (Figure 8-12).
7. Drain the old compressor into a graduated cylinder and record amount drained (Figure 8-13).
8. Position the new or rebuilt compressor.
9. Install the mounting bolt(s) into the compressor bracket(s) and brace(s).
10. Position and reinstall the belt(s).
11. Replace the clutch lead wire.
12. Using new gaskets or O-rings, replace the suction and discharge lines in the reverse order as removed in step 2.

    NOTE: Steps 12 through 14 procedures are found in Chapter 6 of this manual.

13. Leak-test the system.
14. Evacuate the system.
15. Charge the system with refrigerant.
16. Complete the system performance test job sheet.

Spade-type connectors are single or multiple electrical connectors that are flat spade shaped and fit inside a matching connector. Modern connectors outside the passenger compartment use weather pack seals.

SERVICE TIP:
A magnetic parts tray to hold nuts and bolts under the hood while servicing a vehicle will save you both time and aggravation.

FIGURE 8-12 Remove the compressor from the vehicle.

FIGURE 8-13 Drain the compressor oil into a graduated container.

INSTALLING INLINE FILTER

After a catastrophic compressor failure there will be small metal particles left in the refrigerant system. Many manufacturers do not recommend chemical flushing of their refrigerant system. Instead of chemical flushing some manufacturers recommend flushing with clean refrigerant (recover, charge, and then recover again). This process is generally ineffective in removing all the particles left behind. A more practical alternative to protect the system is to install an inline auxiliary filter into the refrigerant system to trap debris and contaminants that may remain in the system after a compressor or desiccant failure.

Auxiliary inline filters may be installed downstream in the liquid line or at the suction port to the compressor inlet or in both locations for maximum protection. The liquid line inline filter is connected in series to the high-pressure refrigerant line between the condenser outlet and the restriction device (i.e., TXV or FOT) to trap small particles of debris (Figure 8-14) that may be left in the system after component failure. Instruction for installing an auxiliary filter may be found in the job sheet at the end of this chapter entitled "Installation of Auxiliary Liquid Line Filter."

An additional inline filter may be installed in low-pressure suction line at the compressor inlet to trap any remaining small particles of debris from entering the air-conditioning compressor (Figure 8-15). This filter screen is mechanically pressed into the suction line with a special service tool kit.

After the air-conditioning system has been operated for a few hours, the filter should be checked and the system retested. The quickest method for testing the filter is to perform a temperature drop test across the inlet and outlet of the filter assembly. A clamp-on thermocouple is the simplest and most accurate method of checking line temperature. If the filter is unrestricted the filter inlet and outlet temperatures should be the same; the allowable range of temperature drop across the filter is 0–6°F (0–3°C). If the temperature drop exceeds 10°F (6°C) the filter will need to be cleaned or replaced.

NOTE: If the filter becomes clogged, air-conditioning system performance will be negatively affected. The restricted filter will also restrict the flow of refrigerant oil through the system, which could cause premature compressor failure.

FIGURE 8-14 Inline filter may also be connected in series to the high-pressure refrigerant line to trap small particles of debris.

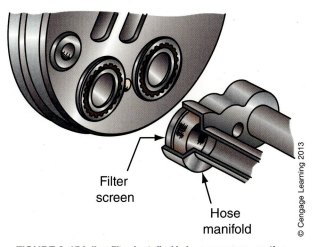

Filter screen

Hose manifold

FIGURE 8-15 Inline filter installed in low-pressure suction line at the compressor inlet will trap small particles of debris from entering the air-conditioning compressor.

NIPPONDENSO

Nippondenso has, perhaps by far, the greatest number of makes and models of compressors available for automotive air-conditioning service. They are used on many car lines, such as Acura, BMW, Chevrolet, Chrysler, Corvette, Dodge, Ford, Honda, Toyota, Lexus, Lincoln, Mazda, Mercedes-Benz, Mercury, Merkur, Mitsubishi, Plymouth, and Porsche.

There are over 25 different models and more than 100 styles of Nippondenso compressors in current use. Other models, such as Ford's FX6 and FX15 and Chrysler's A590 and C171, are also manufactured by Nippondenso. Other models no longer in current production have been replaced, generally by improved design models. Most compressors, which are designated for particular applications, cannot be interchanged. A replacement compressor, then, is best identified by giving the supplier information such as:

- OEM number from identification tag
- Make, model, and year of vehicle
- Grooves in pulley: 1, 2, 4, 5, or 6
- Accessories, such as power steering
- Series and date of manufacture (VIN)

Compressors must be identified by model number, not just by appearance.

Before attempting any installation, it is always wise to make a visual inspection to determine if the supplied replacement compressor is comparable to the defective unit. If the supplied compressor is without a clutch assembly, pull the clutch assembly from the defective compressor to make a comparison. Check the shaft size and length. Also, determine if it is a splined or keyed shaft. Many clutches are not interchangeable.

SERVICING THE NIPPONDENSO COMPRESSOR

The procedures for servicing the Nippondenso compressor are given in three sections: replacing the shaft seal, checking and adding oil, and servicing the clutch. These procedures assume that the compressor has been removed from the vehicle and that the services are to be performed "on the bench."

REPLACING THE SHAFT SEAL

Seal replacement for the Nippondenso compressor is somewhat different from most other compressors in that the front head assembly must first be removed. See Photo Sequence 12 for a typical procedure for removing and replacing the Nippondenso shaft seal assembly.

CHECKING AND ADDING OIL TO THE NIPPONDENSO COMPRESSOR

The Nippondenso compressor is factory charged with 13 oz. (384 mL) of refrigeration oil. It is not recommended that the oil level be checked as a matter of routine unless there is evidence of a severe loss.

The following procedure assumes that the suction and discharge service valves have been removed from the compressor.

Draining the Compressor

1. Drain the compressor oil through the suction and discharge service ports into a graduated container.
2. Rotate the crankshaft one revolution to be sure that all oil is drained.
3. Note the quality and quantity of the oil drained. Inspect the drained oil for brass or metallic particles, which indicate a compressor failure. Record (in ounces or milliliters) the amount of oil removed.
4. Discard the old oil as required by local regulations.

SPECIAL TOOLS
Bearing remover/pulley installer
External snapring pliers
Graduated container
Hub remover
Three-jaw puller
Shaft key remover
Shaft protector
Shaft seal seat installer
Shaft seal seat remover

Classroom Manual
Chapter 6, page 207

Do not attempt to remove the service valves before the refrigerant has been removed from the system.

TYPICAL PROCEDURE FOR REMOVING AND REPLACING THE NIPPONDENSO COMPRESSOR SHAFT SEAL ASSEMBLY

P12-1 After removing the clutch and coil assemblies, use the shaft key remover to remove the shaft key.

P12-2 Remove the felt oil absorber and retainer from the front head cavity.

P12-3 Clean the outside of the compressor with pure mineral spirits and air dry. Do not submerge the compressor into mineral spirits. Drain the oil into a graduated measure.

P12-4 Remove the six through bolts from the front head. Use the proper tool; some require a 10 mm socket and others require a 6 mm Allen wrench. Discard the six brass washers (if so equipped) and retain the six bolts.

P12-5 Gently tap the front with a plastic hammer to free it from the compressor housing. Remove and discard the head-to-housing O-ring and the head-to-valve plate gasket.

P12-6 Place the front head on a piece of soft material, such as cardboard, cavity side up. Use the shaft seal seat remover to remove the seal seat.

P12-7 Using both hands, remove the shaft seal cartridge.

P12-8 Liberally coat all seal parts, compressor shaft, head cavity, and gaskets with clean refrigeration oil. Carefully install the shaft seal cartridge, making sure to index the shaft seal on the crankshaft slots.

P12-9 Install the seal seat into the front head using the seal seat installer.

P12-10 Install the head-to-valve plate gasket over the alignment pins in the compressor housing. Install the head-to-housing O-ring. Carefully slide the head onto the compressor housing, making sure that the alignment pins engage in the holes in the head.

P12-11 Using six new brass washers (if required), install the six compressor through bolts. Using a 10 mm socket or a 6 mm Allen wrench, as required, tighten the bolts to a 260 in.-lb. (29.4 N·m) torque. **SERVICE TIP:** Use an alternate pattern when torquing the bolts.

P12-12 Replace the oil with clean refrigeration oil. Install the crankshaft key using a drift. Align the ends of the felt and its retainer and install them into the head cavity. Be sure the felt and retainer are fully seated against the seal plate. Replace the clutch and coil assembly.

Refilling the Compressor

1. Add oil, as follows:
   a. If the amount of oil drained was 3 oz. (89 mL) or more, add an equal amount of clean refrigeration oil.
   b. If the amount of oil drained was less than 3 oz. (89 mL), add 5–6 oz. (148–177 mL) of clean refrigeration oil.
2. If the compressor is to be replaced, drain all of the oil from the new or rebuilt compressor and replace the oil as outlined in step 1 (a or b, as applicable).

SERVICING THE NIPPONDENSO COMPRESSOR CLUTCH

The Nippondenso compressor may be equipped with either a Nippondenso or Warner clutch assembly. Though these two clutches are similar in appearance, their parts are not interchangeable. Complete clutch assemblies are, however, interchangeable on this compressor.

The apparent difference in the two clutches is that the Nippondenso pulley (Figure 8-16) has two narrow single-row bearings that are held in place with a wire **snapring**. The Warner clutch (Figure 8-17) has a single wide double-row bearing that is staked or crimped in place.

Be sure to use the proper type and grade of oil to ensure refrigerant compatibility.

SERVICE TIP: Oil is added into the suction or discharge port(s). Rotate the compressor crankshaft at least five revolutions by hand after adding oil.

Classroom Manual Chapter 8, page 263

A **snapring** is a spring steel ring used to secure and retain a component to another component.

SERVICE TIP: The shaft/hub key need not be removed. Take care not to lose the shim washer(s).

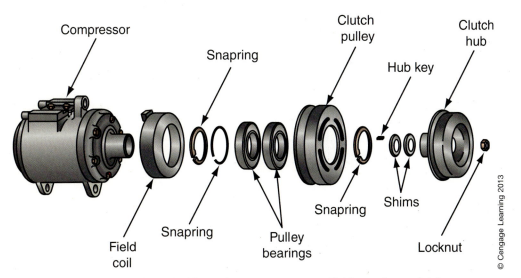

FIGURE 8-16 Exploded view of a Nippondenso compressor with Nippondenso clutch.

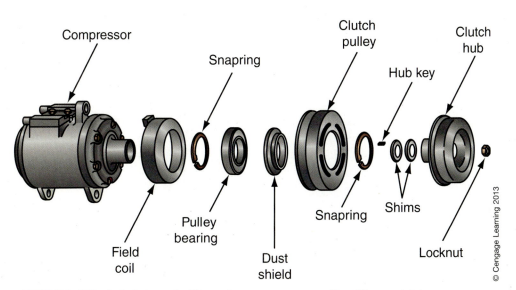

FIGURE 8-17 Exploded view of a Nippondenso compressor with a Warner clutch.

© Cengage Learning 2013

Removing the Clutch

1. Remove the hub nut.
2. Use the hub remover and remove the clutch hub (Figure 8-18).
3. Use the snapring pliers to remove the pulley retainer snapring.
4. With the shaft protector in place (Figure 8-19), remove the pulley and bearing assembly with the three-jaw puller.
5. Use the snapring pliers to remove the field coil retaining snapring (Figure 8-20).
6. Note the location of the coil electrical connector and lift the field coil from the compressor.

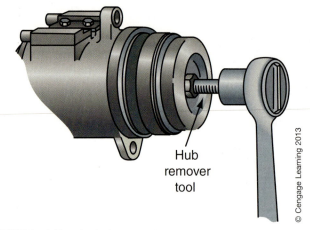

Hub remover tool

© Cengage Learning 2013

FIGURE 8-18 Use the hub removal tool to remove the clutch hub.

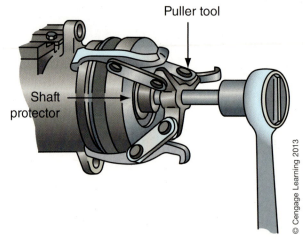

Puller tool

Shaft protector

© Cengage Learning 2013

FIGURE 8-19 With the shaft protector in place, use a three-jaw puller to remove the pulley and bearing assembly.

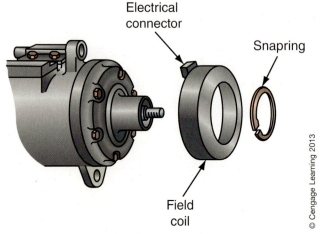

Electrical connector

Snapring

Field coil

© Cengage Learning 2013

FIGURE 8-20 Use a snapring pliers to remove the clutch coil assembly.

Replacing the Pulley Bearing

1. Support the pulley with the proper clutch pulley support.
2. Drive out the bearing(s) using a hammer and bearing remover (Figure 8-21).
3. Lift out the dust shield and retainer or leave them in place. Make sure that the dust shield is in place *before* installing the bearing(s).
4. Install the new bearing(s) using the bearing installer and the hammer (Figure 8-22). The bearing(s) must be fully seated in the rotor.
5. Replace the wire snapring if the clutch is a Nippondenso. If it is a Warner clutch, stake the bearing in place using the prick punch and the hammer.

Installing the Clutch

1. Install the field coil. Be sure the locator pin on the compressor engages with the hole in the clutch coil.
2. Install the snapring. Be sure the bevel edge of the snapring faces out.
3. Slip the rotor/bearing assembly squarely on the head. Using the bearing remover/pulley installer tool, *gently tap* the pulley onto the head (Figure 8-23).
4. Install the rotor/bearing snapring. The bevel edge of the snapring must face out.

SERVICE TIP:
Before reassembly, use pure mineral spirits to clean all parts, including the pulley bearing surface and the compressor front head.

If index pins are not engaged properly, misalignment occurs.

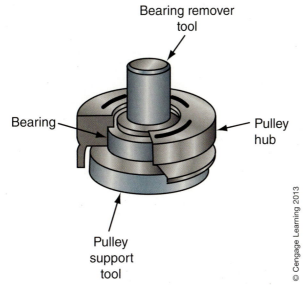

© Cengage Learning 2013

FIGURE 8-21 Use a hammer (not shown) and bearing remover tool to drive out the bearing after placing the hub on a pulley support.

© Cengage Learning 2013

FIGURE 8-22 Use a bearing installer and hammer to drive the new bearing(s) into the pulley hub.

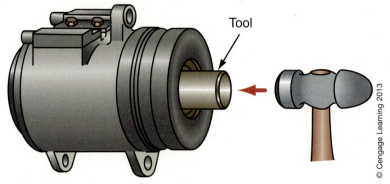

Tool

FIGURE 8-23 Gently tap the pulley assembly onto the compressor head.

A snapring installed backward will not hold properly.

Hub damage may occur if the key is not properly aligned.

CAUTION: Do not drive (hammer) the hub on; to do so will damage the compressor.

5. Install shim washers or be sure they are in place. Check the shaft/hub key to ensure proper seating.
6. Align the hub keyway with the key in the shaft. Press the hub onto the compressor shaft using the hub replacer tool (Figure 8-24).
7. Using a nonmagnetic feeler gauge, check the air gap between the hub and rotor (Figure 8-25). The air gap should be 0.021–0.036 in. (0.53–0.91 mm).

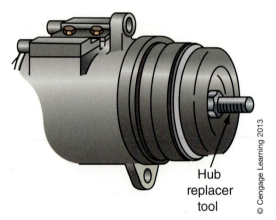

Hub replacer tool

FIGURE 8-24 Press the hub onto the shaft sing the hub replacer tool.

Feeler gauge

FIGURE 8-25 Check the air gap between the hub and rotor using a nonmetallic feeler gauge.

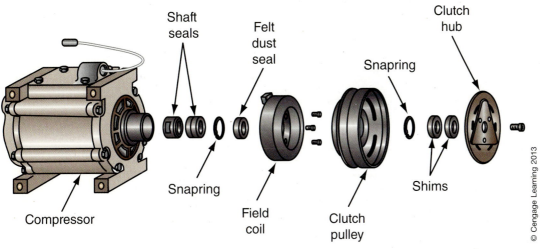

FIGURE 8-26 Panasonic's rotary vane compressor.

Classroom Manual
Chapter 8, page 261

⚠️

CAUTION:
The use of proper refrigerant is important. Do not mix refrigerants.

The rotary compressor is not sensitive to an overcharge of oil. An overcharge, however, may reduce overall system capacity and efficiency.

CAUTION:
Use the proper oil and the correct amount of oil. Too much oil will reduce the system capacity, and too little oil will result in insufficient lubrication.

Oil removed from an air-conditioning system or component must be discarded in accordance with EPA regulations.

SPECIAL TOOLS
External snapring pliers
Internal snapring pliers

8. Turn the shaft (hub) one-half turn and recheck the air gap. Change the shim(s) as necessary to correct the air gap.
9. Install the locknut and tighten to 10–14 ft.-lb. (13.6–19.0 N·m).
10. Recheck the air gap. See steps 7 and 8.

PANASONIC (MATSUSHITA)

The Panasonic rotary vane-type compressor (Figure 8-26) was first introduced in 1993 by Ford Motor Company on the Probe. The compressor, which is belt driven off the engine, uses HFC-134a as a refrigerant.

SERVICING THE PANASONIC VANE-TYPE COMPRESSORS

The main components of the Panasonic vane-type compressor are the rotor with three vanes, a sludge control valve, a discharge valve, and a thermal protector. The only service that may be accomplished is checking and adjusting oil, servicing the clutch, replacing the shaft seal, and servicing the thermal protector, sludge control, and discharge valve.

All service procedures assume that the compressor has been removed from the vehicle and is being worked with on the bench.

CHECKING AND ADJUSTING COMPRESSOR OIL LEVEL

A new Panasonic Rotary compressor contains 6.78 oz. (200 mL) of a special paraffin-base refrigeration oil, designated as YN-9.

It is necessary to adjust the oil any time the compressor is serviced or when being replaced, as outlined in the following procedure.

Procedure

1. Drain the oil from the defective compressor into a calibrated container and note the amount removed. Allow the compressor to drain thoroughly.
2. Drain the oil from the replacement compressor into a second calibrated container. Allow the compressor to drain thoroughly.
3. Add the same amount of clean refrigeration oil to the replacement compressor that was removed from the defective compressor.
4. Add an additional 0.68 oz. (20 mL) of oil.

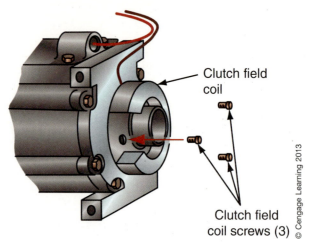

Clutch field coil

Clutch field coil screws (3)

© Cengage Learning 2013

FIGURE 8-27 Using a driver remove/install the three clutch field coil screws.

SERVICING THE CLUTCH ASSEMBLY

Removing the Clutch and Coil

1. Using an Allen wrench, remove the clutch armature Allen bolt.
2. Remove the clutch armature.
3. Remove the shim(s) and set aside.
4. Using **internal snapring** pliers, remove the clutch rotor/pulley snapring.
5. Remove the clutch rotor/pulley.
6. Using a screwdriver, remove the three clutch field coil screws. Remove the clutch field coil (Figure 8-27).

> An **internal snapring** is a retaining device used to retain or hold a component inside a cavity or case.

Replacing the Coil and Clutch

1. Replace the clutch field coil and secure with three screws (see Figure 8-27).
2. Replace the clutch rotor/pulley and secure with the internal snapring.
3. Replace the shim(s).
4. Replace the clutch armature and secure with the Allen bolt.

SERVICING THE COMPRESSOR SHAFT SEAL

To service the compressor shaft seal, proceed as follows:

Removing the Shaft Seal

1. Remove the clutch. It is not necessary to remove the clutch coil for seal service.
2. Remove the felt dust seal from the seal cavity (Figure 8-28).
3. Using internal snapring pliers, remove the shaft seal snapring (Figure 8-29).
4. Using the seal remover, remove the seal seat (Figure 8-30).
5. Using the seal remover/installer, remove the shaft seal (Figure 8-31).

Installing the Shaft Seal

1. Coat all seal parts with clean refrigeration oil.
2. Install the shaft seal, using the remover/installer tool.
3. Install the shaft seal, using the seal remover tool.
4. Replace the shaft seal snapring.
5. Replace the felt dust seal.
6. Replace the clutch assembly.

> A seal, if installed backward, is almost impossible to remove.

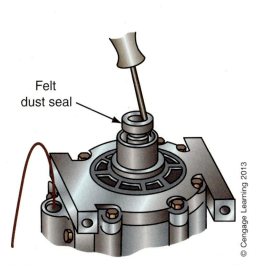

FIGURE 8-28 Remove/replace the felt dust seal.

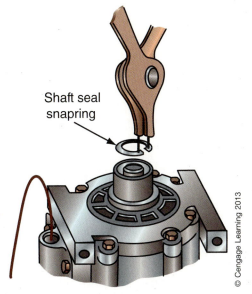

FIGURE 8-29 Remove/replace the shaft seal snapring.

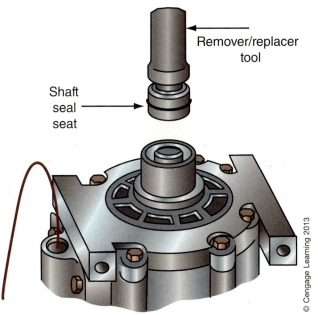

FIGURE 8-30 Remove/install the shaft seat.

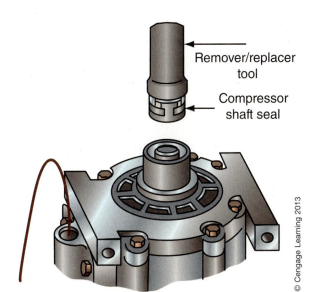

FIGURE 8-31 Remove/install the shaft seal.

SERVICING THE COMPRESSOR

The thermal protector, sludge control, and discharge valve are the only components that may be serviced in the Panasonic rotary compressor. It is not necessary to remove the clutch assembly or the shaft seal assembly for this service. For the disassembly of the Panasonic vane rotary compressor to service the internal parts, follow the procedure listed below. The procedure may be followed in reverse order for reassembly.

Before reassembly, liberally coat all parts with clean refrigeration oil that is compatible with the refrigerant requirements of the system.

Procedure

1. Replace the discharge valve and stopper.
2. Secure the discharge valve and stopper with the two bolts removed in disassembly.
3. Install the thermal protector and secure it with the snapring.
4. Replace the thermal protector housing with a new gasket and secure it with four capscrews.
5. Secure the thermal protector hold-down bracket with the screw previously removed.
6. Replace the two compression springs and spring stoppers.
7. With a new gasket in place, install the oil control valve with three bolts.
8. Install the housing cover. Secure the housing cover with two Allen bolts and six hex nuts.
9. Replace the refrigeration oil.

SANDEN

There are about 20 models of the Sanden compressor, formally known as Sankyo. These compressors are used by Chevrolet, Chrysler, Dodge, Fiat, Ford, Honda, Jeep, Mazda, Peugeot, Renault, Subaru, and Volkswagen. The considerations for selecting a replacement Sanden compressor are:

- Head style: horizontal or vertical O-ring; horizontal or vertical pad; vertical flare
- Clutch diameter: from 3.8 in. (96.5 mm) to 5.6 in. (142 mm)
- Number of grooves in clutch rotor: either 1 or 2 V-groove or 4, 5, 6, 7, or 10 poly-groove
- Mounting boss measurement, front to rear, outside to outside: from 2.85 in. (72.4 mm) to 4.41 in. (112 mm)
- Type of refrigerant: R-12 or R-134a

 WARNING: **Do not mix refrigerants.**

SERVICING THE SANDEN (SANKYO) COMPRESSOR

Servicing the Sanden/Sankyo compressor is limited in this manual to: replacing the shaft oil seal, checking and adjusting the proper oil level, and servicing the clutch. This procedure assumes that the compressor has been removed from the vehicle and is being serviced on the bench.

REPLACING THE COMPRESSOR SHAFT OIL SEAL

The following procedures may be followed when replacing the compressor shaft seal.

Removing the Shaft Seal

1. Using a ¾ in. hex socket and spanner wrench (Figure 8-32), remove the crankshaft hex nut.
2. Remove the clutch front plate (Figure 8-33) using the clutch front plate puller.
3. Remove the shaft key and spacer shims and set them aside.
4. Using the snapring pliers (Figure 8-34), remove the seal retaining snapring.
5. Remove the seal seat using the seal seat remover and installer (Figure 8-35).
6. Remove the seal (Figure 8-36) using the seal remover tool.
7. Remove the shaft seal seat O-ring (Figure 8-37) using the O-ring remover.
8. Discard all parts removed in steps 5, 6, and 7.

FIGURE 8-32 Remove the crankshaft hex nut.

FIGURE 8-33 Remove the clutch front plate.

FIGURE 8-34 Remove the seal seat snapring.

FIGURE 8-35 Remove the seal seat.

FIGURE 8-36 Remove the seal.

FIGURE 8-37 Remove the O-ring.

FIGURE 8-38 Use the air gap gauge to check the rotor-to-hub clearance.

Installing the Shaft Seal

1. Clean the inner bore of the seal cavity by flushing it with clean refrigeration oil.
2. Coat the new seal parts with clean refrigeration oil.
3. Install the new shaft seal seat O-ring. Make sure it is properly seated in the internal groove. Use the remover tool to position the O-ring properly.
4. Install the seal protector on the compressor crankshaft. Liberally lubricate the part with clean refrigeration oil.
5. Place the new shaft seal in the seal installer tool, and carefully slide the shaft seal into place in the inner bore. Rotate the shaft seal clockwise (cw) until it seats on the compressor shaft flats.
6. Rotate the tool counterclockwise (ccw) to remove the seal installer tool.
7. Remove the shaft seal protector.
8. Place the shaft seal seat on the remover/installer tool and carefully reinstall the shaft seal in the compressor seal cavity.
9. Replace the seal seat retainer.
10. Reinstall the spacer shims and shaft key.
11. Position the clutch front plate on the compressor crankshaft.
12. Using the clutch front plate installer tool, a small hammer, and an air gap gauge, reinstall the front plate (Figure 8-38).
13. Draw down the front plate with the shaft nut. Use the air gap gauge for go at 0.016 in. (0.4 mm) and no-go at 0.031 in. (0.79 mm).
14. Using the torque wrench, tighten the shaft nut to a torque of 25–30 ft.-lb. (33.0–40.7 N·m).

CHECKING COMPRESSOR OIL LEVEL

The compressor oil level should be checked at the time of installation and after repairs are made when it is evident that there has been a loss of oil. The Sankyo compressor is factory charged with 7 fl. oz. (207 mL) of oil. A special angle gauge and dipstick are used to check the oil level. The oil chart (Figure 8-39) compares the oil level with the inclination angle of the compressor.

This procedure may also be followed with the compressor in the vehicle after first ensuring that the refrigerant has been recovered.

CAUTION:
Do not touch the carbon ring face with your fingers. Normal body acids will etch the seal and cause early failure.

Do not discard parts that may be reused, such as shaft keys, spacers, nuts, bolts, clamps, snaprings, and so on.

Some seal faces are made of a ceramic material. These seals may be damaged if care is not taken during handling.

Use nonmetallic feeler gauges to determine clutch air gap.

Inclination Angle In Degrees	Acceptable Oil Level In Increments
0	6–10
10	7–11
20	8–12
30	9–13
40	10–14
50	11–16
60	12–17

© Cengage Learning 2013

FIGURE 8-39 Dipstick reading vs. inclination angle.

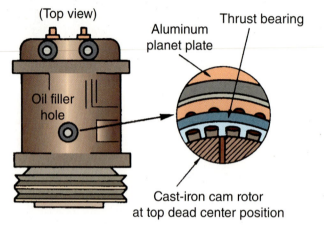

(Top view)

Thrust bearing

Aluminum planet plate

Oil filler hole

Cast-iron cam rotor at top dead center position

© Cengage Learning 2013

FIGURE 8-40 Position the rotor to top dead center (TDC).

Do not attempt to remove the oil plug until after ensuring that the refrigerant has been removed from the system.

TDC is an acronym for "top dead center."

SERVICE TIP: The dipstick tool for this procedure is marked in eight increments. Each increment represents 1 oz. (29.57 mL) of oil.

An overcharge of oil has basically the same effect as an overcharge of refrigerant.

Preparing the Compressor

1. Position the angle gauge tool across the top flat surfaces of the two mounting ears.
2. Center the bubble and read the inclination angle.
3. Remove the oil filler plug. Rotate the clutch front plate to position the rotor at TDC (Figure 8-40).
4. Face the front of the compressor. If the compressor angle is to the right, rotate the clutch front plate ccw by 110 degrees. If the compressor angle is to the left, rotate the plate cw by 110 degrees (Figure 8-41).

Checking the Oil Level

1. Insert the dipstick until it reaches the stop position marked on the dipstick.
2. Remove the dipstick and count the number of increments of oil.
3. Compare the compressor angle and the number of increments with the table (see Figure 8-39).
4. If necessary, add oil to bring the oil to the proper level. *Do not overfill.* Use only clean refrigeration oil of the proper grade.

SERVICING THE CLUTCH

Although this procedure presumes that the compressor is removed from the vehicle, if ample clearance is provided in front of the compressor for clutch service, it need not be removed for this service.

FIGURE 8-41 Rotate the clutch front plate.

Removing the Clutch

1. Use a ¾ in. hex socket and spanner wrench to remove the crankshaft hex nut as shown in Figure 8-32.
2. Remove the clutch front plate, using the clutch front plate puller as shown in Figure 8-33.
3. Using the snapring pliers, remove the internal and external snaprings (Figure 8-42 and Figure 8-43).
4. Using the pulley puller (Figure 8-44), remove the rotor assembly.
5. If the clutch coil is to be replaced, remove the three retaining screws and the clutch field coil. Omit this step if the coil is not to be replaced.

Replacing the Rotor Bearing

1. Using the snapring pliers, remove the bearing retainer snapring.
2. From the back (compressor) side of the rotor, knock out the bearing using the bearing remover tool and a soft hammer.
3. From the front (clutch face) side of the rotor, install the new bearing using the bearing installer tool and a soft hammer. Take care not to damage the bearing with hard blows of the hammer (Figure 8-45).
4. Reinstall the bearing retainer snapring.

FIGURE 8-42 Remove the internal snapring.

FIGURE 8-43 Remove the external snapring.

FIGURE 8-44 Remove the rotor assembly.

FIGURE 8-45 Install the rotor bearing.

FIGURE 8-46 Drive the front plate onto the shaft.

Replacing the Clutch

1. Reinstall the field coil (or install a new field coil, if necessary) using the three retaining screws.
2. Align the rotor assembly squarely with the front compressor housing.
3. Using the rotor two-piece installer tools and a soft hammer, carefully drive the rotor into position until it seats on the bottom of the housing.
4. Reinstall the internal and external snaprings using the snapring pliers.
5. Align the slot in the hub of the front plate squarely with the shaft key.
6. Drive the front plate on the shaft using the installer tool and a soft hammer (Figure 8-46). *Do not use unnecessary hard blows.*
7. Check the air gap with go and no-go gauges.
8. Replace the shaft nut and tighten it to a torque of 25–30 ft.-lb. (33.9–40.7 N·m) using the torque wrench.

 WARNING: **Careful handling of all seal parts is important. The carbon seal face and the steel seal seat must not be touched with the fingers because of the etching effect of the acids normally found on the fingers.**

ELECTRICALLY-DRIVEN AIR-CONDITIONING COMPRESSOR

Many hybrid electric vehicles are equipped with a high-voltage three phase alternating current air-conditioning compressor which has specific safety and service requirements. This section is meant only as an overview for system service. Always refer to specific vehicle manufacturer service information for specific service procedures.

An electrically-driven air-conditioning compressor (Figure 8-47) does not have a drive belt but is instead connected to the high-voltage harness by an orange colored wire and connector harness. The compressor requires special ND-11 insulating refrigerant oil (Figure 8-48). Avoid oil cross contamination during system evacuation and recharging by using hoses designed for ND-11. Using incorrect lubricant can set diagnostic trouble codes (DTC) and may cause damage to the electrically-driven compressor and may result in the leakage of electrical power.

CAUTION: Alignment is essential in order to prevent damage to mating surfaces.

A BIT OF HISTORY

Tecumseh was one of the first manufacturers of automotive air-conditioning compressors. A large, heavy, cast-iron flywheel-pulley compressor was used from the late 1940s through the late 1950s. Known as Tecumseh's Model HH, it was discontinued in 1958 in favor of a smaller, somewhat lighter model LB. The model LB, which was also made of cast iron, was soon discontinued, however and was replaced by the popular HA and HG series, which are still found on limited applications.

A **carbon seal face** is made of a carbon composition rather than another material, such as steel or ceramic.

Etching is the unintended erosion of a metal surface generally caused by acid exposure.

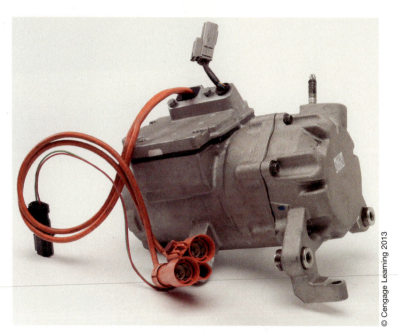

FIGURE 8-47 Typical electronic inverter air-conditioning compressor found on hybrid electric vehicles.

⚠ **CAUTION REFRIGERANT UNDER HIGH PRESSURE**

Improper service methods may cause injury. Air-conditioning system should be serviced by qualified personnel. See Repair Manual.

┌─ Refrigerant ─┐   ┌─ Oil ─┐
HFC134a Max. 0.58kg(1.29lbs.)   ND-OIL   **11** SAE
USE ONLY Min. 0.48kg(1.06lbs.)   OR EQUIVALENT   J-639

MFD, BY DENSO MANUFACTURING MICHIGAN INC.

FIGURE 8-48 Typical refrigerant system label for electronic inverter air-conditioning compressor system.

Classroom Manual

Chapter 8, page 277

Oil must be discarded in a manner consistent with the Environmental Protection Agency (EPA) guidelines.

An overcharge of oil will reduce system capacity.

Crayon or chalk temporarily marks component locations. A scribe or file is used to permanently mark the location.

Electronic Inverter Compressor Removal and Service

1. Disconnect the negative cable from high-voltage battery pack.
2. Remove high-voltage service plug.
3. Attach R/R/R equipment and recover refrigerant from the system.
4. Disconnect discharge hose and subassembly from compressor (Figure 8-49).
5. Seal the opening on the discharge hose using vinyl tape to prevent foreign matter and moisture from entering the system.
6. Disconnect suction hose and subassembly from compressor (Figure 8-50).
7. Again seal the opening on the suction hose using vinyl tape to prevent foreign matter and moisture from entering the system.
8. Wearing insulated lineman gloves remove electric inverter compressor connector and 3 harness hold-down clamps. Release the green-colored lock (1) and disconnect the connector (2) (Figure 8-51). Insulate the end of the vehicle harness connector with vinyl tape.
9. Remove the three mounting bolts and remove compressor from vehicle (Figure 8-52).
10. Check for debris in the discharge port of the compressor and drain the old compressor and inspect condition of the oil. If any particles or debris is present in either the oil or the discharge port, the dryer must also be replaced.

FIGURE 8-49 Electronic inverter air-conditioning compressor discharge hose.

FIGURE 8-50 Electronic inverter air-conditioning compressor suction hose.

11. Ready the new compressor for installation. Gradually remove the service valves from the suction and discharge ports to release the inert gas (helium) from the new compressor. The new compressor is shipped charged with 4.05 fl. oz. (115 cc) of new ND-11 oil. To determine the amount of oil to remove from the new compressor follow the following formula:

115 cc − (Amount drained from old compressor) = (Amount of oil to remove from new compressor before installation)

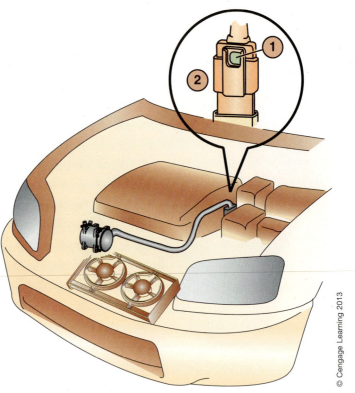

FIGURE 8-51 Electronic inverter air-conditioning compressor electrical connector and harness.

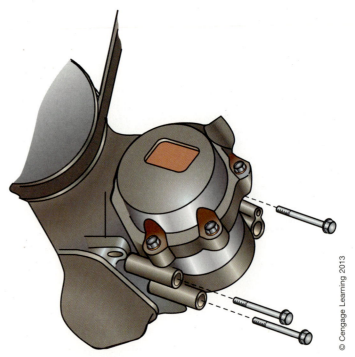

FIGURE 8-52 Electronic inverter air-conditioning compressor mounting points.

12. If a new compressor is installed without adjusting the oil volume the refrigerant system will be overcharged with oil reducing the effectiveness of the heat exchangers and may cause refrigerant system failure and abnormal vibration.

13. Install the new compressor mounting it with only 2 bolts (bolts 1 and 2) as indicated in Figure 8-53 first. Torque bolts to 18 ft. lbs. (25 N·m).

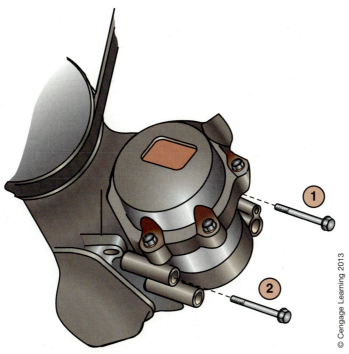

FIGURE 8-53 Mount electronic inverter air-conditioning compressor with only 2 of the 3 bolts.

© Cengage Learning 2013

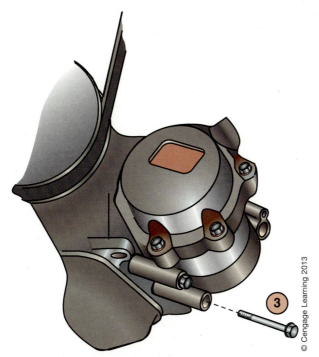

FIGURE 8-54 Mount electronic invert air-conditioning compressor 3 bolt and torque to specification.

© Cengage Learning 2013

14. Install remaining bolt (3) (Figure 8-54) and torque bolt to 18 ft. lbs. (25 N·m).
15. Wearing insulated lineman gloves reinstall electric inverter compressor connector and three harness hold-down clamps and lock the green-colored lock.
16. Install new o-rings lubricated with ONLY new ND-11 oil on both discharge and suction hose assemblies and reconnect hoses. Torque retaining bolts to 87 in. lbs. (9.8 N·m).

17. Install the high-voltage service plug.
18. Reconnect the negative cable from high-voltage battery pack.
19. Review manufacturer service information to see if a system initialization needs to be preformed after battery disconnect.
20. Recharge the system with the proper amount of refrigerant.
21. Warm-up compressor.
22. Inspect for refrigerant leaks and perform a system performance test.

CASE STUDY

A customer brings her vehicle into the shop because the air conditioner does not work. Initial inspection reveals that the clutch does not energize when the air conditioner is turned on, but the blower motor is operative.

The technician checks all of the fuses and circuit breakers that may affect the clutch circuit and finds that all are good. Next, the technician uses a voltmeter to check the available voltage at the clutch coil connector. The test indicates that 12.6 volts are available.

The technician then performs a voltage drop test across the ground side of the clutch coil. The voltmeter indicates 12.6 volts. The conclusion is that the ground provision of the clutch coil is defective and must be repaired.

After the ground wire, which is bonded to body metal, is cleaned and reconnected, the clutch functions properly and the air conditioner is fully operational.

TERMS TO KNOW

Carbon seal face

Etching

Flange

Internal snapring

Mounting boss

Running design change

Shaft seal

Snapring

ASE-STYLE REVIEW QUESTIONS

1. A slipping compressor clutch may be due to all of the following, *except*:
   A. Incorrect clutch air gap
   B. Internal compressor problem
   C. Excessively high refrigerant pressure
   D. Low refrigerant charge

2. The electromagnetic clutch is being discussed:
   Technician A says that the clutch will slip if the air gap is too close.
   Technician B says that the clutch will slip if the belt is close.
   Who is correct?
   A. A only
   B. B only
   C. Both A and B
   D. Neither A nor B

3. There are signs of oil spray on the compressor clutch hub and nearby underhood areas.
   Technician A says that a leaking compressor shaft seal could be the cause.
   Technician B says that a faulty compressor clutch bearing could be the cause.
   Who is correct?
   A. A only
   B. B only
   C. Both A and B
   D. Neither A nor B

4. *Technician A* says that it is virtually impossible to insert the seal seat backward when using the appropriate tool.

   Technician B says that it is almost impossible to remove a seal seat that has been installed backward. Who is correct?

   A. A only C. Both A and B
   B. B only D. Neither A nor B

5. Most air-conditioning systems that have R-134a as the refrigerant also use _____ as a lubricant.

   A. Polyalkaline glycol (PAG)
   B. Polyol ester (POE)
   C. Mineral oil
   D. Alkylbenzene oil

6. The compressor clutch circuit is being tested.

   Technician A says if the compressor clutch coil voltage drop is equal to source voltage (battery) then the clutch coil is defective.

   Technician B says if the test lamp does not light when connected across the clutch's harness terminals, then the problem is probably a shorted clutch coil.

   A. A only C. Both A and B
   B. B only D. Neither A nor B

7. O-rings are being discussed:

   Technician A says that R-134a O-rings are usually black.

   Technician B says that R-12 O-rings are blue or green. Who is correct?

   A. A only C. Both A and B
   B. B only D. Neither A nor B

8. *Technician A* says that the seal cavity may be flushed with clean refrigeration oil.

   Technician B says that the seal cavity may be flushed with clean mineral spirits.

   Both agree that the residue must be disposed of in a proper manner. Who is correct?

   A. A only C. Both A and B
   B. B only D. Neither A nor B

9. To measure and adjust compressor clutch air gap, a _____ feeler gauge should be used.

   A. nonmetallic C. carbon steel
   B. soft steel D. metalic

10. A vehicle returns to the shop for a repeat failure of the compressor system due to compressor clutch slippage and overheating. Which of the following is the most likely cause?

    A. A loose drive belt
    B. A low refrigerant charge level.
    C. A faulty compressor clutch diode.
    D. High resistance in the compressor clutch electrical circuit.

ASE CHALLENGE QUESTIONS

1. During the compression stroke of an air-conditioning compressor, the suction valve is closed by the:

   A. Discharge valve. C. Valve spring.
   B. Discharge pressure. D. Suction pressure.

2. If the compressor oil has been drained and it shows signs of metallic particles, all of the following should be done, *except*:

   A. All system inlet screens should be cleaned after flushing.
   B. The receiver-drier or accumulator should be replaced after flushing.
   C. The entire system should be flushed.
   D. A full charge of heavier refrigeration oil should be added after flushing.

3. The angle of the swash plate in a variable displacement compressor is controlled by:
   A. Suction pressure differential in the crankcase.
   B. Pressure differential between the low and high sides.
   C. Both A and B
   D. Neither A nor B

4. The shims shown in the illustration are used to:
   A. Align the clutch pulley.
   B. Secure the hub key.
   C. Back up the locknut.
   D. Adjust the air gap.

5. The purpose of the compressor is to:
   A. Pump low-pressure vapor to a high-pressure vapor.
   B. Pump low-pressure vapor to a low-pressure liquid.
   C. Pump low-pressure liquid to a high-pressure liquid.
   D. None of the above

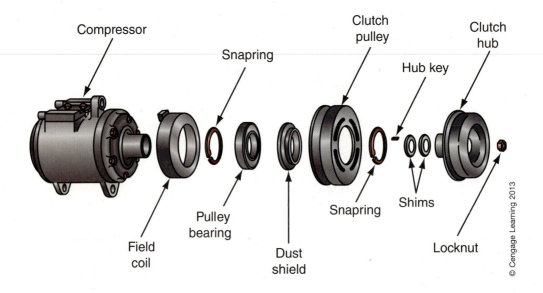

Compressor · Snapring · Clutch pulley · Clutch hub · Hub key · Field coil · Pulley bearing · Dust shield · Snapring · Shims · Locknut

© Cengage Learning 2013

Name _____ Date _____

COMPRESSOR IDENTIFICATION

Upon completion of this job sheet, you should be able to identify the different types of compressors used in automotive air-conditioning systems.

NATEF Correlation

HEATING AND AIR CONDITIONING: A/C System Diagnosis and Repair; *Locate and interpret vehicle and major component identification numbers (VIN, vehicle certification labels, calibration decals).* **(P-1)**

Tools and Materials

Several vehicles equipped with air-conditioning systems

Describe the vehicle being worked on.

Vehicle 1

Year _____ Make _____ Model _____

VIN _____ Engine type and size _____

Vehicle 2

Year _____ Make _____ Model _____

VIN _____ Engine type and size _____

Vehicle 3

Year _____ Make _____ Model _____

VIN _____ Engine type and size _____

Vehicle 4

Year _____ Make _____ Model _____

VIN _____ Engine type and size _____

Procedure

1. Visually inspect the air-conditioning system and describe its overall condition.

_____

_____

Vehicle	One	Two	Three	Four
Compressor:				
Type	_____			
Cylinders	_____			
Refrigerant:				
Type	_____			
Charge: Lb./oz.	___/___	___/___	___/___	___/___
mL	_____	_____	_____	_____
Lubricant:				
Type	_____			
Charge: oz.	___/___	___/___	___/___	___/___
mL	_____	_____	_____	_____
Belt Type	_____			

Instructor's Response _____

_____

REFRIGERANT OIL RETURN OPERATION AND COMPRESSOR OIL SELECTION/REPLACEMENT

Name _____ **Date** _____

Upon completion of this job sheet, you should be able to determine the approved lubricant for the refrigerant system and add the proper amount. Lubricant is added to the refrigerant system when a component is replaced or after a large leak has been repaired. It is important to maintain the specified amount of lubricant in the system and not to overfill it. A lubricant return operation should be performed prior to replacing any major refrigerant system component or compressor assembly. The intent of this operation is to allow the majority of the system oil to be returned to the air-conditioning compressor assembly.

NATEF Correlation

HEATING AND AIR CONDITIONING: A/C System Diagnosis and Repair; *Locate and interpret vehicle and major component identification numbers (VIN, vehicle certification labels, calibration labels).* **(P-1)**

HEATING AND AIR CONDITIONING: A/C System Diagnosis and Repair; *Inspect the condition of discharged oil; determine necessary action.* **(P-2)**

HEATING AND AIR CONDITIONING: A/C System Diagnosis and Repair; *Determine recommended oil for system application.* **(P-1)**

Tools and Materials

Late-model vehicle

Service manual or information system

Safety glasses or goggles

Hand tools, as required

Refrigerant recovery/recycling/charge station

Describe the vehicle being worked on.

Year _____ Make _____ Model _____

VIN _____ Engine type and size _____

Procedure

In order to perform the refrigerant return operation, the air-conditioning system must be operating and there must be no evidence of a large amount of refrigerant oil loss.

1. Start the engine and allow it to idle at 1,500 rpm.
2. Turn on the air-conditioning system and set it to MAX or select the Recirculation mode.
3. Set the blower motor speed to HI.

CAUTION: If excessive refrigerant oil loss has occurred, never perform the lubricant return operation because compressor damage may result.

4. Allow the engine to idle at 1,500 rpm for 10 minutes.

5. Turn off the engine.

6. Recover the refrigerant and record the amount of refrigerant oil removed from the system by the refrigerant recovery equipment. Add this amount of refrigerant oil back into the system along with the amount specified with the specific component being replaced.

   a. Record amount of oil removed _____

   b. Record amount of refrigerant removed _____

7. Replace the system component and evacuate and recharge the system according to manufacturer's service procedures. Add the specified amount of refrigerant oil for the component being replaced, plus the amount of refrigerant oil recorded during the recovery process.

   a. Amount of oil removed in step 6a _____

   b. Amount of oil specified to be added to the system for the component replaced _____

   c. Total amount of refrigerant to be added to the system prior to charging (a + b) = _____

After replacing the evaporator, condenser, or refrigerant storage container (receiver-drier/accumulator), add the manufacturer-specified amount of refrigerant oil.

An example of the amount of lubricant to be added is listed in the following table.

Part Replaced	Refrigerant Oil to Be Added to System
Condenser	1.2 fl. oz. (35 mL)
Refrigerant storage container (receiver-drier/accumulator)	0.3 fl. oz. (10 mL)
Evaporator	2.5 fl. oz. (75 mL)

Instructor's Response _____

_____

_____

Name _____ Date _____

CHECK AND CORRECT COMPRESSOR OIL LEVEL

Upon completion of this job sheet, you should be able to check and correct compressor lubricant levels.

NATEF Correlation

HEATING AND AIR CONDITIONING: A/C System Diagnosis and Repair; *Select oil type; measure and add oil to the A/C system as needed.* **(P-1)**

Tools and Materials

An air-conditioning system compressor

Service manual

Selected air conditioning tools

Lubricant, if required

Describe the vehicle being worked on.

Year _____ Make _____ Model _____

VIN _____ Engine type and size _____

Procedure

_____

Task Completed

1. What type of refrigerant is the compressor designed for?
   _____

2. What type of lubricant is the compressor designed for?
   _____

3. What is the lubricant capacity?
   oz. _____ mL _____

4. Following procedures outlined in the service manual, drain the lubricant from the compressor. ☐

5. How much lubricant was drained from the compressor?
   oz. _____ mL _____

6. When refilling, how much clean, fresh lubricant should be added to the compressor? oz._____ mL _____

7. Should the lubricant removed in step 4 be reused?

_____

Why?

_____

_____

Instructor's Response _____

_____

_____

Name _____ **Date** _____

REPLACE A/C COMPRESSOR ASSEMBLY

Upon completion of this job sheet, you should be able to recover refrigerant from an A/C system, remove and replace a compressor assembly, evacuate and recharge a system, as well as perform a leak test of the system.

NATEF Correlation

A/C System Diagnosis and Repair; *Leak test A/C system; determine necessary action.* **(P-1)**

A/C System Diagnosis and Repair; *Inspect the condition of discharged oil; determine necessary action.* **(P-2)**

A/C System Diagnosis and Repair; *Determine recommended oil for system application.* **(P-1)**

REFRIGERATION SYSTEM COMPONENT DIAGNOSIS AND REPAIR; Compressor and Clutch; *Remove and reinstall A/C compressor and mountings; measure oil quantity; determine necessary action.* **(P-1)**

Tools and Materials

Test vehicle

Refrigerant recovery and recycling unit

Manifold and gauge set

Refrigerant leak detector

Hand tools, as required

Safety glasses

Describe the vehicle being worked on.

Year _____ Make _____ Model _____

VIN _____ Engine type and size _____

Procedure

_____

Task Completed

1. Connect the gauge set to the system. Note the high-side and low-side pressures with the system off.
   a. High-side pressure _____

   _____

   b. Low-side pressure_____

   _____

2. Connect the manifold gauge set to the recovery unit. Start the unit following the unit manufacturer's instructions.

☐

3. Wait 5 minutes and note if pressure is detected (if pressure is present, rerun the recovery process). The system should hold a vacuum for 2 minutes. Did it? _____

☐ 4. Follow the steps "Removing and Replacing the Compressor" listed in Chapter 5, and refer to the manufacturer's recommended procedure for the removal and replacement of the component.

5. Evacuate the system. Did the system hold a vacuum? _____

6. List the amount and type of oil used:

   a. Type of oil: _____

   b. Amount of oil required: _____

7. Recharge the system.

   a. Record the amount of refrigerant added _____

☐ 8. Check the system for leaks using a halon leak detector.

9. Record the system running pressure.

   a. High-side pressure _____

   _____

   b. Low-side pressure _____

   _____

Instructor's Response _____

_____

_____

Name _____ Date _____

COMPRESSOR CLUTCH AMPERAGE DRAW AND RESISTANCE TEST

Upon completion of this job sheet, you should be able to inspect air-conditioning compressor clutch amperes draw and resistance and compare it to a calculated value, as well as perform a voltage drop test across the coil assembly. You should also be able to diagnose unusual operating noises in the air-conditioning system and determine necessary action.

NATEF Correlation

A/C SYSTEM DIAGNOSIS AND REPAIR: *Diagnose unusual operating noises in the A/C system; determine necessary action.* **(P-2)**

OPERATING SYSTEMS AND RELATED CONTROLS DIAGNOSIS AND REPAIR: *Inspect and test A/C-heater blower, motors, resistors, switches, relays, wiring, and protection devices; perform necessary action.* **(P-1)**

Tools and Materials

Late-model vehicle

Service manual or information system

Safety glasses or goggles

Hand tools, as required

Digital multimeter (DMM)

T-pins

Describe the vehicle being worked on.

Year _____ Make _____ Model _____

VIN _____ Engine type and size _____

Procedure

Follow procedures outlined in the service manual. Give a brief description of your procedure, following each step. Ensure that the engine is cold and wear eye protection.

1. Visually check the coil for loose connections or cracked insulation. Were any faults detected?

_____

_____

2. Inspect the clutch plate and hub assembly. Check for signs of looseness between the plate and hub. Were any faults detected?

_____

_____

3. Check the rotor and bearing assembly.

4. Check the bearing for signs of excessive noise, binding, or looseness. Replace the bearing if necessary. Were any faults detected?

_____

_____

5. Locate the air-conditioning compressor clutch wiring diagram for the vehicle you are servicing, as well as the amperage draw and resistance specifications for the compressor clutch coil.

    a. Record resistant specifications.

    b. List amperage draw specifications.

6. Ensure the ignition switch is in the OFF position. Remove the A/C compressor clutch relay.

7. Using a digital multimeter (DMM) set to the DC ampere scale, connect the meter leads in series across the contact point circuit of the relay connector (these are the larger cavities). On an ISO relay connector, this would be terminal #30 and terminal #87.

8. Start the engine and select A/C. Record amperage (I) draw and compare it to specifications.

    a. Record Amperes (I) (Current)

    b. Is this recorded reading within specifications?

9. Turn the engine off and replace the relay.

10. Set the DMM to ohms and disconnect the connector from the compressor clutch.

11. Measure the resistance of the compressor clutch coil.

    a. Ω Resistance (R)

    b. Is this recorded reading within specifications?

12. Reconnect the A/C compressor clutch connectors. Set the DMM to read DC volts. Restart the vehicle and measure the battery voltage at the compressor clutch by back probing the connector. Shut the vehicle off.

    a. Voltage (E)

        1. Using Ohm's law, calculate the current, $I = E/R =$ _____ vs. _____ Actual (reading from 8a).

13. Were the actual and the calculated results the same? If not, why?_____

14. Using Ohm's law, calculate the resistance of the A/C compressor clutch, $E/I = R$.

    a. Using Ohm's law, calculate the current, $R = E/I =$ _____ vs._____ Actual (reading from 11a).

_____

15. Were the actual and the calculated results the same? If not, why?_____

_____

_____

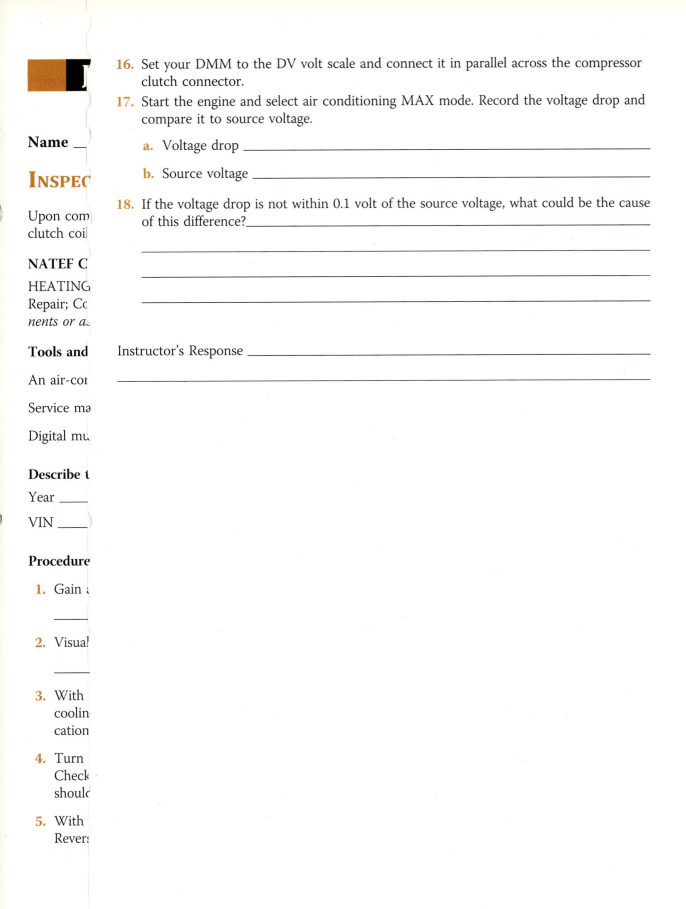

Name _____

16. Set your DMM to the DV volt scale and connect it in parallel across the compressor clutch connector.

17. Start the engine and select air conditioning MAX mode. Record the voltage drop and compare it to source voltage.

   a. Voltage drop _____

   b. Source voltage _____

18. If the voltage drop is not within 0.1 volt of the source voltage, what could be the cause of this difference?_____

_____

_____

_____

Instructor's Response _____

_____

INSPEC

Upon com
clutch coi

NATEF C

HEATING
Repair; Co
nents or a

Tools and

An air-cor

Service ma

Digital mu

Describe t

Year _____

VIN _____

Procedure

1. Gain

   _____

2. Visual

   _____

3. With
   coolin
   cation

4. Turn
   Check
   shoul

5. With
   Rever

6. Select the diode test setting on the DMM. First, connect the leads to the diode, placing the positive lead on one side and the negative lead on the opposite side of the diode; note the reading. Then reverse the leads and note the reading.

Forward bias reading _____

Reverse bias reading _____

NOTE: The DMM should read 0.7–0.5 volt in one direction (forward bias) and infinity or out of limits in the other direction (reverse bias) if the diode is functioning correctly.

7. Based on the inspection above and tests, what is your recommendation? _____

_____

_____

_____

Instructor's Response _____

_____

_____

JOB SHEET

Name _____ Date _____

REMOVING AND REPLACING A COMPRESSOR CLUTCH

Upon completion of this job sheet, you should be able to remove and replace a typical compressor clutch.

NATEF Correlation

REFRIGERATION SYSTEM COMPONENT DIAGNOSIS AND REPAIR: Compressor and Clutch; *Inspect A/C compressor drive belts; replace and adjust as needed.* **(P-2)**

REFRIGERATION SYSTEM COMPONENT DIAGNOSIS AND REPAIR: Compressor and Clutch; *Inspect, test, and replace A/C compressor clutch components or assembly.* **(P-2)**

Tools and Materials

Compressor with clutch

Service manual

Selected air conditioning tools

Clutch components, as required

Describe the vehicle being worked on.

Year _____ Make _____ Model _____

VIN _____ Engine type and size _____

Procedure

Following procedures outlined in the service manual, perform the following task and write a short summary of each step in the space provided:

1. Remove the clutch hub and plate assembly.
2. Visually inspect the hub and plate assembly. Note any problems.

   _____

   _____

3. Remove the pulley and bearing assembly.
4. Carefully inspect the bearing for signs of wear or roughness. Note any problems.

   _____

   _____

5. Make an electrical resistance check of the coil.
   Coil resistance should be: _____ ohms_____   Coil resistance is: _____ ohms

6. Replace parts, as needed. List additional parts required. _____

_____

7. Reassemble the clutch.

Instructor's Response _____

_____

_____

Name _____ **Date** _____

INSTALLATION OF AUXILIARY LIQUID LINE FILTER

Upon completion of this job sheet, you should be able to determine the need for an air-conditioning system filter and perform the necessary action. The focus of this job sheet is on the installation of an auxiliary filter to trap debris and contaminants that may remain in a system after a compressor or desiccant failure.

NATEF Correlation

HEATING AND AIR CONDITIONING: Refrigeration System Component Diagnosis and Repair; Evaporator, Condenser, and Related Components; *Determine need for A/C system filter; perform necessary action.* **(P-3)**

Tools and Materials

Late-model vehicle

Service manual or information system

Safety glasses or goggles

Hand tools, as required

Describe the vehicle being worked on.

Year _____ Make _____ Model _____

VIN _____ Engine type and size _____

Procedure

Follow procedures outlined in the service manual. Give a brief description of your procedure, following each step. Ensure that the engine is cold, and wear eye protection.

Contaminants can lodge in screens and valves, causing a restriction. Any time a major component, such as a compressor and/or receiver-drier/accumulator assembly, is replaced an auxiliary filter should be installed. The following procedure will outline the steps for installing a filter in the liquid line.

1. Recover the refrigerant from the system. Follow the procedures outlined in job sheet "Recover and Recycle Refrigerant," and replace any other component that needs service at this time, following the appropriate job sheet for each operation.
   a. List amount of refrigerant recovered.
   b. List amount of oil removed during recovery process.

2. Locate the liquid refrigerant line between the condenser outlet and the metering device. Find an accessible area with a straight section of line longer than the filter's assembly. Following the instructions accompanying the filter, determine the length of line that needs to be removed, and mark the tubing. How much line must be removed?

_____

_____

☐

3. Using a tubing cutter, remove the section of line and debur using the deburring tool.
4. Lubricate the O-rings and ferrule fitting that accompany the filter kit assembly and install according to the directions in the kit. Never reuse O-rings. Outline the instructions. _____

_____

_____

_____

_____

_____

5. Tighten the ferrule fittings to the specified torque. List torque specifications.

_____

☐

6. Evacuate and recharge the system.
7. Perform system performance check to verify proper system operation and recheck the system for leaks.
   a. Were any leaks detected?
   b. List high-side and low-side pressure readings.
   c. List duct discharge temperatures.

Instructor's Response _____

_____

_____

Chapter 9

CASE AND DUCT SYSTEMS

UPON COMPLETION AND REVIEW OF THIS CHAPTER, YOU SHOULD BE ABLE TO:

- Remove and replace the blower motors.

- Remove and replace the blower motor resistor or power module.

- Remove and replace the heater core.

- Remove and replace the evaporator.

- Perform odor control treatment of the case and duct system.

- Test the vacuum system.

- Perform temperature control door adjustment.

- Replace the cabin air filter.

- Troubleshoot, service, and adjust the operation of the in-vehicle mode circuits such as vent, HI-LO, MAX (cool/heat), and defrost.

A maze of ducts, vents, motors, wiring, and vacuum hoses makes up the typical automotive air-conditioning case and duct system in today's modern vehicle (Figure 9-1). The somewhat inaccessibility of most of its components adds to the mystique of this often neglected component of the air-conditioning system.

While it is true that there are literally hundreds of variations, troubleshooting and servicing are not difficult if one is familiar with the system.

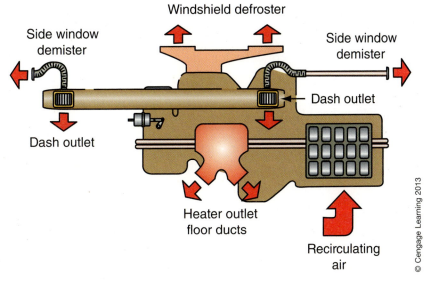

FIGURE 9-1 A typical case/duct system.

FIGURE 9-2 Clean debris from the fresh air inlet.

FRESH AIR INLET

Classroom Manual
Chapter 9, page 285

Most of the heater and air conditioning mode functions are performed with some outside air, except when MAX is selected. Though not generally noticeable, the quality of the air-conditioning system can be affected if the fresh air inlet screen is blocked with leaves or other **debris**. In time, if neglected, this debris can deteriorate and be pulled into the heater/evaporator case where it can cause serious airflow blockage through the evaporator and heater core. The fresh air inlet (Figure 9-2) is often concealed by the hood and is therefore overlooked during preventive maintenance. Cleaning this area should be a part of a periodic preventive maintenance schedule.

Debris is foreign matter, such as the remains of something broken or deteriorated.

COMPONENT REPLACEMENT

The greatest problem arises from the lack of data necessary to properly service any particular unit or system. It is necessary to have the manufacturer's service manuals for specific step-by-step procedures. For example, consider the replacement of a HI/LO **actuator** motor. The "big three" automakers differ for one year model, as follows.

HI/LO is also referred to as BI-LEVEL.

Chrysler

An **actuator** is a device that transfers a vacuum or electric signal to a mechanical motion. An actuator typically performs an on/off or open/close function.

1. Remove the left and right underpanel silencer ducts.
2. Remove the floor console.
3. Remove the center floor heat adaptor duct.
4. Remove the rear seat heat forward adaptor duct.
5. Loosen the center support bracket; pry rearward to gain access to the actuator.
6. Remove the actuator retaining screws.
7. Remove the actuator (Figure 9-3).
8. Remove the electrical connections from the actuator.
9. To reinstall, reverse the order of removal.

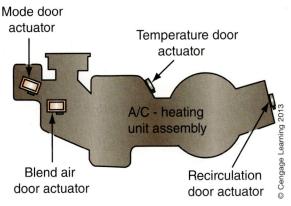

FIGURE 9-3 Typical Chrysler mode door actuator location.

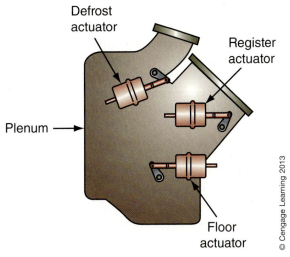

FIGURE 9-4 Typical Ford mode door actuator details.

Ford

1. Disconnect the vacuum hose.
2. Remove the retaining screws.
3. Remove the actuator from the linkage.
4. Remove the actuator (Figure 9-4).
5. For installation, reverse the removal procedures.

General Motors

1. Disable the air bag deployment system.
2. Remove the battery negative (–) cable and fuse.
3. Remove the instrument panel.
4. Remove the floor outlet assembly.
5. Remove the windshield defroster vacuum hoses.
6. Remove the windshield defroster air distribution assembly.
7. Remove the vacuum hose from the upper/lower mode valve actuator.
8. Remove the retaining nuts or screws.
9. Remove the actuator (Figure 9-5).
10. To install the new actuator, reverse the preceding procedure.

General Motors calls its air bag deployment system a "Supplemental Inflatable Restraint (SIR)."

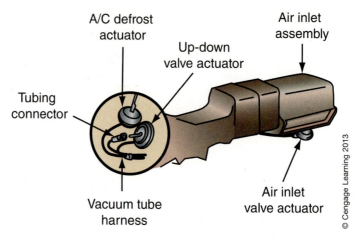

FIGURE 9-5 Typical General Motors mode door actuator details.

A/C defrost actuator

Up-down valve actuator

Air inlet assembly

Tubing connector

Vacuum tube harness

Air inlet valve actuator

© Cengage Learning 2013

Classroom Manual

Chapter 9, page 304

This comparison is not to suggest that Ford's procedure is the simplest or that General Motors' procedure is the most difficult. The procedures vary considerably for all year/model applications. The example, which was randomly selected, is intended to provide a general comparison of what may be expected in the day-to-day service of air-conditioning systems and to express the importance of having an appropriate service manual at hand.

BLOWER MOTOR

Blower motor replacement is generally a little more straightforward than some of the other case/duct components (Figure 9-6). See Photo Sequence 13. This procedure, however, should be considered typical for any type vehicle.

For reassembly or the installation of a new blower motor assembly, reverse P14-9 through P14-1. If the gasket, P14-7, was damaged or destroyed, replace it with a new gasket or seal the mating surfaces with a suitable caulking material.

Silicone rubber, available in tube form, is ideal for sealing mating surfaces.

REPLACING THE POWER MODULE OR RESISTOR

 WARNING: This component may be very hot. Take care before touching it with bare hands.

SPECIAL TOOLS

Coolant recovery system, if applicable

Hose clamp pliers, if applicable

Classroom Manual

Chapter 10, page 327

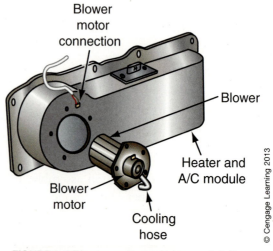

Blower motor connection

Blower

Heater and A/C module

Blower motor

Cooling hose

© Cengage Learning 2013

FIGURE 9-6 Blower motor and plenum details.

TYPICAL PROCEDURE FOR REMOVING A BLOWER MOTOR

P13-1 Disconnect the battery ground (−) cable.

P13-2 Disconnect or remove any wiring, brackets, or braces that hamper blower motor service.

P13-3 Disconnect the blower motor lead(s).

P13-4 Disconnect the blower motor ground (−) wire.

P13-5 Disconnect the blower motor cooling tube (if applicable).

P13-6 Remove the attaching screws.

P13-7 Remove the blower and motor assembly. The sealing gasket often acts as an adhesive. If this is the case, carefully pry the blower flange away from the case.

P13-8 Remove the shaft nut or clip, if applicable.

P13-9 Remove the blower wheel. Do not lose the space; use it on the replacement motor. Reverse the sequence to install the new motor.

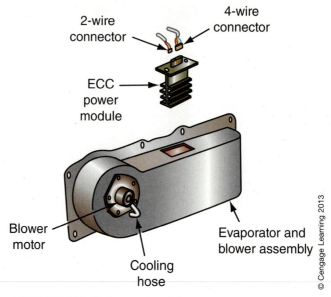

FIGURE 9-7 Typical power module.

1. Remove the brace(s) or cover(s) that may restrict access to the **power module** or resistor (Figure 9-7).
2. Remove the electrical connector(s).
3. Remove the retainer(s), if equipped.
4. Remove the retaining screws or nuts.
5. Remove the power module or resistor.
6. For replacement, reverse the procedure.

REPLACING THE HEATER CORE

Access to the heater core is gained by following directions given in specific service manuals. This procedure is typical and assumes the procedure for access to the heater core is available.

1. Drain the cooling system into a clean container. The coolant may be reused, reclaimed, or discarded in a manner consistent with Environmental Protection Agency (EPA) guidelines.
2. Disconnect the battery ground (−) cable.
3. Disconnect the heater hoses at the bulkhead. This is a good opportunity to inspect the heater hoses and replace any that show signs of deterioration.
4. Gain access to the heater core as outlined in the appropriate service manual.
5. Remove the retaining screws, brackets, or straps.
6. Remove the core from the case (Figure 9-8).

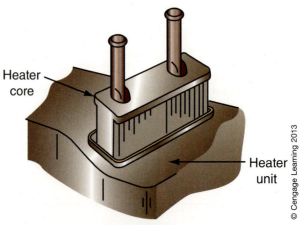

FIGURE 9-8 Removing the heater core from the case.

REPLACING THE EVAPORATOR CORE

To illustrate the importance of proper service manuals for this service, it may be noted that the 1994 Ford Taurus/Mercury Sable manual instructs the technician, "Using a **hot knife**, cut the top of the air-conditioning evaporator housing between the raised outline." An illustration is included in the manual to show the area to be cut (Figure 9-9). Some service manuals simply say to remove the blower motor and assembly screws before separating the case halves and removing the evaporator (Figure 9-10).

The following procedure assumes that access to the evaporator has been determined.

1. Recover the refrigerant.
2. Drain the radiator if the heater hose(s) has to be removed to gain access.
3. Remove the heater hose(s), if necessary.
4. Remove any wiring harness, heat shields, brackets, covers, and braces that may restrict access to the evaporator core.
5. Remove the liquid line at the thermostatic expansion valve (TXV) or fixed orifice tube (FOT).
6. Remove the suction line at the evaporator or accumulator outlet.
7. Gain access to the evaporator core as outlined in the service manual.
8. Lift the evaporator from the vehicle.
9. Drain the oil from the evaporator into a calibrated cup.
10. For replacement, reverse the preceding procedure. First, replace the oil with the same amount and type as drained in step 9.

A **hot knife** is a tool that has a heated blade. It is used for separating objects, for example, evaporator cases.

Drain coolant into a clean container so it may be reused.

Replace any heater hoses found to be brittle or damaged.

Discard all O-rings. They should be replaced with new O-rings on reassembly.

Dispose of used oil in accordance with local ordinances.

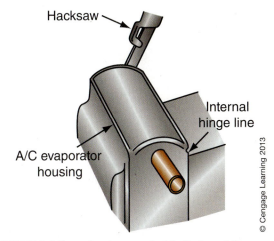

FIGURE 9-9 Use a hacksaw or hot knife to cut the case.

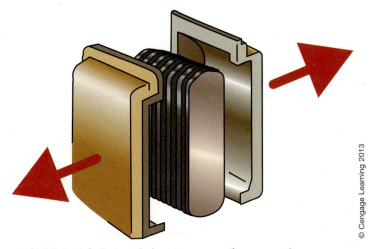

FIGURE 9-10 Split case halves to remove the evaporator.

SPECIAL TOOLS

Flare nut wrench set
Torque wrench

REMOVING AND REPLACING THE EVAPORATOR

This procedure is given in four parts:

1. Aftermarket
2. Factory or Dealer Installed (Domestic)
3. Factory or Dealer Installed (Import)
4. Rear Heating/Cooling Unit

Prior to servicing the evaporator, the air-conditioning system refrigerant must be removed by the recovery process. This procedure should be performed prior to step 1 outlined in either part. On completion of this service, the air-conditioning system should be properly evacuated and charged with the appropriate refrigerant as outlined in Chapters 5 and 6 of this manual.

Part 1. Aftermarket

1. Remove the liquid line from the metering device.
2. Remove the suction line from the evaporator.
3. Remove and discard O-rings, if equipped.
4. Remove the electrical lead wire(s). Also, disconnect the ground wire.
5. Remove the mounting hardware; remove the evaporator from the vehicle.
6. Check to see if there is a measurable amount of lubricant in the evaporator.

   NOTE: Add an equivalent amount of proper, clean, and fresh lubricant to the replacement evaporator.

7. To install the evaporator, reverse steps 1, 2, 4, and 5. Install new O-rings, if applicable.

Part 2. Factory or Dealer Installed (Domestic) (Figure 9-11)

1. Remove the liquid line from the metering device inlet.

   NOTE: If an H-valve metering device, skip to step 5.

2. Remove and discard the O-ring, if equipped.
3. Remove the suction line from the evaporator.

H-valve systems are not equipped with an accumulator, but instead have a receiver-drier.

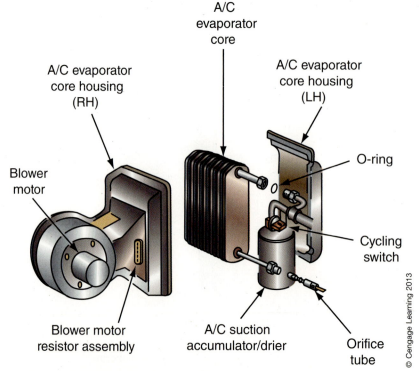

FIGURE 9-11 Exploded view of an evaporator assembly.

© Cengage Learning 2013

4. Remove and discard the O-ring, if equipped.

   NOTE: If not an H-valve system, skip steps 5 and 6 and proceed with step 7.

5. Remove the suction/liquid lines from the H-valve.
6. Remove and discard the gasket or O-rings, as applicable.
7. Remove the accumulator, if equipped, following procedures as outlined in this chapter.
8. Remove any mechanical linkages or vacuum lines from the evaporator controls.
9. Remove mounting bolts and hardware, as applicable, from the evaporator housing.
10. Separate the housing to gain access to the evaporator core.
11. Carefully lift the evaporator assembly from the vehicle.

    NOTE: Do not force the assembly.

12. If there is a measurable amount of lubricant in the evaporator, add an equivalent amount of proper, clean, and fresh lubricant to the replacement evaporator.
13. Install a replacement evaporator by reversing steps 1, 3, and 7–11 or 5 and 7–11, as applicable.

> Use new gaskets and O-rings when required.

Part 3. Factory or Dealer Installed (Import) (Figure 9-12)

1. Remove the liquid line and suction line from the evaporator.
2. Immediately cap the fittings (evaporator and hose) to keep moisture and debris out of the system.

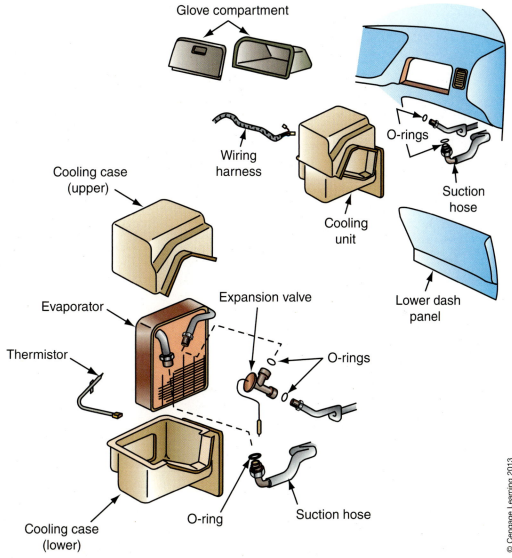

FIGURE 9-12 Typical import cooling unit.

© Cengage Learning 2013

3. Remove and discard O-rings or gaskets, as applicable.
4. Remove any obstructions, such as dash panels or the glove box to gain access to the cooling unit.
5. Remove mechanical linkages, electrical wires, or vacuum lines from the cooling unit.
6. Remove mounting bolts and hardware, as applicable, from the cooling unit.
7. Lift the cooling unit assembly from the vehicle. There should be no need to force the assembly.
8. Separate the housing to gain access to the evaporator core.
9. Install a replacement evaporator core by reversing steps 1 and 4–8.

   NOTE: Use new O-rings or gaskets to replace those removed in step 3.

10. If there was a measurable amount of lubricant in the evaporator, add an equivalent amount of proper, clean, and fresh lubricant to the replacement evaporator.

Part 4. Rear Heating/Cooling Unit (Figure 9-13)

There are many variations on procedures to service the rear heating/cooling unit. The manufacturer's service manual should be followed for any particular procedure. The following procedure is given as typical only.

1. Drain the coolant from the radiator. It may not be necessary to drain all of the coolant, however.

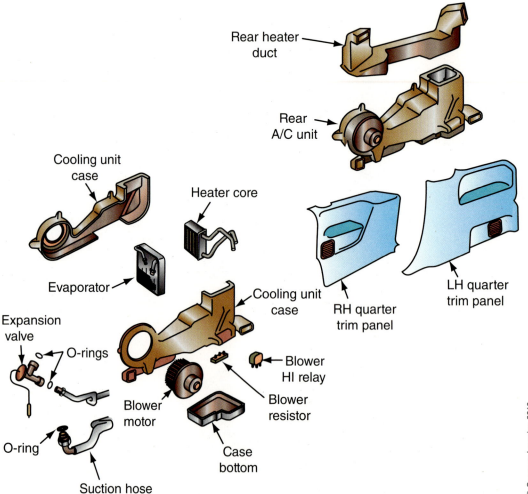

FIGURE 9-13 Typical rear heating/cooling unit.

2. Remove the liquid line and suction line from the evaporator.

3. Immediately cap the fittings (evaporator and hose) to keep moisture and debris out of the system.

4. Remove and discard O-rings or gaskets, as applicable.

5. Remove the coolant hoses from the heater core.

6. Remove the heater grommet, if applicable.

7. Remove any obstructions, such as seats, panels, trim, or controls, to gain access to the heating/cooling unit.

8. Remove any mechanical linkages, electrical wires, or vacuum lines from the heating/cooling unit.

9. Remove mounting bolts and hardware, as applicable, from the unit.

10. Lift the heating/cooling assembly from the vehicle. There should be no need to use force.

11. Separate the case to gain access to the evaporator and heater cores as well as to the metering device.

   NOTE: If when replacing the evaporator there is a measurable amount of lubricant, add an equivalent amount of proper, clean, and fresh lubricant to the replacement evaporator.

12. Install a replacement component by reversing steps 1, 2, and 5–11.

   NOTE: Use new O-rings or gaskets to replace those removed in step 4.

ODOR PROBLEMS

An odor emitting from the air-conditioning system ducts may be caused by a leaking heater core or hose inside the heater/evaporator case. An odor may also be caused by refrigeration oil leaking into the heater/evaporator case due to a leaking evaporator. The remedy is to repair or replace the leaking parts.

A musty odor is usually due to water leaks, a clogged evaporator drain tube, or mold and mildew on the evaporator core. Mold and mildew, which are fungi, are most common in air-conditioning systems in vehicles operated in hot and humid climates. The odor, generally noted during startup, may be caused by debris in the heater/evaporator case or by microbial fungi growth on the evaporator core.

Odor Control Treatment

When a musty odor develops in the HVAC system, the only remedy is to eradicate the odor causing microscopic mold, bacteria, and mildew. The most common practice is to apply a liquid antimicrobial deodorizer and disinfectant product that is commercially available from both parts stores and vehicle manufacturers.

For best results, follow the directions that come with the product. The following is a typical process for eliminating HVAC odor and applying the product to specific locations in the air duct system. This procedure could vary from one vehicle to another due to system control features. Always consult the manufacturer's recommendations.

 WARNING: **Always read and follow all instructions and warnings that come with disinfectant products, and only use these products in a well-ventilated area with vehicle doors and windows open and wearing eye protection. Do not use these products around flames or other sources of ignition.**

1. Connect a siphon-type air-conditioning disinfectant sprayer to a 12 oz. (354.88 mL) bottle of disinfectant solution.

2. Remove the cabin air filter (if equipped) and reinstall the cover.

Classroom Manual
Chapter 9, page 299

SPECIAL TOOLS

Siphon-type air-conditioning disinfectant sprayer

Liquid antimicrobial disinfectant 12 fluid ounces (354.88 mL)

3. Spray approximately 4 oz. (118.29 mL) of the solution into the fresh air inlet with the blower on high speed, the mode door in the fresh air position, and the temperature control set to cold.
4. Next, locate the recirculated air inlet on the passenger side of the interior compartment and set the HVAC control panel to recirculation mode. Spray approximately 2 oz. (59.15 mL) of the solution into the air recirculation intake with the blower on high speed.

   NOTE: In some instances, it may be necessary to remove the blower motor resistor in order to direct the disinfectant sprayer toward the evaporator core.

5. Next, repeat step 3, but this time place the temperature control on full hot position.
6. Finally, turn the system and blower off and spray the remainder of the product into each of the air outlets in the system (defroster, floor vent, panel discharge, and side vents).
7. Reinstall the cabin air filter if one was removed in step 2.
8. Allow the vehicle to sit for at least 30 minutes with the windows open. Operate the system before returning it to the customer.

 WARNING: This procedure should only be performed on a cold vehicle to prevent the disinfectant from coming in contact with hot engine components. Take extreme care not to get disinfectant in eyes, on hands, or on clothing. Wash thoroughly with soap and water immediately after handling.

 WARNING: If the disinfectant gets into the eyes, hold the eyelids open and flush with a steady, gentle stream of water for 15 minutes. Immediately seek professional medical attention.

Delayed Blower Control

An aftermarket delayed blower control may be installed in many systems to reduce the probability of a recurrence of odors caused by mold and mildew. It is installed following the instructions included with the package or as given in a manufacturer's **technical service bulletin (TSB)**. The delayed blower control is used to dry out the evaporator and air distribution system. After the air-conditioning compressor has been in operation for 4 or 5 minutes, the control will cause the blower motor to run after the ignition switch is placed in the OFF position. The delayed blower control will operate the blower motor at high speed for 5 minutes to clear the evaporator core of accumulated condensate, thereby reducing the recurrence of odors caused by mold and mildew.

TESTING THE VACUUM SYSTEM

The first step in troubleshooting the vacuum system is to ensure that manifold vacuum is available at the selector switch. Vacuum diagrams are generally provided in the manufacturer's service manual to help identify color coding and connections (Figure 9-14). The following is a typical procedure to quickly check the vacuum control system for improper or erratic direction of airflow from the outlets.

1. Are other vacuum-operated devices operational?
2. Do other vacuum motors operate properly?
   a. If yes, there is a vacuum source.
   b. If no, proceed with step 3.

CAUTION:
Do not permit the coils of the blower resistor to become grounded to any metal surface.

Technical service bulletins (TSB) contain updated information provided by the vehicle manufacturer regarding vehicle problems and offering solutions to problems encountered in their vehicles.

Classroom Manual
Chapter 9, page 304

In some vehicles, the vacuum connection will be at the base of the carburetor.

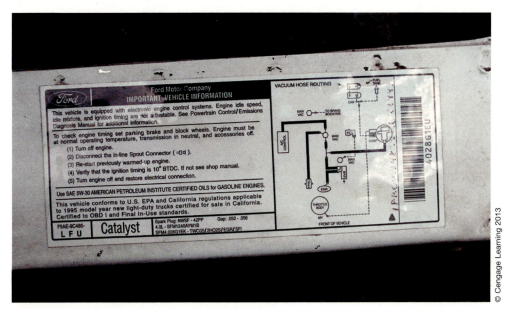

FIGURE 9-14 An under hood vacuum system decal.

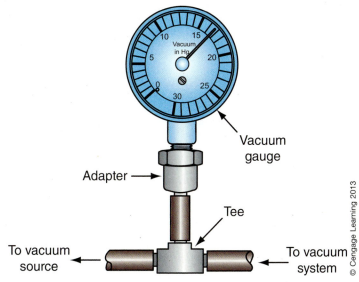

FIGURE 9-15 Check the vacuum source.

3. Disconnect the hose at the manifold inlet.
4. Connect a vacuum gauge, a short hose, and a vacuum tee in line with the vacuum source and system (Figure 9-15).
5. Is there now a vacuum signal?
   a. If yes, check for a defective check valve or hose(s).
   b. If no, check for blockage or restriction at the manifold fitting and correct it as required.

Vacuum Switch

The vacuum control provides a vacuum passage for selected circuits in the control system. To test for a defective vacuum control:

1. Disconnect the hose from the inoperative vacuum motor at the switch.
2. Connect a vacuum gauge to the vacant port.

SPECIAL TOOLS

Vacuum pump
Vacuum gauge
Vacuum hose

FIGURE 9-16 **A typical vacuum check valve.**

3. Move the switch through all of its positions.
4. Is there a vacuum at the port in either position?
   a. If yes, the switch is probably all right.
   b. If no, the switch may be defective; proceed with step 5.
5. Is there a vacuum signal at any of the other ports?
   a. If yes in step 5 but no in step 4, the switch is defective and should be replaced.
   b. If no, the problem may be a defective restrictor, check valve, hose, or reserve tank.

Check Valve

A check valve (Figure 9-16) allows flow in one direction and blocks (checks) the flow in the other direction. See Photo Sequence 14 for testing a check valve.

Reserve Tank

Using the same setup as for testing the check valve, insert the hose onto the vacuum reserve tank instead of the check valve.

1. Start the vacuum pump.
2. Observe the vacuum gauge.
3. Turn off the vacuum pump. If there is a vacuum and it holds for 5 minutes, the tank may be considered all right. If there is little or no vacuum or it does not hold for 5 minutes, the tank is defective.

Leaks in vacuum tanks may usually be repaired by using a fiberglass-reinforced resin.

Hose

A vacuum hose is often made of synthetic rubber or nylon. Deterioration, cracking, and splitting are problems found with vacuum hoses. The best way to determine the condition of vacuum hoses is by visual inspection. If a hose shows signs of deterioration, it should be replaced.

Restrictor

A restrictor is generally a porous bronze filter whose purpose is to prevent minute particles of dust and debris from entering the vacuum system where they could restrict control circuits or cause component damage. The restrictor also slows the mode door operation to reduce the noise from mode door operation. It is not practical to clean a restrictor. If in doubt, the simplest remedy is to replace it.

Check valves are used to prevent a loss of vacuum during acceleration and after engine shutdown.

Vacuum reserve tanks may be made of plastic or metal.

SPECIAL TOOL

Heat gun

Classroom Manual
Chapter 10, page 326

TYPICAL PROCEDURE FOR TESTING A CHECK VALVE

P14-1 Remove the check valve from the vehicle.

P14-2 Attach a vacuum source, such as a vacuum pump. The direction of flow should be away from the pump.

P14-3 Turn on the pump and observe the gauge. If there is a vacuum, the check valve is good. Proceed with step 4. If there is no vacuum, the check valve is defective and must be replaced.

P14-4 Turn off the pump.

P14-5 Disconnect the check valve.

P14-6 Reverse and reconnect the check valve.

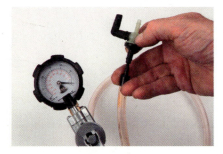

P14-7 Turn on the vacuum pump and observe the gauge. If there is a vacuum, the check valve is defective and must be replaced. If there is now no vacuum but the pump held a vacuum for step 3, the check valve is good and may be returned to the vehicle.

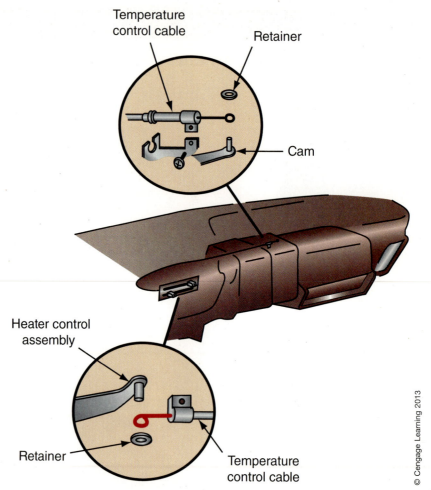

FIGURE 9-17 Temperature door adjustment details.

© Cengage Learning 2013

TEMPERATURE DOOR CABLE ADJUSTMENT

Remove access panels or components to gain access to the **temperature door** (Figure 9-17) and proceed as follows:

1. Loosen the cable and attach the fastener at the heater case assembly.
2. Make sure that the cable is properly installed and routed to ensure no binding and freedom of movement.
3. Place the temperature control lever in the full cold position and hold it in place.
4. Tighten the cable fastener that was loosened in step 1.
5. Move the temperature control lever from full cold to full hot to full cold positions.
6. Repeat step 5 several times and check for freedom of movement.
7. Recheck the position of the door. If it is loose or out of position, repeat steps 2 through 7. If it is still in position and secure, replace the access panels and covers.

MODE SELECTOR SWITCH

Classroom Manual

Chapter 9, page 287

The mode selector switch provides an electrical or vacuum signal to the mode doors or control module assembly. The mode selector switch determines mode door position and air discharge location selected by drive (i.e., dash vent, defrost). Photo Sequence 15 illustrates a typical procedure for removal and replacement of the mode selector switch.

REMOVING AND REPLACING THE MODE SELECTOR SWITCH

The following is a typical procedure for removing a mode selector switch. The switch is replaced by reversing the procedure given. For specific procedures, consult the appropriate manufacturer's service manual.

P15-1 Disconnect the battery ground-cable.

P15-2 Remove the instrument panel finish appliqué.

P15-3 Remove the screws holding the control assembly to the instrument panel.

P15-4 Sufficiently pull the control assembly from the instrument panel to gain access to the rear electrical connector, control cable, and mode switch.

P15-5 Depress the latches of the electrical connector to disengage the connector from the control assembly.

P15-6 Disconnect the temperature control connector from the control assembly.

P15-7 Remove the knob from the mode selector switch.

P15-8 Remove the mode selector switch attaching screw(s) and remove the switch.

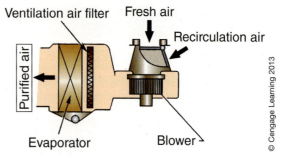

FIGURE 9-18 Some systems today have a cabin air filter.

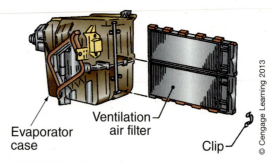

FIGURE 9-19 Remove the air filter from the evaporator case.

Classroom Manual

Chapter 9, page 300

CABIN AIR FILTER

The procedure for cleaning or replacing the filter varies from vehicle to vehicle, so it is important to follow manufacturer's recommended procedures. If the vehicle is equipped with an air filter (Figure 9-18), instructions may be found in the owner's manual or on a label inside the glove box. The following is a typical procedure:

1. Remove the dash undercover.
2. Remove the glove box.
3. Remove instrument reinforcement from the instrument panel.
4. Remove the filter retaining clip.
5. Remove the air filter from the case/duct (Figure 9-19).
6. Clean and install a new filter.
7. Replace components in reverse order used to remove them.

CUSTOMER CARE: Both the vehicle owner and the service technician often neglect the cabin air filter. As the technician, you need to educate the consumer on the advantages of frequent service of the cabin air filter. A properly maintained system will result in improved airflow and air quality for the passenger compartment.

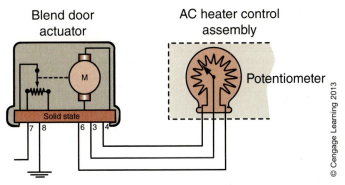

Blend door
actuator

AC heater control
assembly

Potentiometer

Solid state

© Cengage Learning 2013

FIGURE 9-20 Typical electronic mode door actuator (motor) wiring diagram for a manual climate control system.

ELECTRIC MODE DOOR ACTUATOR

Many automatic and manual climate control air-conditioning systems use electronic mode door actuators to control mode door operation. On an automatic climate control system the actuator motors are sent commands from the climate control module or body computer. On manual air-conditioning systems a control knob operates a potentiometer (variable resistor) which varies a voltage signal to the mode door motor (Figure 9-20). This is how the temperature knob (hot/cold) activates the electronic blend door actuator on a manual system. A control module is not used and the control voltage is supplied directly from the A/C-Heater control head assembly.

The following is a typical testing procedure (refer to Figure 9-20), it is not meant to replace manufacturer service information:

1. Disconnect mode door actuator connector.
2. Turn ignition on.
3. Test for battery voltage on terminal 7.
4. Test for battery ground on terminal 8.
5. Test for continuity between terminals 3 and 4 as the potentiometer (temperature control) knob is turned. The resistance should change from low to high as the knob is turned in one direction and from high to low in the opposite direction.
6. Test for continuity between terminals 3 and 6 as the potentiometer (temperature control) knob is turned. The resistance should change from low to high as the knob is turned in one direction and from high to low in the opposite direction.
7. Test for battery voltage between terminals 7 and 8.
8. Reconnect mode door actuator connector.
9. Test for near battery voltage on terminal 4.
10. Test for voltage on terminal 3 as the potentiometer (temperature control) knob is turned.

Diagnosis:

- If voltage or ground is missing in steps 3 and 4, check fuse and wires for shorts and opens.
- If continuity was not present in step 5 or 6, check wiring; if ok, replace potentiometer. If signal dropped out, replace potentiometer.

If all the above readings are within range remove mode door actuator and check the door for mechanical binding or looseness or for a broken shaft or door. If door movement checks out ok, then replace the mode door actuator.

TABLE 9-1 TYPICAL CASE/DUCT SYSTEM PROBLEMS

Vacuum System	Motor System
No vacuum-to-air conditioner master control	No power to air-conditioner mode selector
Air conditioner control leaks vacuum	High resistance connection in air conditioner control
Damaged, kinked, or pinched vacuum hose	Broken, loose, or disconnected electrical wiring
Damaged or leaking vacuum motor	Damaged or defective actuator motor
Actuator arm disconnected at door crank	Actuator linkage disconnected at door crank
Damaged or leaking vacuum reserve tank	Defective fuse, circuit breaker, or fusible link
Damaged or leaking check valve	Defective diode or component in programmer

PROBLEMS ENCOUNTERED

Block or ladder diagrams, covering several pages, are often used in manufacturers' service manuals to troubleshoot problems with automotive air-conditioning duct systems. It is therefore recommended that the appropriate service manuals be consulted for specific troubleshooting procedures. Table 9-1 lists the problems that typically are encountered in the case/duct system.

Always follow a manufacturer's recommended procedures and heed its cautions when troubleshooting any control system. The unintentional grounding of some circuits can cause immediate and permanent damage to delicate electronic components. The use of a testlight, powered or non-powered, is not recommended for underdash service. The battery in a powered testlight or the added resistance on a non-powered test lamp may be sufficient to cause failure to the delicate balance of solid-state electronic circuits.

These circuits are susceptible to damage by electrostatic discharge (ED) merely by touching them. Electrostatic discharge is a result of static electricity, which "charge" a person simply by their sliding across a seat, for example. To provide an extra margin of safety, the technician should wear a grounding bracelet (Figure 9-21), an electrical conducting device that surrounds the wrist and attaches to a known ground source. This device ensures that the body will not store damaging static electricity by providing a path to ground for it to be discharged.

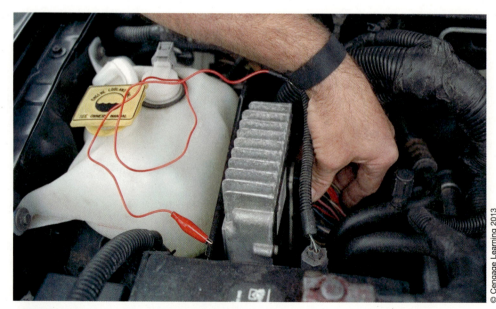

© Cengage Learning 2013

FIGURE 9-21 Wear a grounding bracelet when working around sensitive electronic components.

Passenger Compartment Heating Performance (Trouble Tree)

The following is an example of a typical diagnostic trouble tree for heating system performance. Always refer to specific manufacturer diagnostic information when troubleshooting system malfunctions.

Step #	Diagnostic Step	YES	NO
1	Start the engine and allow it to idle for 10 minutes. Does the engine reach operating temperature?	Go to step 2.	Go to step 8.
2	a. Select the FLOOR discharge mode. b. Set the blower to the lowest speed setting. c. Set the temperature control to maximum heat setting. d. Feel the heater core inlet and outlet hose temperature. e. Does the inlet hose feel hotter than the outlet hose?	Go to step 6.	Go to step 3.
3	a. Install a thermometer in the center panel vent. b. Connect a clamp-on thermocouple or other contact thermometer to the heater core outlet hose. c. Select the PANEL discharge mode. d. Select the highest blower speed setting. e. Set the temperature control to maximum heat setting. f. Record the temperature at: 　Center panel vent _____ 　Heater core outlet hose _____ g. Compare the recorded temperatures. h. Are the two temperature readings about equal?	Go to step 4.	Go to step 5.
4	Inspect and repair the following areas for cold air leaking into the passenger compartment or duct system: ☐ Cowl ☐ Recirculation door ☐ HVAC case assembly Were repairs completed?	Go to step 9.	—
5	a. Inspect the temperature control door operation. b. Perform necessary repairs. Were repairs completed?	Go to step 9.	—
6	a. Turn the engine off. b. Backflush the heater core. c. Start the engine. d. Select the FLOOR discharge mode. e. Set the blower to lowest speed setting. f. Set the temperature control to maximum heat setting. g. Feel the heater core inlet and outlet hose temperature. h. Does the inlet hose feel hotter than the outlet hose?	Go to step 7.	Go to step 9.
7	Replace the heater core.	Go to step 9.	—
8	Repair the low engine temperature condition. Refer to the diagnostic section if the engine fails to reach operating temperature.	Go to step 9.	—
9	Operate the system to verify the repair. Was the original system complaint corrected and is the system functioning as designed?	System is operating as designed.	Go to step 1.

CASE STUDY

A customer brings a late-model vehicle into the shop with the complaint that air does not come out of the dash outlets regardless of the mode selected. Before attempting to check the under-dash air distribution system, it is noted that the vehicle is equipped with an air bag system. The service manual cautions that the air bag system should be disarmed before performing any underdash service. Following service manual proce-dures, the technician disarms the air bag. In this case, the technician disconnects and tapes the neg-ative (–) battery terminal, removes the fuses, dis-connects the wiring harness and, finally, removes the air bag module from the vehicle. By taking time to heed the service manual warnings, possible air bag deployment and injury are avoided.

ASE-STYLE REVIEW QUESTIONS

1. *Technician A* says that a slight amount of conditioned air is made available at the defroster duct outlet at all times to prevent windshield fogging.
   Technician B says that a positive in-vehicle pressure is maintained at all times to prevent exhaust gas infil-tration. Who is correct?
   A. A only
   B. B only
   C. Both A and B
   D. Neither A nor B

2. *Technician A* says that a vacuum reserve tank helps maintain a vacuum in the system at all times only when the engine is running.
   Technician B says that the check valve prevents vac-uum loss when the engine is stopped. Who is correct?
   A. A only
   B. B only
   C. Both A and B
   D. Neither A nor B

3. *Technician A* says that a check valve prevents vacuum flow in either direction.
   Technician B says that a check valve permits vacuum flow in either direction. Who is correct?
   A. A only
   B. B only
   C. Both A and B
   D. Neither A nor B

4. *Technician A* says cleaning the fresh air inlet should be part of a periodic preventative maintenance schedule.
   Technician B says debris in the fresh air intake can contaminate the refrigerant.
   Who is correct?
   A. Technician A only
   B. Technician B only
   C. Both A and B
   D. Neither A nor B

5. An odor from the evaporator may be caused by all of the following, *except:*
   A. Mold
   B. Mildew
   C. Oil
   D. All of the above

6. An odor emitting from the case and duct system may be caused by all of the following, *except:*
   A. A leaking heater core
   B. Refrigerant oil leak
   C. Mold and mildew
   D. Refrigerant gas leak

7. It is necessary to drain the engine coolant before replacing which component?
   A. The vacuum diaphragm
   B. The blower motor
   C. The heater core
   D. All of the above

8. *Technician A* says mode doors may be vacuum operated.
   Technician B says mode doors may be electrically operated.
   Who is correct?
   A. Technician A only
   B. Technician B only
   C. Both A and B
   D. Neither A nor B

9. *Technician A* says when a musty odor develops the odor causing mold and mildew must be eradicated.
   Technician B says that a liquid antimicrobial deodor-izer and disinfectant is used to eliminate odors in duct systems.
   Who is correct?
   A. Technician A only
   B. Technician B only
   C. Both A and B
   D. Neither A nor B

10. *Technician A* says a block or ladder diagrams are used to aid in troubleshooting air-conditioning duct system faults.
    Technician B says the use of test lights is not recom-mended for under dash service.
    Who is correct?
    A. Technician A only
    B. Technician B only
    C. Both A and B
    D. Neither A nor B

1. All of the following statements about a typical air-conditioning system set to MAX cooling are true, *except:*
   A. The heater coolant flow control valve is open.
   B. The compressor clutch coil is energized.
   C. The blower motor is running.
   D. The outside/recirculate door is positioned to recirculate.

2. When should the cabin air filter be replaced if the vehicle is operated in dusty or high pollution conditions?
   A. Every 3 months or 3000 miles
   B. Once a year or every 12,000 miles
   C. Every 15,000 miles
   D. Every 30,000 miles

3. The *least* likely cause of an inoperative vacuum motor is:
   A. A split hose
   B. A defective check valve
   C. A defective vacuum switch
   D. A kinked hose

4. A customer complains of the passenger compartment temperature always being too hot, cold, or not changing as the temperature range is changed on the control panel. On an electronic mode door actuator (motor) system all of the following could cause this problem, *except:*
   A. Faulty actuator
   B. Faulty potentiometer
   C. Open or shorted wire
   D. Mode door moving too freely

5. During normal comfort control operation with the windows closed, harmful gases are not allowed to enter the vehicle because:
   A. The vehicle is airtight when the windows are closed
   B. They are removed by natural convection in the ambient airstream
   C. They are carried away by the force of the ram air
   D. Of a slight in-vehicle positive pressure

Name _____   **Date** _____

CASE/DUCT SYSTEM DIAGNOSIS

Upon completion of this job sheet, you should be able to make basic checks of the vacuum system of an air-conditioning case/duct system.

NATEF Correlation

HEATING AND AIR CONDITIONING: Operating Systems and Related Controls Diagnosis and Repair; *Diagnose malfunctions in the vacuum and mechanical components and controls of the heating, ventilation, and A/C (HVAC) system; determine necessary action.* **(P-2)**

HEATING AND AIR CONDITIONING: Operating Systems and Related Controls Diagnosis and Repair; *Inspect and test A/C-heater ducts, doors, hoses, cabin air filters, and outlets; perform necessary action.* **(P-3)**

Tools and Materials

Vehicle with manually controlled, factory-installed air-conditioning system

Service manual

Chapter 8 of Classroom Manual

Selected air-conditioning system tools

Vacuum pump

Describe the vehicle being worked on.

Year _____ Make _____ Model _____

VIN _____ Engine type and size _____

Procedure

Disconnect the vacuum source hose and connect a vacuum pump to the vacuum reserve tank. It will not be necessary to run the engine for a vacuum source. Gain access to the vacuum motors of the mode doors and determine the vacuum signal applied to each for the following air delivery conditions.

Use the abbreviations:
  fv for full vacuum
  pv for partial vacuum
  nv for no vacuum

Air Delivery Condition	Defroster	Actuator Motor (Pot)		
		A/C	Bi-level	Air Inlet
1. MAX	_____	_____	_____	_____
2. NORM	_____	_____	_____	_____
3. Bi-level (B/L)	_____	_____	_____	_____
4. VENT	_____	_____	_____	_____
5. Heat (HTR)	_____	_____	_____	_____
6. BLEND	_____	_____	_____	_____
7. Defog	_____	_____	_____	_____
8. OFF	_____	_____	_____	_____

Instructor's Response _____

_____

_____

Name _____ **Date** _____

AIR DELIVERY SELECTION

Upon completion of this job sheet, you should be able to trace the air delivery in each of the six basic air delivery modes.

NATEF Correlation

HEATING AND AIR CONDITIONING: Operating Systems and Related Controls Diagnosis and Repair; *Check operation of automatic and semi-automatic heating, ventilation, and air-conditioning (HVAC) control systems; determine necessary action.* **(P-3)**

Tools and Materials

Service manual

Blue and red pencil

Procedure

In each of the diagrams below, show the position of the mode doors using a red pencil. Show the airflow using a blue pencil.

1. A typical dual-zone duct system with passenger-side full hot selected and driver-side full cold selected. Both driver and passenger selected panel air.

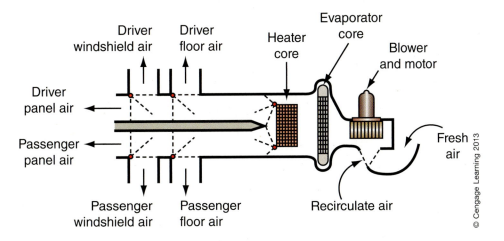

2. A typical dual-zone duct system with passenger-side full cold selected and driver-side warm selected. Passenger selected panel air and driver selected floor air.

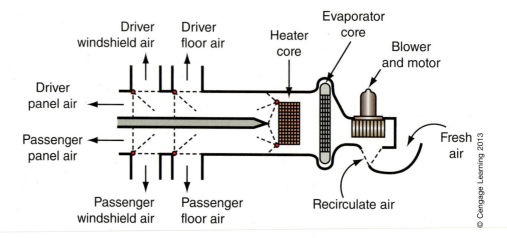

3. Airflow when DEFROST is selected from both passenger and driver outlets and full heat is selected by both driver and passenger.

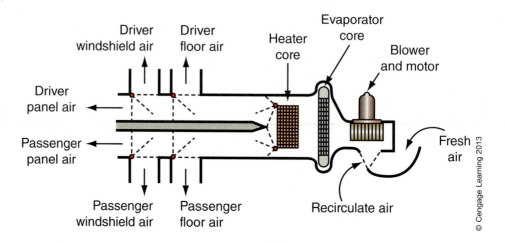

4. Both driver and passenger select MAX cooling with panel air.

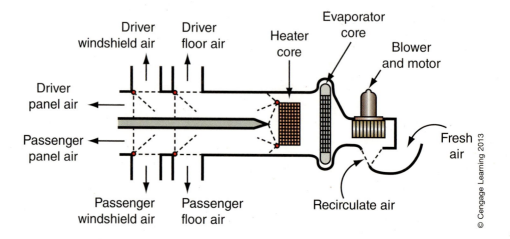

5. Airflow in the cooling mode when BI-LEVEL is selected. Both driver and passenger select bi-level cooling with panel air discharge, but the driver wants full cold while the passenger wants warm air.

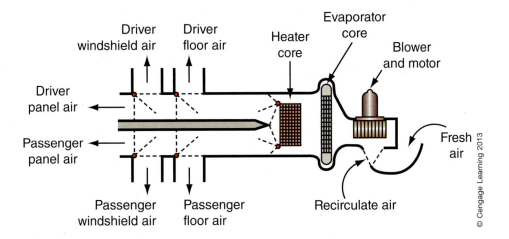

Instructor's Response _____

_____

_____

Name _____ Date _____

REPLACE CASE AND DUCT SYSTEM COMPONENTS

Upon completion of this job sheet, you should be able to remove and replace case and duct system components.

NATEF Correlation

HEATING AND AIR CONDITIONING: Operating Systems and Related Controls Diagnosis and Repair; *Inspect and test A/C-heater ducts, doors, hoses, cabin air filters, and outlets; perform necessary action.* **(P-3)**

Tools and Materials

Vehicle with air-conditioning system in need of case/duct service

Service manual

Appropriate tools

Describe the vehicle being worked on.

Year _____ Make _____ Model _____

VIN _____ Engine type and size _____

Procedure

1. Determine what component part or assembly is in need of replacement. Describe your _____ procedure and how you arrived at your decision.

   Procedure _____

   _____

   Component _____

   _____

   ⚠️

   CAUTION:
   If equipped with air bag(s), follow specific manufacturer's service procedures for replacing defective components.

2. Look up the manufacturer's procedures in the appropriate service manual.
   Service manual (title) _____ Year _____

   Section _____ Page _____ Component _____

3. Following the recommended procedures, remove necessary components to gain access to and remove the defective part.

   Procedure _____

   _____

4. Obtain the replacement part. Compare it with the part removed. Are they the same or is _____ the new part improved?

Conclusion _____

_____

5. Install the new part.

Procedure _____

_____

6. If possible, check the new component for proper operation.

Procedure _____

_____

7. Replace all components removed in step 3 to gain access.

Procedure _____

_____

8. Write a brief summary of any problems encountered with this repair.

_____

_____

Instructor's Response _____

_____

_____

Name _____ **Date** _____

ADJUST A DOOR CABLE

Upon completion of this job sheet, you should be able to adjust a case/duct system door cable.

NATEF Correlation

HEATING AND AIR CONDITIONING: Operating Systems and Related Controls Diagnosis and Repair; *Inspect and test A/C-heater control cables and linkages; perform necessary action.* **(P-3)**

Tools and Materials

Vehicle with air-conditioning system

Service manual

Hand tools, as required

Describe the vehicle being worked on.

Year _____ Make _____ Model _____

VIN _____ Engine type and size _____

Procedure

1. Following procedures outlined in the service manual, gain access to the control cable.

   Procedure _____

   _____

2. Adjust the cable.

   Procedure _____

   _____

3. Check cable operation. Readjust, if necessary.

   Procedure _____

   _____

4. In reverse order, replace components removed in step 1.

Procedure _____

_____

5. What specific problems, if any, were encountered during this procedure?

_____

_____

Instructor's Response _____

_____

_____

Name _____ **Date** _____

HVAC ODOR CONTROL TREATMENT

Upon completion of this job sheet, you should be able to eradicate the odor-causing microscopic mold, bacteria, and mildew that may develop in an HVAC system.

NATEF Correlation

REFRIGERATION SYSTEM COMPONENT DIAGNOSIS AND REPAIR: Evaporator, Condenser, and Related Components; *Inspect evaporator housing water drain; perform necessary action.* **(P-3)**

Tools and Materials

Late-model vehicle

Service manual or information system

Safety glasses or goggles

Hand tools, as required

Siphon-type air-conditioning disinfectant sprayer

Liquid antimicrobial disinfectant, 12 fluid ounces (354.88 mL)

Describe the vehicle being worked on.

Year _____ Make _____ Model _____

VIN _____ Engine type and size _____

Procedure

For the best results, follow the directions that come with the product. The following is a typical process for eliminating HVAC odor and applying the product to specific locations in the air duct system. This procedure may vary from one vehicle to another due to system control features. Always consult the manufacturer's recommendations. Give a brief description of your procedure following each step. Ensure that the engine is cold, and wear OSHA-approved eye protection.

1. Connect a siphon-type air-conditioning disinfectant sprayer to a 12 oz. (354.88 mL) bottle of disinfectant solution.
2. Inspect the evaporator housing water drain; perform the necessary actions if it is determined to be restricted. Was a restriction found?

   _____

3. Spray approximately 4 oz. (118.29 mL) of the solution into the fresh air inlet with the blower on high speed, the mode door in the fresh air position, and the temperature control set to cold.

4. Next, locate the recirculated air inlet on the passenger side of the interior compartment and set the HVAC control panel to recirculation mode. Spray approximately 2 oz. (59.15 mL) of the solution into the air recirculation intake with the blower on high speed.

5. Next, repeat step 3, but this time place the temperature control on full hot position.

6. Finally, turn the system and blower off and spray the remainder of the product into each of the air outlets in the system (defroster, floor vent, panel discharge, and side vents). Did you have enough product to complete this procedure?

7. Allow the vehicle to sit for at least 30 minutes with the windows open. Operate the system before returning it to the customer. Did the odor control treatment remove the objectionable small?

_____

_____

Instructor's Response _____

_____

_____

Name _____ **Date** _____

TEMPERATURE CONTROL DIAGNOSIS

Upon completion of this job sheet, you should be able to diagnose temperature control problems in the heater/ventilation system and determine the necessary action.

NATEF Correlation

HEATING AND AIR CONDITIONING: Heating, Ventilation, and Engine Cooling Systems Diagnosis and Repair; *Diagnose temperature control problems in the heater/ventilation system; determine necessary action.* **(P-2)**

Tools and Materials

Late-model vehicle

Service manual or information system

Safety glasses or goggles

Hand tools, as required

Thermometer

Describe the vehicle being worked on.

Year _____ Make _____ Model _____

VIN _____ Engine type and size _____

Procedure

Task Completed

The temperature control regulates the temperature of the air inside the vehicle. The blue region is for cooler temperatures and the red areas are for warmer temperatures. Blended air is achieved by mixing the amount of cooled (blue area) air with warmed (red area) air. The following are typical diagnostic procedures for temperature control operation; for specific information, refer to the vehicle manufacturer's diagnostic information.

1. First, verify that the engine coolant has reached operating temperature and that the air-conditioning system is functioning correctly. ☐

2. Check the heater core inlet and outlet temperature of the heater hoses. Both hoses should be hot. ☐

3. Verify that the fresh air intake is not obstructed with debris and that the cabin air filter is clean, if the cabin is equipped with one. ☐

4. Place the temperature control selector in the full hot position and select the panel vent mode. Place the thermometer in the center panel vent. The outlet temperature should be approximately 60°F (15.5°C) above ambient air temperature.

    a. Ambient air temperature _____

    b. Outlet temperature _____

5. Place the temperature control in the full cold position and select panel vent mode. The thermometer should still be in the center panel vent. The outlet temperature should be approximately 40°F (4.4°C) below ambient air temperature.

   a. Ambient air temperature _____

   b. Outlet temperature _____

☐   6. If the outlet temperature failed either steps 4 or 5, check the temperature control mode door actuator operation.

   a. If the temperature control mode door is cable actuated, go to Job Sheet 55, Adjust a Door Cable.

7. If the system passed both steps 4 and 5, the system is functioning correctly. Always refer to the manufacturer's information, when available, to determine the proper system functioning parameters.

Instructor's Response _____

_____

_____

JOB SHEET

61

Name _____ Date _____

ELECTRONIC ACTUATOR CONTROL DIAGNOSIS

Upon completion of this job sheet, you should be able to diagnose failures in the electrical controls of heating, ventilation, and A/C (HVAC) systems and to determine the necessary action.

NATEF Correlation

HEATING AND AIR CONDITIONING: Operating Systems and Related Controls Diagnosis and Repair; *Diagnose failures in the electrical controls of heating, ventilation, and A/C (HVAC) systems; determine necessary action.* **(P-2)**

Tools and Materials

Late-model vehicle

Service manual or information system

Safety glasses or goggles

Hand tools, as required

Digital multimeter

Describe the vehicle being worked on.

Year _____ Make _____ Model _____

VIN _____ Engine type and size _____

Procedure

Task Completed

With the various designs of electronic HVAC control heads and actuators in use today, it is necessary to refer to the manufacturer's specific diagnosis and troubleshooting information. Proceed with extreme care when diagnosing and servicing the underdash electrical systems. One improper test point could cause serious damage to one of the onboard computers. The following is meant to be a general procedure for inspecting the electronic actuator and control head operation for an electronically controlled temperature control door. The temperature control regulates the temperature of the air inside of the vehicle. The blue region is for cooler temperatures, and the red areas are for warmer temperatures. Blended air is achieved by mixing the amount of cool (blue area) air with warmed (red area) air. The following are typical diagnostic procedures for temperature control operation; for specific information, refer to the vehicle manufacturer's diagnostic information.

Extreme caution must be exercised while working on underdash components. Failure to do so could inadvertently trigger the inflatable restraint system.

1. First, verify that the engine coolant has reached operating temperature and that the air-conditioning system is functioning correctly. ☐

2. Check the heater core inlet and the outlet temperature of the heater hoses. Both hoses should be hot. ☐

3. Verify that the fresh air intake is not obstructed with debris and that the cabin air filter is clean, if the cabin is equipped with one.

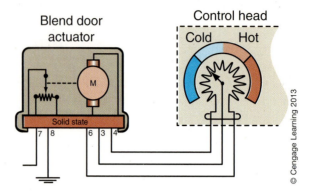

Blend door actuator

Control head

Cold   Hot

M

Solid state

7 8   6 3 4

© Cengage Learning 2013

4. Place the temperature control selector in the full hot position and select the panel vent mode. Place the thermometer in the center panel vent. The outlet temperature should be approximately 60°F (15.5°C) above ambient air temperature.

   a. Ambient air temperature _____

   b. Outlet temperature _____

5. Place the temperature control in the full cold position and select panel vent mode. The thermometer should still be in the center panel vent. The outlet temperature should be approximately 40°F (4.4°C) below ambient air temperature.

   a. Ambient air temperature _____

   b. Outlet temperature _____

☐

6. If the outlet temperature failed either steps 4 or 5, check the temperature control mode door actuator operation.

☐   a. First, verify that 12 volts are available at the actuator power supply wire.

☐   b. Next, verify that the actuator ground wire is intact and functioning properly.

☐   c. Then, obtain specific diagnostic information for the system and check both input and output signals.

   d. Most systems allow for testing of the actuator through the scan tool or climate control panel. Actuators that are duty cycled are more accurately diagnosed through this method.

7. If the system passed both steps 4 and 5, the system is functioning correctly. Always refer to the manufacturer's information, when available, to determine the proper system functioning parameters.

Instructor's Response _____

_____

_____

Name _____ Date _____

REMOVE AND REPLACE THE HEATER CORE

Upon completion of this job sheet, you should be able to remove and reinstall the heater core assembly.

NATEF Correlation

HEATING AND AIR CONDITIONING: Heating, Ventilation, and Engine Cooling Systems Diagnosis and Repair; *Remove and reinstall the heater core.* **(P-3)**

Tools and Materials

Late-model vehicle

Service manual or information system

Safety glasses or goggles

Hand tools, as required

Describe the vehicle being worked on.

Year _____ Make _____ Model _____

VIN _____ Engine type and size _____

Procedure

Access to the heater core is gained by following the procedures outlined in the appropriate service manual. The following procedure is typical and assumes the procedure for access to the heater core is available. Give a brief description of your procedure following each step. Ensure that the engine is cold, and wear OSHA-approved eye protection.

CAUTION: Do not use undue force when connecting the heater hoses. Damage to the new heater core may occur if care is not taken.

1. Drain the cooling system into a clean container. The coolant may be reused, reclaimed, or discarded in a manner consistent with Environmental Protection Agency (EPA) guidelines.
   How much coolant and what type was removed?

2. Disconnect the battery ground (–) cable. A "Keep Alive" device may need to be used so volatile memory is not lost in modules.

3. Disconnect the heater hoses at the bulkhead. This is a good opportunity to inspect the heater hoses and replace any that show signs of deterioration.
   Do any hoses require replacement?

4. Gain access to the heater core as outlined in the appropriate service manual.
   How long does the labor time list this entire service procedure should take (Flat Rate)?

5. Remove the retaining screws, brackets, or straps.

6. Remove the core from the case. Can you see the cause of the failure?

7. Install the new heater core.

8. Reverse disassembly procedures.

9. Refill the cooling system. How much coolant was required? What type of coolant was required?

10. Did you complete the job in the "Flat Rate" time specified in the labor time guide?

Instructor's Response _____

_____

_____

Name _____ **Date** _____

REMOVE AND REPLACE THE EVAPORATOR

Upon completion of this job sheet, you should be able to remove and reinstall an evaporator assembly.

NATEF Correlation

HEATING AND AIR CONDITIONING: Heating, Ventilation, and Engine Cooling Systems Diagnosis and Repair; *Remove and reinstall an evaporator; measure the oil quantity; determine the necessary action.* **(P-3)**

Tools and Materials

Late-model vehicle

Service manual or information system

Safety glasses or goggles

Hand tools, as required

Describe the vehicle being worked on.

Year _____ Make _____ Model _____

VIN _____ Engine type and size _____

Procedure

Access to the evaporator assembly is gained by following the procedures outlined in the appropriate service manual. The following procedure is typical and assumes the procedure for access to the evaporator is available. Give a brief description of your procedure following each step. Ensure that the engine is cold, and wear OSHA-approved eye protection.

1. Recover the refrigerant. How much refrigerant was removed? Was any refrigerant oil removed, if so how much?

2. If the heater hose(s) and heater core must be removed to gain access, drain the cooling system into a clean container. The coolant may be reused, reclaimed, or discarded in a manner consistent with Environmental Protection Agency (EPA) guidelines. Was it necessary to drain the cooling system?

3. Disconnect the battery ground (–) cable. A "Keep Alive" device may need to be used so volatile memory is not lost in control modules.

4. Disconnect the heater hoses at the bulkhead, if necessary. This is a good opportunity to inspect the heater hoses and replace any that show signs of deterioration. Was this step necessary? Do any hoses require replacement?

5. Remove any wiring harness, heat shields, brackets, covers, and braces that may restrict access to the evaporator core.

6. Remove the liquid line at the thermostatic expansion valve (TXV) or fixed orifice tube (FOT).

CAUTION:
Do not use undue force when connecting the heater hoses. Damage to the heater core may occur if care is not taken.

7. Remove the suction line at the evaporator or accumulator outlet.
8. Gain access to the evaporator core as outlined in the service manual.
9. Lift the evaporator from the vehicle.
10. Drain the oil from the evaporator into a calibrated cup.
    How much if any was removed?
11. For replacement, reverse the preceding procedure. First, replace the oil with the same amount and type as drained in step 9, if greater than the amount specified by the manufacturer.

    How much oil was added to the system? Why?_____

    _____

12. What was the labor time listed for this complete procedure in the labor time guide?

    _____

    _____

Instructor's Response _____

_____

_____

JOB SHEET

Name _____ **Date** _____

TEST A BLOWER MOTOR

Upon completion of this job sheet, you should be able to test a blower motor circuit and determine proper operation of the blower resistor or power module.

NATEF Correlation

HEATING AND AIR CONDITIONING: Operating Systems and Related Controls Diagnosis and Repair; *Inspect and test A/C-heater blower, motors, resistors, switches, relays, wiring, and protection devices; perform necessary action.* **(P-2)**

Tools and Materials

Late-model vehicle

Shop manual

Voltmeter

Fused jumper wire

Safety glasses or goggles

Hand tools, as required

Describe the vehicle being worked on.

Year _____ Make _____ Model _____

VIN _____ Engine type and size _____

Procedure

After each of the following steps, write a brief description of your procedure, followed by your findings.

Test the blower motor:

1. With the ignition switch in the ON position, turn the blower control to:

   a. HIGH _____

   b. MED-HI _____

   c. MED-LO _____

   d. LOW _____

   Did the blower run in any speed? Explain: _____

   _____

2. Turn the blower control and ignition switch OFF. Disconnect the blower motor and connect the voltmeter: one lead to ground and the other lead to the disconnected wire. Is there a voltage? _____ Why? _____

_____

3. Reconnect the blower motor connector. Using T-pins back probe the blower motor power and ground wires. Connect the DMM red lead to the power feed wire and connect the DMM black lead to the ground wire (or blower motor housing, in some applications).

4. Turn the ignition switch ON. While observing the voltmeter, turn the blower control to:

   a. HIGH _____

   b. MED-HI _____

   c. MED-LO _____

   d. LOW _____

   Was there voltage noted in any speed position? Explain: _____

   _____

   e. Turn the blower control and ignition switch OFF.

5. Connect one end of a fused jumper wire to the battery's positive terminal. While wearing OSHA-approved safety glasses or goggles, carefully connect the other end of the jumper wire to the blower motor terminal. Was there a "spark"? ____ Did the fuse "blow"? _____ Did the motor "run"? _____ Describe what happened. _____

   _____

6. Conclusion. Write a brief summary of your findings.

   _____

   _____

   _____

   _____

Instructor's Response _____

_____

_____

Name _____ **Date** _____

REMOVE AND REPLACE A BLOWER MOTOR

Upon completion of this job sheet, you should be able to remove and replace a blower motor.

NATEF Correlation

HEATING AND AIR CONDITIONING: Operating Systems and Related Controls Diagnosis and Repair; *Inspect and test A/C-heater blower, motors, resistors, switches, relays, wiring, and protection devices; perform necessary action.* **(P-2)**

Tools and Materials

Late-model vehicle

Shop manual

Safety glasses or goggles

Hand tools, as required

Describe the vehicle being worked on.

Year _____ Make _____ Model _____

VIN _____ Engine type and size _____

Procedure

Follow the procedures outlined in the manufacturer's shop manual. These procedures are given as a typical guideline for the task. After each step, write a brief summary of your procedure.

1. While wearing safety glasses or goggles, carefully remove the battery ground cable. If this is required, a "Keep Alive" memory saving device may need to be used. Was it used?

2. Disconnect the BCM or PCM (if applicable) following procedures given in the service manual. Was this necessary?

3. Remove any components, such as coolant reservoir, that may prevent blower motor removal. Was it necessary to remove additional components?

   _____

   _____

4. Remove electrical connector(s) and ground wire(s), if applicable. What connection need to be removed?

   _____

5. Remove all retaining screws and fasteners.

6. If necessary, use a sharp utility knife to cut through any gasket material that may restrict the removal of the blower motor assembly.

7. Lift the blower and motor from the case/duct.

8. Remove the retaining nut or clip from the motor shaft and remove the blower, if applicable.

9. Slide the blower onto the new motor shaft and secure it with a nut or clip.
   Was it necessary to reuse the old blower motor turbine (squirrel cage)?

10. Reverse the removal procedure and replace the blower and motor. Replace any gasket material cut in step 6 with black weatherstrip adhesive. Do not use RTV.
    What did you use to seal the unit?

11. Make electrical connections, reversing the order of steps 1, 2, and 4.

12. Replace any components removed in step 3.

13. Test blower motor operation and check for noises. Is it operating as designed and is it noise free?

14. What is the labor time listed for this service procedure in the labor time guide? _____

_____

Instructor's Response _____

_____

_____

Name _____ Date _____

INSPECT AND REPLACE CABIN AIR FILTER

Upon completion of this job sheet, you should be able locate, inspect, and replace the cabin air filter.

NATEF Correlation

HEATING AND AIR CONDITIONING: Operating Systems and Related Controls Diagnosis and Repair; *Inspect and test A/C-heater ducts, doors, hoses, cabin filters, and outlets; perform necessary action.* **(P-1)**

Tools and Materials

Late-model vehicle

Service manual or information system

Safety glasses or goggles

Hand tools, as required

Cabin air filter

Describe the vehicle being worked on.

Year _____ Make _____ Model _____

VIN _____ Engine type and size _____

Procedure

The procedure for cleaning or replacement of the cabin air filter varies from manufacturer to manufacturer. Location of the access panel also varies among vehicle models, so it is important to follow manufacturers' recommended procedures. Typical replacement intervals for the cabin air filter are every 15,000 miles. The following is a typical procedure for replacement of a cabin air filter element with an access panel located in the passenger compartment.

1. Remove the dash undercover (kick panel).
2. Remove the glove box assembly.
3. Remove the access panel for the cabin air filter located on the duct case.
4. Remove the filter retaining clip.
5. Remove the cabin air filter from the case/duct.

6. Clean or install a new filter element.
7. Replace components in the reverse order used to remove them.

NOTE: Access to some cabin air filters is in the engine compartment under the cowl cover. Always refer to vehicle service information.

Instructor's Response _____

_____

_____

Chapter 10

DIAGNOSIS AND SERVICE OF SYSTEM CONTROLS

BASIC TOOLS

Basic mechanic's tool set
Fender cover
Digital multimeter
Jumper leads
Fused jumper leads

UPON COMPLETION AND REVIEW OF THIS CHAPTER, YOU SHOULD BE ABLE TO:

- Discuss the methods used to diagnose fuse and circuit breaker defects.

- Recognize and identify the components of the climate control system.

- Understand and practice the methods used to diagnose compressor clutch malfunctions.

- Identify and troubleshoot the different types of pressure- and temperature-actuated controls.

- Understand the function of and be able to troubleshoot the components of an automatic temperature control system.

- Identify and test sensors and actuators on an automatic climate control system.

- Activate the self-diagnostic test mode on an automatic climate control system.

- Use a scan tool to diagnose an automatic climate control system and verify the operation of the CAN bus line.

The control system of an automotive air-conditioning system, at first, may seem to be very complex. And, indeed, it is a complex system of many single wires. Compare the control system schematic to a road map. As one looks at a road map and notes the many highways and byways, it also looks complex. There is, however, only one route that is of interest at any one time. All the other routes are unimportant for any particular journey. For the most part, the same is true when diagnosing any control system or subsystem; although the "map" may seem very complex, most of it will prove to be of no interest.

The schematic in Figure 10-1 is a composite of several car line schematics and, while it may be representative of several make or model automobiles, it should not be considered typical for any specific make or model. For specific information, manufacturers' shop and service manuals must be consulted.

Refer to the schematics of this text as you are led through a systematic approach to diagnosis, troubleshooting, and repair procedures for today's modern automotive air-conditioning systems.

CUSTOMER CARE: Climate control systems today are very complicated and not for the do-it-yourselfer. There are many similarities among all the systems, but the terminology may differ from manufacturer to manufacturer, and the specific values for sensor diagnosis may differ. In order to be successful in diagnosing automatic climate control systems, you must have access to a high-quality information system, such as manufacturer service information or systems like ALLDATA or Mitchell on Demand.

An electrical schematic often requires several pages in a service manual.

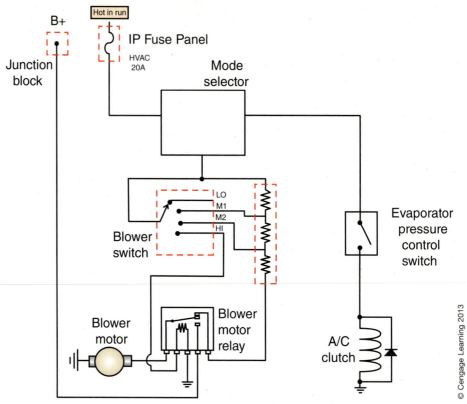

FIGURE 10-1 A typical automotive air-conditioning electrical system schematic.

FUSES AND CIRCUIT BREAKERS

Note that there are several fuses, a circuit breaker, and a **fusible link** in the schematic. The purpose of these devices is to provide optimum protection to all of the circuits at all times. A fuse or fusible link is generally used in circuits that are hot all the time; that is, circuits that are not interrupted when the ignition switch is **open**. This provides a positive non-restorable interruption of power should an **overload** occur when the vehicle is unattended. A circuit breaker, on the other hand, is generally used only in circuits that are interrupted when the ignition switch is open (off).

There are several methods that may be used to check a fuse or circuit breaker: in-vehicle testing with a **voltmeter** or non-powered test lamp and out-of-vehicle testing with an **ohmmeter** or powered test lamp.

To test a fuse or circuit breaker in the vehicle, use a voltmeter or test lamp as follows:

1. Connect one lead of the test lamp or voltmeter to body ground (−).
2. Touch the other lead to the positive (hot) side of the fuse or circuit breaker in the fuse block or holder (Figure 10-2). If the lamp does not light or if voltage is not indicated, power is not available and the problem is elsewhere. If the lamp lights or if voltage is indicated, power is available. Proceed with step 3.
3. Touch the lead to the other side of the fuse or circuit breaker (Figure 10-3). If the lamp does not light or if voltage is not indicated, the fuse is blown or the circuit breaker is defective. Proceed with step 4. If the lamp lights or voltage is indicated, the problem is elsewhere and further testing is necessary.
4. Test protected components for **shorts** or overloads, then replace the fuse or circuit breaker.

FIGURE 10-2 Touch the lead to the hot side of the fuse block or holder.

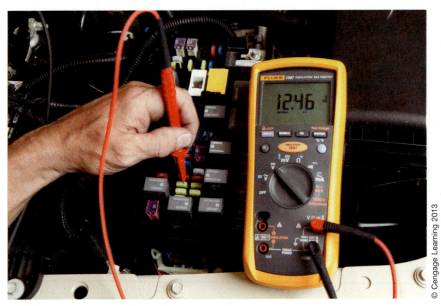

FIGURE 10-3 Touch the lead to the other side of the fuse.

To test a fuse or circuit breaker that has been removed from the vehicle, follow this procedure:

1. Set the omm to the ohmmeter scale, touch the leads together, and zero the meter or make sure that the test lamp battery is good.
2. Touch the two leads of the ohmmeter or test lamp to either side of the fuse or circuit breaker (Figure 10-4). If the ohmmeter indicates a low resistance or if the test lamp lights, the fuse or circuit breaker is good. If there is no resistance indicated on the ohmmeter or if the test lamp does not light, the fuse is blown or the circuit breaker is defective.

SPECIAL TOOLS

Nonpowered test lamp

Voltmeter

Make sure that the test lamp is not burned out.

SPECIAL TOOLS

Powered test lamp

Ohmmeter

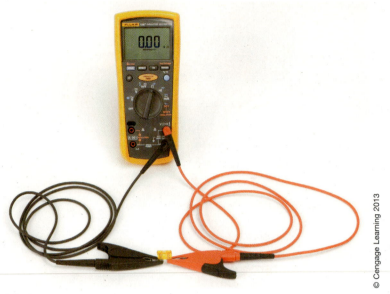

FIGURE 10-4 Touch the two leads to either side of the fuse.

Classroom Manual Chapter 10, page 319

SPECIAL TOOLS

Powered test lamp

Ohmmeter

THERMOSTAT

The thermostat (Figure 10-5) cycles the air-conditioning compressor electromagnetic clutch on and off as determined by a preset temperature. There are two types of thermostat: fixed and variable. Testing either type of thermostat is a relatively simple matter if it has been removed from the vehicle. Proceed as follows.

Variable-Type Thermostat

The variable-type thermostat is generally found on aftermarket air-conditioning systems.

1. Connect an ohmmeter or powered test lamp to the two terminals of the thermostat (Figure 10-6).
2. While observing the ohmmeter or test lamp, rotate the thermostat from fully clockwise (cw) to fully counterclockwise (ccw). If a low resistance is noted or if the test lamp lights, the thermostat is probably all right. If no resistance is noted or if the test lamp does not light, the thermostat is defective.
3. Repeat step 2 several times to ensure stable and consistent results.

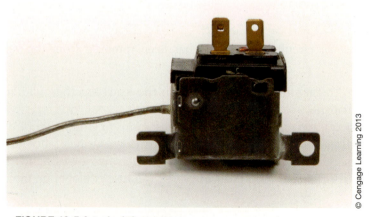

FIGURE 10-5 A typical thermostat.

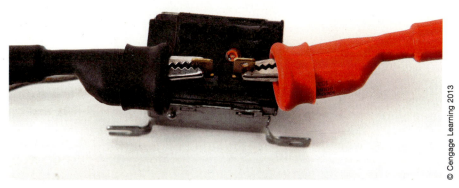

FIGURE 10-6 Connect an ohmmeter to the two terminals of a thermostat.

Fixed-Type Thermostat

A fixed-type thermostat that has no provisions for temperature adjustment is generally found on factory-installed air-conditioning systems. Two beakers of water are required: one cooled with ice (32°F or 0°C), and the other heated to about 120°F (49°C). The thermostat is tested as follows:

1. Connect the ohmmeter or test lamp in the same manner as in the adjustable thermostat test.
2. Is low resistance noted or is the lamp lit? Generally, at ambient temperature, the thermostat will be closed. If the answer is yes, proceed with step 3. If the answer is no, proceed with step 5.
3. Immerse the capillary tube end or remote bulb into the ice bath (Figure 10-7).
4. Did the resistance increase or the lamp go out? A reduction in temperature below the set point should open the thermostat contacts. If the answer is yes, proceed with step 5. If the answer is no, the contacts are stuck closed and the thermostat is defective.
5. Immerse the capillary tube in the hot bath.
6. Did the resistance decrease (Figure 10-8) or the lamp light? If the answer is yes, the thermostat is probably all right. If the answer is no, the thermostat is probably defective with contacts stuck open.

SPECIAL TOOLS

Powered test lamp

Ohmmeter

Systems with a fixed thermostat usually maintain the desired in-car temperature by tempering cooled and heated air in the plenum section of the duct system.

CAUTION:
Technicians need to wear a static discharge wrist strap when servicing electrical devices containing solid-state components such as control modules. Static electric discharge may damage sensitive circuits.

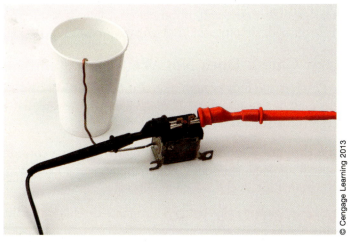

FIGURE 10-7 Immerse the cap tube into an ice bath.

FIGURE 10-8 Did the resistance decrease?

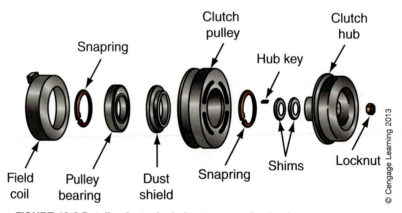

FIGURE 10-9 Details of a typical electromagnetic clutch.

Classroom Manual

Chapter 10, page 329

ELECTROMAGNETIC CLUTCH

The electromagnetic clutch (Figure 10-9) starts compressor action when wanted and stops it when it is not wanted. The clutch either works or it does not work. It may be noisy when it works, which is a sign that it needs attention before it fails. If it does not work, the problem may be that it is burned, slipping, will not engage, or will not disengage.

The Compressor Clutch Does Not Work

If the compressor clutch is the only component that does not work or if it works intermittently, the problem may be the clutch. It is more likely, however, that it is in the clutch electrical circuit (Figure 10-10).

Testing the Clutch Circuit. The following procedure should be considered a typical procedure only. For specific procedures, follow the manufacturer's instructions outlined in the appropriate service manual for safely testing an electrical circuit.

1. Touch the test lamp leads to the battery terminals to check the integrity of the fuse and bulb (Figure 10-11).
2. Disconnect the clutch coil from the wiring harness.

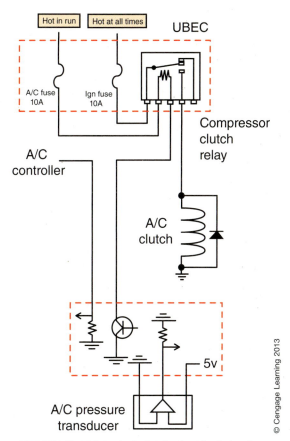

FIGURE 10-10 A typical clutch electrical circuit.

FIGURE 10-11 Touch the fused test lamp leads to the vehicle battery to test for integrity.

3. Turn the ignition switch to ON and place the air-conditioning system controls in any COOL position.
4. Connect a fused test lamp from ground to the positive (+) terminal of the disconnected wiring harness.
   a. If the test lamp does not light, the fuse, clutch relay, or wiring may be defective. Troubleshooting procedures are similar to those outlined for the blower motor circuit. Repair or replace components as necessary.

   NOTE: The problem could be in the power train control module (PCM). Consult the manufacturer's service manual for troubleshooting procedures. Photo Sequence 16 illustrates a typical procedure for testing PCM wiring and circuits.

   b. If the test lamp lights, proceed with step 5.
5. Turn the control back and forth from minimum (MIN) to maximum (MAX) cooling several times while observing the test lamp.
   a. If the test lamp flickers or goes out, the PCM may be defective. Further testing of the PCM is indicated.
   b. If the test lamp remains on, proceed with step 6.
6. Remove the clutch and clutch coil and bench test the individual parts as follows:
   a. Visually inspect the clutch rotor and armature assembly.
   b. If the rotor or armature is heavily scored, as in Figure 10-12, or shows signs of overheating, replace the assembly.
7. Bench test the clutch coil.
   a. Connect a jumper wire from the clutch coil frame (Figure 10-13A) or ground wire (Figure 10-13B) to the battery ground (−) cable.
   b. Connect a fused test lamp from the battery positive (+) terminal to the clutch coil lead wire.
   c. If the test lamp does not light, the clutch coil is defective (open) and must be replaced.
   d. If the test lamp lights, the clutch coil is not defective.

   NOTE: If the clutch coil is shorted, the test lamp will also light. If suspected of being shorted, hold the resistance test as outlined in Job Sheet 49.

   e. Reinspect the ground wire, rotor, and armature to determine the problem.
   f. Correct as necessary.

FIGURE 10-12 Inspect the clutch rotor and armature surfaces.

Procedure for Testing PCM-Controlled Air Conditioners

P16-1 Locate the power train control module (PCM) and gain access to its wiring harness. Disconnect the PCM.

P16-2 Check for a poor connection at the PCM.

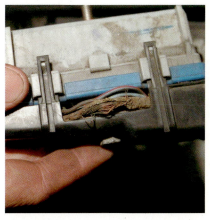

P16-3 Inspect the wiring harness for damage.

P16-4 Connect a digital voltmeter to the relay driver circuit at the PCM harness connector.

P16-5 Turn the ignition switch to ON. Do not start the engine.

P16-6 Observe the voltmeter while moving connectors and wiring harness relating to the relay.

P16-7 Note any change in voltage while moving the relay driver wiring harness. Change indicates a wiring harness fault.

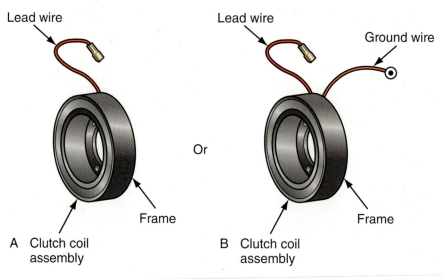

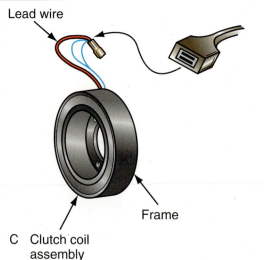

FIGURE 10-13 Clutch coil electrical connection: (A) grounded through the frame, (B) grounded through the ground wire, (C) grounded through the connector.

Noisy

Problem	Remedy
1. Slipping belt	Tighten belt
2. Misaligned belt	Align belt
3. Clutch slipping	See "Clutch Will Not Engage"
4. Rotor/pulley snapring missing	Replace snapring
5. Rotor/pulley snapring improperly installed	Properly install new snapring
6. Rotor-to-armature air gap too small	Properly adjust air gap
7. Improper field coil snapring	Install proper snapring
8. Field coil snapring installed improperly	Reinstall snapring
9. Damaged bearing	Replace clutch assembly

Burned Clutch

A burned clutch is often noted by charred paint, blued steel, melted bearing seals, broken springs, or a charred field coil. To prevent a recurrence, all of the problems leading to a

© Cengage Learning 2013

burned clutch should be addressed before replacing it. These problems and the recommended remedies are:

Problem	Remedy
1. Compressor shaft seal leak	Replace shaft seal
2. Compressor thru bolt leak	Repair as required
3. Oil leak: engine, power steering, transmission	Repair or replace as required
4. Missing rotor/pulley snapring	Make sure snapring is installed
5. Improperly installed rotor/pulley snapring	Make sure snapring is properly installed
6. Improper field coil snapring	Make sure snapring is proper
7. Improperly installed field coil snapring	Make sure field coil snapring is installed properly
8. Mismatched components	Replace with matched components

Clutch Will Not Engage

Problem	Remedy
1. Excessive air gap	Adjust air gap
2. Poor electrical connection(s)	Repair as required
3. Undersized wiring	Use minimum 18-gauge wire
4. Damaged wiring	Repair as required
5. Defective clutch relay	Replace relay
6. Electrical component failure	Replace defective component
7. Shorted field coil	Replace field coil
8. Open field coil	Replace field coil

Many component malfunctions are due to a defective ground connection.

Clutch Will Not Disengage

Problem	Remedy
1. Improper air gap	Adjust air gap
2. Rotor/pulley snapring not installed	Install snapring
3. Rotor/pulley snapring improperly installed	Reinstall snapring
4. Electrical problem	Correct problem as required

PRESSURE SWITCHES AND CONTROLS

There are many types of pressure switch controls used in the automotive air-conditioning system. These are the low-pressure cutoff switch, high-pressure cutoff switch, compressor discharge pressure switch, and pressure cycling switch.

Replacement is rather simple. Remove the old pressure switch and install a new one. A word of caution, however. If it is not known if the switch is equipped with a Schrader-type service port, the refrigerant must first be removed from the system.

At atmospheric pressure, a low-or high-pressure switch should be normally closed (nc). If there is a low-pressure switch, a vacuum pump may be used to see at what low pressure (if any) it opens. Similarly, a nitrogen source may be substituted for the vacuum pump to test the operation of a high-pressure switch. A test lamp is used to determine if and when the pressure switch opens or closes.

Classroom Manual
Chapter 10, page 330

When replacing a pressure switch, make certain that the new replacement is the same pressure range as the old defective one.

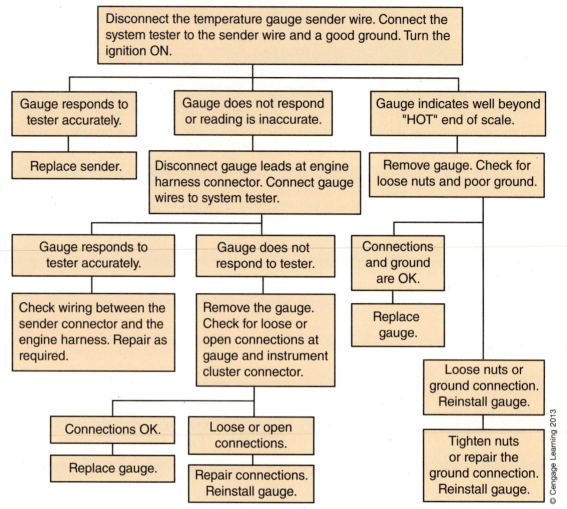

TEMPERATURE GAUGE INACCURATE OR INOPERATIVE

Disconnect the temperature gauge sender wire. Connect the system tester to the sender wire and a good ground. Turn the ignition ON.

Gauge responds to tester accurately.

Replace sender.

Gauge does not respond or reading is inaccurate.

Disconnect gauge leads at engine harness connector. Connect gauge wires to system tester.

Gauge indicates well beyond "HOT" end of scale.

Remove gauge. Check for loose nuts and poor ground.

Gauge responds to tester accurately.

Check wiring between the sender connector and the engine harness. Repair as required.

Gauge does not respond to tester.

Remove the gauge. Check for loose or open connections at gauge and instrument cluster connector.

Connections and ground are OK.

Replace gauge.

Connections OK.

Replace gauge.

Loose or open connections.

Repair connections. Reinstall gauge.

Loose nuts or ground connection. Reinstall gauge.

Tighten nuts or repair the ground connection. Reinstall gauge.

© Cengage Learning 2013

FIGURE 10-14 A diagnostic chart for a typical temperature gauge/sending unit test.

COOLANT TEMPERATURE WARNING SWITCHES

Classroom Manual
Chapter 10, page 333

There are two types of coolant temperature warning systems: warning lamps and gauge systems. Testing the warning lamp system is rather straightforward. If the lamp(s) is/are good and the wiring is sound, an inoperative system is generally due to a defective sending unit. It is a relatively simple matter to substitute a new sending unit if a defective unit is suspected. The gauge system requires the use of a tester, however. A diagnostic chart of a typical temperature gauge/sending unit test is shown in Figure 10-14.

VACUUM SWITCHES AND CONTROLS

Classroom Manual
Chapter 10, page 335

Vacuum-controlled actuators, often called motors, are used to position the A/C-defog valve, up-mode valve, down-mode valve, and the inside air valve (Figure 10-15).

Sensor Testing with a Scan Tool

Most scan tools will display the voltage values or switch position of many sensors. Access to this information differs, depending on the scan tool used. For example, when using the Tech2 (Figure 10-16), if the tool display reads BODY COMP MENU, select STATE DISPLAY. The display will change to BODY COMP STATE. By selecting SENSORS, the

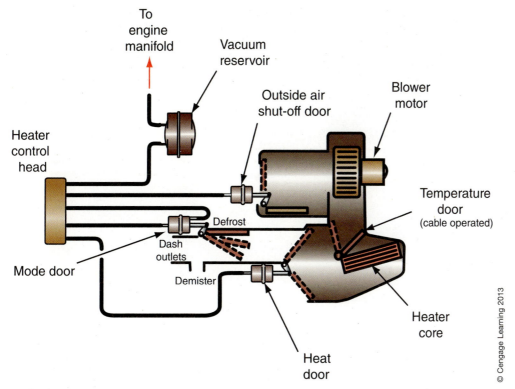

FIGURE 10-15 A typical vacuum control schematic.

FIGURE 10-16 The Tech2 scan tool used by General Motors and Isuzu.

value of selected sensors can be viewed. If the technician is interested in a switch position, select INPUTS/OUTPUTS, and the display will indicate the various positions of various switches used as inputs to the computer.

BREAKOUT STATE

A **breakout box** (Figure 10-17) is a device that, when connected between the module and the wiring harness, allows the technician to "see" the exact information the computer is receiving and sending.

The breakout box taps directly into the sensor or actuator circuit, providing the technician with the exact voltage signal being sent or received. A breakout box connected into the system allows a digital multimeter (DMM) to be used to measure the voltage signals and resistance values of the circuit (Figure 10-18). The diagnostic manual provided with

A **breakout box** is a tool that is connected to the vehicle wiring harness and is used for pinpoint testing circuits with a voltmeter or ohmmeter.

FIGURE 10-17 A breakout provides test points for voltmeter and ohmmeter connections.

FIGURE 10-18 Using the breakout box to test a circuit.

the breakout box should be used as a guide through a series of test procedures. Comparing the test results with specifications will lead to the problem area.

AUTOMATIC (ELECTRONIC) TEMPERATURE CONTROLS

The two basic automatic temperature control (ATC) systems are those that use their own microprocessor (climate control module) and those that incorporate the controls of the system into the BCM. Always refer to published manufacturer diagnostic information and technical service bulletins (TSBs) before attempting any repair. As with any other electrical system diagnosis and repair procedure, you will need to refer to circuit schematics and specifications for the vehicle you are diagnosing.

Even ATC systems of the same manufacturer will have different service and diagnostic procedures based on the specific model and year of the vehicles. It cannot be overemphasized that specific make, model, and year vehicle service information is required to accurately diagnose and troubleshoot ATC systems. That being said, there are similarities among systems and the fundamental sensor operation and diagnosis are the same.

With the complexity and interconnectedness of today's computer and network-controlled systems, you must always be thinking about the interrelatedness of various vehicle systems and subsystems. Prior to attempting to diagnose any ATC system complaint, first verify that there are no power train codes (P0XXX or P1XXX) or network codes (U0XXX) present. A P0115—Engine Coolant Temperature Circuit Malfunction—may not allow the compressor clutch to be engaged.

Programmer

The programmer is generally identified by the electrical and vacuum lines attached to it. It primarily contains a small circuit board that controls a small reversible DC motor that adjusts the air mix valve to blend cold and warm air. It also contains four vacuum **solenoids** that control the various vacuum mode actuators. The programmer also provides data for blower speed selection and operation to provide selected in-vehicle temperature conditions.

Typically, to remove the programmer (Figure 10-19), gain adequate access by removing the right-side sound barrier and glove box. Then:

1. Remove the brackets and covers to gain access to the programmer.
2. Remove the threaded rod from the programmer.
3. Remove the vacuum connector retaining nut.
4. Remove the vacuum and electrical connectors from the programmer.
5. Remove the programmer from the vehicle.
6. To replace, reverse the preceding procedure.

Blower Control

For any given ECC signal, the blower control module has a predetermined blower motor voltage value. A signal from the BCM to the programmer causes a variable voltage signal to be sent to the power module.

Variable Resistor Test. (Figure 10-20) The control assembly variable resistor, also referred to as a sliding resistor or blower speed control, can be tested in or out of the vehicle by the following procedure:

1. Disconnect the electrical connector to the variable resistor.
2. Connect an ohmmeter's test leads across the terminals.
3. Set the comfort control lever to the 65 selection and note the resistance. The resistance in this position should be less than 390 ohms (390 Ω).

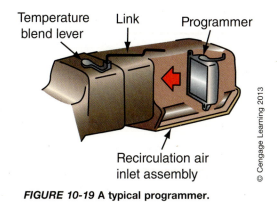

Temperature blend lever   Link   Programmer

Recirculation air inlet assembly

© Cengage Learning 2013

FIGURE 10-19 A typical programmer.

Classroom Manual
Chapter 10, page 339

Solenoids are electromagnetic devices controlled remotely by electrically energizing or de-energizing a coil.

SERVICE TIP: If you suspect an intermittent electrical problem with the climate control module, grasp the wiring harness near the module connection and shake it. Next, lightly tap the module assembly to see if the problem is affected by the vibrations. This will aid in locating troublesome intermittent internal connections or board failures.

The variable resistor assembly is also referred to as a sliding resistor.

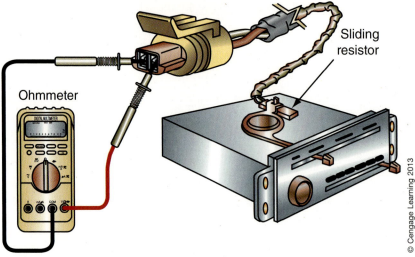

FIGURE 10-20 Testing the control assembly slide resistor with a DMM on the Ω setting. The resistor values should be within specifications and the transition should be smooth as the level is moved.

4. Slowly move the comfort control lever toward the right while observing the ohmmeter. When the ohmmeter indicates 930 Ω, the lever should be near the 75 setting.

   NOTE: The ohmmeter should have indicated a smooth increase in resistance.

5. Move the lever to the 85 setting. The resistance value should increase smoothly to at least 1,500 Ω.

If the resistance values are not within these specifications or the increase in resistance is not smooth, the control assembly must be replaced.

Clutch Control

The compressor clutch is controlled by the PCM from inputs to the PCM, such as engine coolant temperature as well as rpm, and to the body computer module (BCM), such as outside temperature, in-car temperature, sun load temperature, and air-conditioning system high-and low-side temperatures/pressures.

The clutch diode (Figure 10-21) is found connected across the electromagnetic clutch coil in many systems. Its purpose is to prevent unwanted electrical **spikes** that could damage delicate minicomputer electrical systems and subsystems as the clutch is engaged and disengaged.

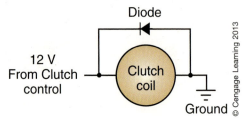

FIGURE 10-21 A clutch diode prevents electrical spikes.

CLIMATE CONTROL SYSTEM SENSORS

Many climate control sensors are similar in design and function to sensors used in other systems, such as engine management and emissions control systems. The service technician needs to have a good understanding of basic electricity and electronics first. The *Today's Technician Automotive Electricity & Electronics* textbook is a good place to start. Today's systems require the use of enhanced diagnostic scan tools and manufacturer service information to accurately diagnose both sensor and software used in climate control systems. Always refer to published diagnostic information and TSBs before attempting any repair. The following is an overview of sensor operation and diagnosis for common automatic climate control systems.

Evaporator Temperature Sensor

The evaporator temperature sensor, also called a fin sensor (Figure 10-22), is an NTC thermistor that monitors the temperature of the evaporator core. The sensor probe is inserted between the evaporator fins to sense changes in core temperature. The sensor modifies a voltage reference signal (Figure 10-23) with a change in evaporator temperature and this signal is sent to the climate control module or BCM/PCM. This sensor may be either a two-wire or a three-wire sensor depending on the make and model of the vehicle. The two-wire design uses a 5 V reference wire and a signal return wire to the control module. The three-wire design has a 5 V reference wire, a signal return wire, and a ground wire all connected to the control module.

The evaporator temperature sensor is used by the control module to turn the compressor clutch ON and OFF. When evaporator temperature typically drops below 33°F (1°C), a command is sent over the data bus to the PCM to interrupt compressor clutch engagement to prevent condensed water from freezing on the evaporator fins and restricting airflow. To prevent rapid compressor cycling, the control module does not request the compressor back on until the evaporator temperature rises to a temperature above the turn-off temperature, generally 37°F (3°C) to 45°F (7°C). By comparing the readings from a DMM, scan tool, and thermometer the sensor can be diagnosed for accurate operation.

As a thermistor's temperature increases, its resistance decreases.

Classroom Manual

Chapter 10, page 353

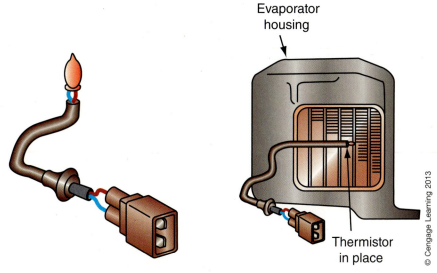

FIGURE 10-22 An evaporator temperature sensor (thermistor).

Evaporator housing

Thermistor in place

© Cengage Learning 2013

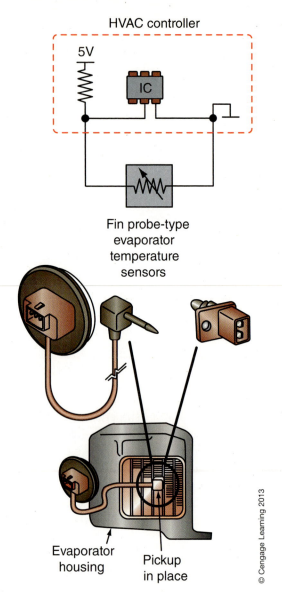

HVAC controller

5V

IC

Fin probe-type
evaporator
temperature
sensors

© Cengage Learning 2013

Evaporator
housing

Pickup
in place

FIGURE 10-23 The evaporator temperature
sensor is an NTC thermistor.

Following is an example of a typical chart for evaporator sensor values at various temperature ranges.

Temperature	Voltage	Resistance in kΩ
50°F (10°C)	1.90	9.95
45°F (7.2°C)	2.06	11.38
40°F (4.4°C)	2.23	13.05
35°F (1.6°C)	2.40	15.00
30°F (−1.1°C)	2.58	17.29
25°F (−3.9°C)	2.76	19.99
20°F (−6.7°C)	2.94	13.19
15°F (−7.2°C)	2.98	23.90

**Classroom
Manual**
Chapter 10,
page 349

Sun Load Sensor

The sun load sensor is located on the top of the dash panel where the sun rays will hit the
sensor in the same manner they would hit the front passengers. The sun load sensor is a

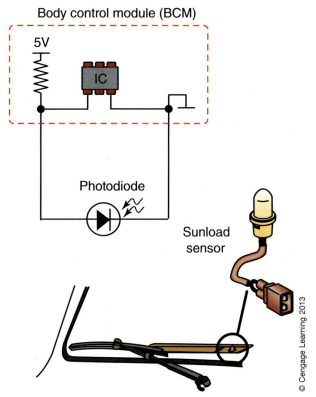

FIGURE 10-24 The sun load sensor is a photodiode that is mounted on the top surface of the instrument panel.

photodiode that modifies a voltage signal by varying sensor resistance with a change in sunlight levels (intensity), not temperature (Figure 10-24). The sun load sensor is not serviceable or adjustable and must be replaced as an assembly if defective. The first step in the diagnostic process is to verify that the sensor is unobstructed:

■ Check that there are no window stickers on the windshield directly above the sensor.
■ Check that there are no items such as parking passes on top of the dashboard in front of sensor.
■ Check that the defroster grille and sun load sensor are properly installed.
■ Check the position of the wiper blades and arms.

When a customer complains of poor system performance on sunny days, especially in the early afternoon, the position of the sun load sensor should be checked. The sun load sensor operation is typically checked using a scan tool and DMM. Refer to vehicle-specific service information and electrical diagrams for diagnostic information. Faults in the sun load sensor circuit will set a diagnostic trouble code.

 WARNING: On vehicles equipped with air bags, disable the air bag system before attempting to diagnose or work on the steering column, instrument panel, and dash assemblies. Failure to follow manufacturer's service warning may result in personal injury or death.

Sun load sensor quick test:

1. Park the vehicle outside on a sunny day.
2. Turn the ignition to the ON position, but do not start the vehicle yet.
3. Connect the scan tool to the vehicle data link connector (DLC).

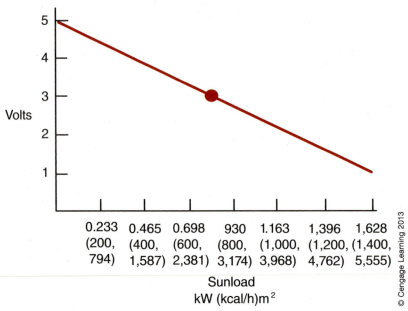

FIGURE 10-25 The sun load sensor is photodiode sensor: As the sun intensity increases, the signal line voltage will decrease.

4. Select "Sensor Data Stream" from the menu.
5. Observe the sun load sensor input value (Figure 10-25).
6. Start the engine and allow it to idle. Turn the air-conditioning system on and select FULL AUTO mode. Set the temperature to approximately 70°F (21°C) on the ATC system. Allow the vehicle to run and the system to stabilize for 10 minutes.
7. Cover the sensor with a dark piece of paper or cloth while observing data values on the scan tool. Cover and uncover the sensor several times, waiting 30–60 seconds between each interval to give the system time to react. The sensor data values should respond quickly to changes in light levels. If the sensor values do not change, further diagnosis will be required. Consult specific vehicle service information for detailed diagnostic steps.

If data steam values are not available with the scan tool you are using, perform the preceding test and pay attention to blower motor speed and center duct output temperature. Both the air volume and temperature should change during the test.

Infrared Temperature Sensor

Classroom Manual

Chapter 10, page 354

The infrared temperature sensor detects and measures thermal radiation emitted by the front seat passenger area and converts this data into a pulse width modulated output signal that is sent to the ATC module. The ATC system software logic uses the input information to adjust airflow temperature and flow rate to compensate for solar heat gain or evaporator heat loss in order to maintain the comfort level selected by the passenger(s). Sensor operation is checked using a diagnostic scan tool. Refer to vehicle-specific service information and electrical diagrams for diagnostic information. The ATC module continuously monitors the infrared temperature sensor circuit and will set diagnostic trouble codes for any circuit or sensor malfunction.

Infrared temperature sensor quick test:

1. Turn the ignition to the ON position, but do not start the vehicle yet.
2. Connect the scan tool to the vehicle DLC.
3. Select "Sensor Data Stream" from the menu.
4. Observe the infrared temperature sensor input value.

5. Start the engine and allow it to idle. Turn the air-conditioning system on and select FULL AUTO mode. Set the temperature to approximately 70°F (21°C) on the ATC system. Allow the vehicle to run and the system to stabilize for 10 minutes.

6. Get two small buckets and two small towels. Fill one bucket with hot water and place a towel in it. Fill the other bucket with cold water and ice and place the other towel in this bucket.

7. Squeeze out the cold towel so as not to drip water on the interior of the car, and allow it to hang directly in front of the sensor while observing data values on the scan tool. Next squeeze out the hot towel so as not to drip water on the interior of the car, and allow it to hang directly in front of sensor while observing data values on the scan tool. Repeat this process from cold to hot several times, waiting 30–60 seconds between each interval to give the system time to react. The sensor data values should respond quickly to changes in heat levels. If the sensor values do not change, further diagnosis will be required. Consult specific vehicle service information for detailed diagnostic steps.

If data steam values are not available with the scan tool you are using, perform the preceding test and pay attention to blower motor speed and center duct output temperature. Both the air volume and temperature should change during the test.

Coolant Temperature Sensor

The engine coolant temperature sensor is an NTC thermistor and is mounted in an engine coolant passage. Like other sensors, it is supplied a 5 V reference voltage that is modified by the resistance of the sensor (Figure 10-26). This modified voltage is sent via the signal line to the PCM/ECM and shared on the data bus line. The sensor resistance can range from 100,000 ohms to a few hundred ohms depending on coolant temperature. An increase

Classroom Manual
Chapter 10, page 354

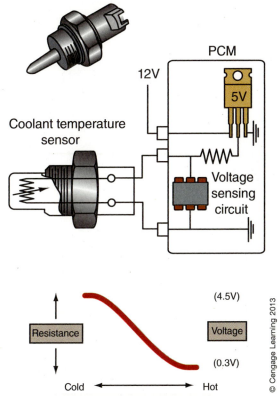

FIGURE 10-26 The engine coolant temperature sensor is an NTC thermistor.

© Cengage Learning 2013

in temperature will cause a decrease in sensor resistance. If a high-temperature condition is detected, the PCM/ECM will deactivate the air-conditioning compressor clutch relay. When the engine returns to normal temperature, the control module will reengage the compressor clutch. The ATC system also uses coolant temperature information as part of its cold engine lockout strategy and will not allow the blower motor to operate until warm coolant is available for heating the air via the heater core.

Coolant temperature sensor quick test:

1. Turn the ignition to the ON position, but do not start the vehicle yet.
2. Connect the scan tool to the vehicle DLC.
3. Select "Sensor Data Stream" from the menu.
4. Observe the coolant temperature value.
5. Using a noncontact infrared thermometer, measure the temperature at or near the coolant temperature sensor. If the data stream value and the measured temperature are the same, proceed to the next step. If the values are not the same, further diagnosis will be required.
6. Start the engine and allow it to idle.
7. For the next 10 minutes monitor the actual sensor temperature and scan tool data value for engine coolant temperature. If they are similar the system is operating normally. If the values are not the same, further diagnosis will be required. Consult specific vehicle service information for detailed diagnostic steps.

Ambient Temperature Sensor

Classroom Manual

Chapter 10, page 350

To test the ambient temperature sensor (ATS), first remove it from its socket and then measure its resistance using an ohmmeter. At an ambient temperature between 70°F and 80°F (21°C and 27°C), the sensor resistance should be between 225 and 235 Ω. If the resistance is not within this range, the ambient sensor is defective and must be replaced.

NOTE: Because ohmmeter battery current flow through the sensor and body heat will affect the readings, do not hold the sensor in your hand or leave the ohmmeter connected for longer than 5 seconds (Figure 10-27).

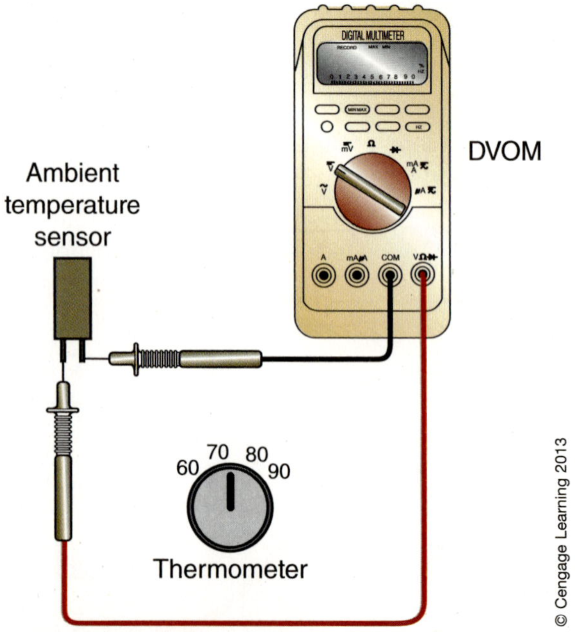

© Cengage Learning 2013

FIGURE 10-27 Test the ambient temperature sensor at room temperature. Avoid touching the sensor during testing. Also, do NOT connect the ohmmeter for longer than 5 seconds.

In-Car Temperature Sensor

The in-car temperature sensor, a thermistor, is located inside an aspirator. To provide an accurate temperature reading, a small sample of air is drawn through the aspirator across the in-car temperature sensor.

The resistive value of the in-car temperature sensor is sent to the BCM and is used by the ECC for calculations to maintain the preselected in-vehicle temperature conditions.

The following procedure may be used to test the in-car temperature sensor:

1. Disconnect its electrical connector. Do not disconnect the aspirator tubes or remove the temperature sensor from the panel.
2. Place a test thermometer into the air inlet grill near the sensor.
3. Set the blower motor speed control to MED.
4. Depress then pull out the NM-A/C button. This will turn off the compressor and close the water valve.
5. Operate the blower while quickly measuring the resistance of the sensor.

NOTE: Do not leave the ohmmeter connected to the sensor terminals for longer than 5 seconds or inaccurate readings will result. Ohmmeter battery current flow through the sensor, a thermistor, and body heat will affect resistance.

Resistance of the in-car temperature sensor should be 1,100–1,800 Ω at an ambient temperature between 70°F and 80°F (21°C and 27°C). The sensor must be replaced if the resistance is not within these specifications.

Aspirator

The **aspirator** is an assembly device that houses the in-car temperature sensor (Figure 10-28). A quick method for testing the aspirator assembly to verify that it is providing enough airflow to the in-car temperature sensor is to set the controls for HI blower speed while in the heat mode of operation. Place a piece of paper, large enough to cover the aspirator

Classroom Manual
Chapter 10, page 352

An **aspirator** is a device that uses a negative pressure (suction) to move air.

To aspirate is to draw by suction.

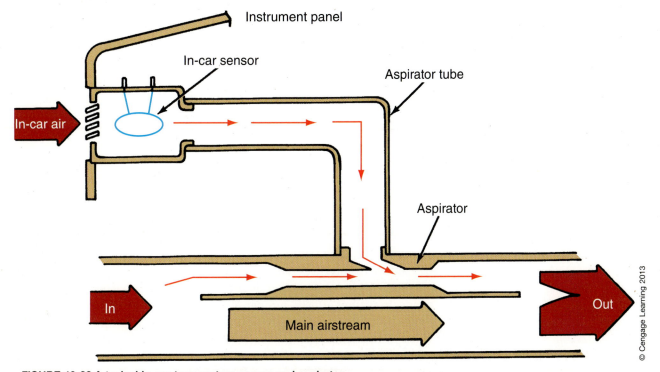

FIGURE 10-28 A typical in-car temperature sensor and aspirator.

© Cengage Learning 2013

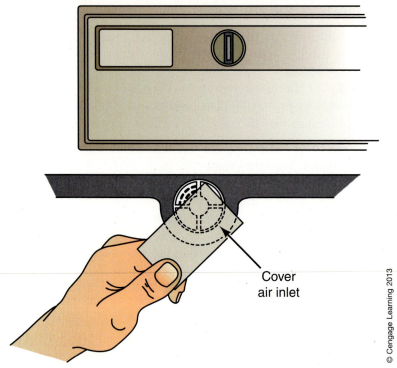

Cover
air inlet

© Cengage Learning 2013

FIGURE 10-29 When performing the aspirator paper test, the vacuum should hold the paper against the grille.

air inlet, over the inlet (Figure 10-29). The suction of the aspirator should be great enough to hold the paper against the inlet grille. If it is not, refer to the aspirator system diagnostic chart in Figure 10-30.

Pressure Transducer

Classroom Manual Chapter 10, page 345

The pressure transducer is mounted on the high-pressure discharge line from the refrigerant compressor and shuts off the compressor if discharge pressure is either too high or too low. Some systems also use the signal to control condenser cooling fan operation. The pressure transducer is a three-wire sensor consisting of a 5-volt reference, ground, and output signal wire (Figure 10-31). The pressure transducer varies voltage on the signal wire in relation to refrigerant system high-side pressure. An increase in pressure will cause an increase in signal voltage. The operating range for the sensor is 0.10 volt for zero (0) psig and 4.90 volts for 450 psig (3,103 kPa). The normal operating range for the sensor is 1.0 V = 75 psig (517 kPa) to 4.4 V = 400 psig (2,758 kPa).

When diagnosing no compressor action you will need to verify the proper operation of the pressure transducer using a scan tool and manifold and gauge set. With the vehicle running and the HVAC system turned on, compare the pressure gauge readings to the pressure displayed on the scan tool data screen to verify proper operation. If the scan tool does not displace the same pressure as that displaced on the gauge diagnosis of the sensor, wiring and processor operation will be required. A quick method for diagnosing sensor wiring and the processor's ability to interpret data is to attempt to set both a high-and a low-voltage signal code. To set a high-voltage code disconnect the sensor harness, and on the harness side of the connector, using a fused jumper lead, connect the 5 V reference wire to the signal wire. If the processor is working correctly a high-voltage (pressure) code will be set. Next, try to set the opposite code for a low-voltage (pressure) condition. Again, disconnect the sensor harness and on the harness side of the connector, using a fused jumper lead, connect the ground wire to the signal wire. If the processor is working correctly a low-voltage (pressure) code will be set.

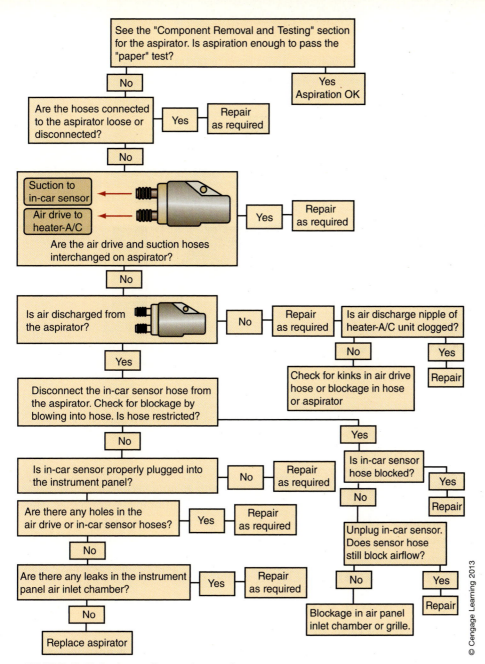

FIGURE 10-30 Aspirator diagnostic test chart.

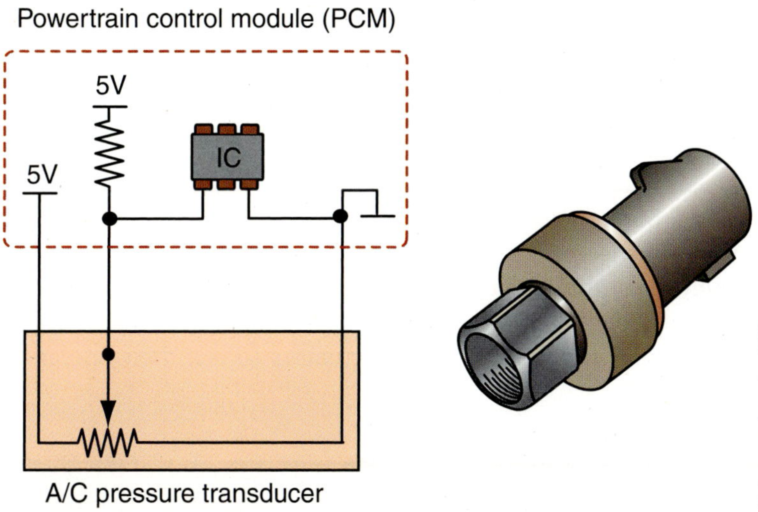

Powertrain control module (PCM)

A/C pressure transducer

FIGURE 10-31 The refrigerant system pressure transducer varies voltage on the signal wire in relation to refrigerant system high-side pressure and sends this information to the control module.

© Cengage Learning 2013

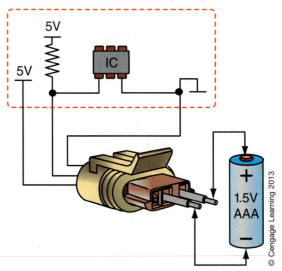

Powertrain control module (PCM)

© Cengage Learning 2013

FIGURE 10-32 The pressure transducer signal line may be test using a 1.5 V battery. Connect a jumper wire from the + battery terminal on the 1.5 V AAA battery to the PCM signal sensing line, and use another jumper lead to connect the negative terminal to the PDM ground to send fixed information to the control module.

SERVICE TIP:
If you suspect an intermittent electrical problem with the climate control system, such as a short or open connection, grasp the wiring harness and shake it, especially near connections and splices, both in the engine compartment and behind the dash panel if access allows. This will aid in locating troublesome intermittent connections.

Classroom Manual

Chapter 9, page 305

A **servomotor** is an electrical motor that is used to control a mechanical device, such as a coolant control valve.

In the event that you do not have access to a scan tool, you can use a 1.5 V AAA battery to fool the computer into thinking it is receiving a valid pressure transducer voltage signal. With the vehicle running and the HVAC system turned on, disconnect the pressure transducer harness connector. Using two jumper leads connect the positive terminal of the 1.5 V battery to the signal wire and connect the ground terminal of the battery to the connector ground wire (Figure 10-32). If the air-conditioning compressor engages, the pressure transducer is faulty; only allow the refrigerant system to operate for less than a minute during this test to avoid the potential for system overpressurization.

TEMPERATURE AND MODE DOOR CONTROL

Heater Flow Control Valve

The heater flow control valve is opened or closed by a signal from the BCM to provide in-vehicle temperature control. If the valve is found to be defective, it must be replaced. Photo Sequence 17 illustrates a typical procedure for replacing a heater flow control valve.

Actuators

An actuator is a device that transforms a vacuum or electrical signal to a mechanical motion. It is the component that performs the actual work commanded by the computer. An actuator may be an electric or vacuum motor, relay, switch, or solenoid that typically performs an on/off, open/close, or push/pull operation.

Testing Actuators. Most systems allow for testing of the actuator through the scan tool or FCC panel while in the correct mode. Actuators that are duty cycled by the computer are more accurately diagnosed through this method. As in the earlier example of retrieving trouble codes from the Chrysler system using the DRB-II scan tool, select ACTUATOR TESTS. This will allow activation of selected actuators to test their operation.

Servomotors

A **servomotor** is a vacuum or electric motor that is used to control the position of the mode and blend air doors in an automotive heating and air-conditioning case/duct system.

TYPICAL PROCEDURE FOR REPLACING A HEATER FLOW CONTROL VALVE

P17-1 Drain the coolant from the system and recycle it.

P17-2 Loosen and remove the inlet hose clamp at the heater core. Slip the heater hoses off their fittings.

P17-3 Loosen and remove the retaining bolt for the bracket of the heater control valve.

P17-4 Remove all other heating control valve retaining bolts and any parts that may interfere with the removal of the control valve.

P17-5 Remove the control valve and inspect the hoses connected to it.

P17-6 Clean the coolant pipes and hoses. Make sure all damaged parts are replaced. Then replace the heater control valve.

P17-7 Install and tighten all heater control valve and valve bracket retaining nuts and bolts.

P17-8 Install new clamps and connect the heater hoses to the heater core.

P17-9 Fill the cooling system with fresh coolant to the correct level. Bleed the system, if necessary.

P17-10 **Pressure test the system and check for leaks. Run the engine and allow it to reach normal operating temperature. Then shut it off and retest for leaks.**

Servomotor Test. The servomotor must be removed from the vehicle for testing. Follow the procedures outlined in the service manual for removing the motor. While operating the motor, check for smooth operation and observe the testlight. If the motor briefly jams, the testlight illumination level will increase. If the testlight flickers while the motor is operating, the motor is not moving in a smooth fashion. The motor must be able to move to the full cw and full ccw position. If the cause of a problem cannot be corrected, the servomotor will have to be replaced.

DIAGNOSING SATC AND EATC SYSTEMS

Semiautomatic or automatic control of the interior (cabin) temperature is made possible through the use of electronic components such as microprocessors, thermistors, and potentiometers that control vacuum and electric actuators. A failure in any of these components will result in inaccurate or no temperature control. Today's technician must possess a basic knowledge and understanding of the operating principles of both the semi-automatic temperature control (SATC) and **electronic automatic temperature control (EATC)** systems and must be proficient at diagnosing and servicing these systems.

These troubleshooting procedures include that portion of the system that controls its operation. Procedures to troubleshoot the other components of the system are found elsewhere in this manual.

EATC stands for electronic automatic temperature control.

SATC System Diagnosis

An SATC system controls both the operating mode and blower fan speed of the air-conditioning system. Faults within the evaporator and heater systems will have an adverse effect on the operation and control of the SATC system.

To properly troubleshoot and service an SATC system, a schematic and specifications for the vehicle being diagnosed are essential. Motors and compressor clutch circuits of the SATC system are tested in the same manner as manual temperature control (MTC) systems.

The air delivery control of SATC systems differs among manufacturers and requires specific diagnostic procedures. Most such systems have specific tests to troubleshoot each particular system, and reference should always be made to each model's service manual. A typical example of one system and its procedures follows.

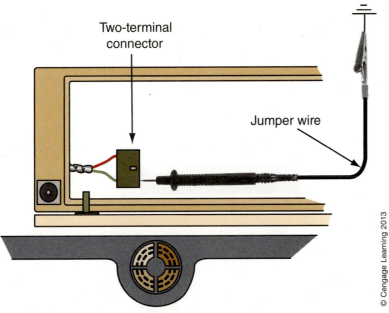

FIGURE 10-33 Jumper wire connections to set the blend door to the minimum position.

Chrysler SATC Troubleshooting. To perform some of the service manual tests on Chrysler's pre-CAN bus network SATC system, it may be necessary to place the blend air door in one of three different positions. When locking the blend door in the full and minimum positions, set the controls for LOW blower speed and BI-LEVEL mode. Follow these procedures for setting the door position:

1. Set the blend door in the full reheat position by disconnecting the in-car sensor and turning the system ON.
2. Obtain the minimum reheat position by connecting a jumper wire between the red terminal wire from the variable resistor and ground (Figure 10-33), and then turning the system ON. DO NOT connect the jumper wire to the sensor side of the red terminal.
3. **a.** Set the blend door in the middle position by first disconnecting the negative battery cable and removing the ground screw on the passenger side cowl.
   b. Next, connect a jumper wire from the blower ground wire (black wire with tracer) to a good ground.
   c. Then reconnect the battery negative cable.
   d. Finally, move the temperature control lever until the blend door moves to the middle position.
   e. Then disconnect the jumper wire.

Refer to the test point diagram in Figure 10-34 for the particular vehicle being serviced. Use a voltmeter to measure the voltage between points A and B and points J and I while the system is placed in any mode other than OFF. The voltage values at these test points should be 11 volts or more.

Before performing the continuity tests (Figure 10-35 and Figure 10-36), disconnect the in-car sensor and the power feed connector. Follow the continuity test procedures to determine any system defects.

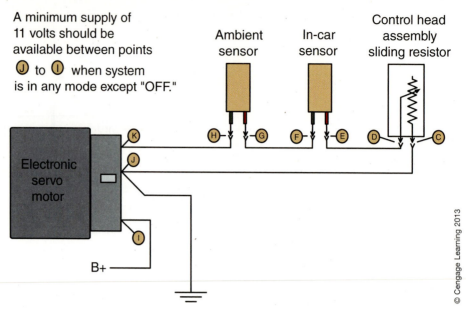

A minimum supply of 11 volts should be available between points Ⓙ to Ⓘ when system is in any mode except "OFF."

Ambient sensor

In-car sensor

Control head assembly sliding resistor

© Cengage Learning 2013

FIGURE 10-34 Chrysler semiautomatic temperature control system electrical test points.

SATC CONTINUITY TEST PROCEDURE

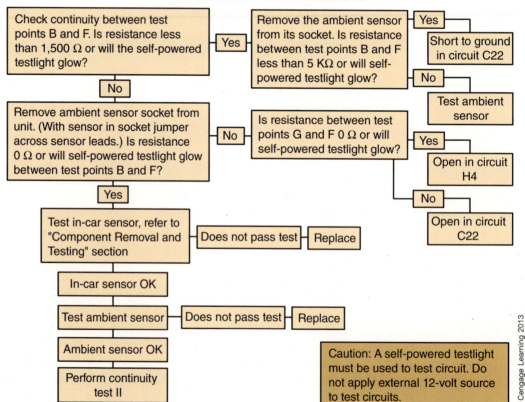

FIGURE 10-35 SATC continuity test procedure (part 1).

© Cengage Learning 2013

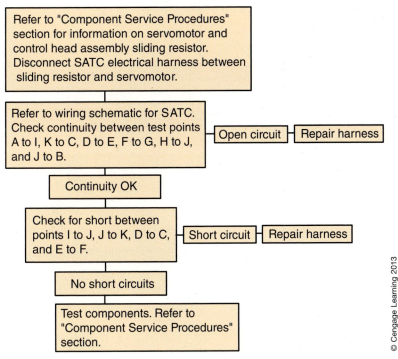

SATC CONTINUITY TEST PROCEDURE

Refer to "Component Service Procedures" section for information on servomotor and control head assembly sliding resistor. Disconnect SATC electrical harness between sliding resistor and servomotor.

Refer to wiring schematic for SATC. Check continuity between test points A to I, K to C, D to E, F to G, H to J, and J to B.

Open circuit — Repair harness

Continuity OK

Check for short between points I to J, J to K, D to C, and E to F.

Short circuit — Repair harness

No short circuits

Test components. Refer to "Component Service Procedures" section.

© Cengage Learning 2013

FIGURE 10-36 **SATC continuity test procedure (part 2).**

EATC System Diagnosis

Proper diagnostics of EATC systems depends on system design. There are two basic system designs: those that use their own microprocessor and those that incorporate the controls of the system into the BCM.

Since the first introduction of CAN networked systems in the early 2000s, system operation has varied depending on vehicle model and year. Some systems allow the retrieval of DTCs and other data using a scan tool. Other systems do not support scan tool diagnostics and still rely on self-diagnostics through the climate control head as in the past. Still others support both scan tool diagnostics and self-diagnostics via the control head.

Scan tool diagnostics operate in the same manner for climate control systems as it does for engine and emissions-related diagnostics. Scan tools give the technician the ability to look at the sensor and actuator date and, in many cases, bidirectional control is possible. Bidirectional control allows the scan tool to function as the control module, allowing the technician to command actuators ON and OFF, such as the compressor clutch relay and mode door motor position. Trouble codes for climate control systems may be in the form of typical power train-related codes (i.e., PXXXX), body-related codes (i.e., BXXXX), and sometimes are network-related fault codes (i.e., UXXXX). It is important never to ignore any diagnostic trouble code when systematically troubleshooting a failure with a climate control system or subsystem. Please refer to the end of Chapter 10 in the Classroom Manual for a list of codes relevant to the climate control system; the list is comprehensive but is in no way meant to substitute for manufacturer's service and diagnostic information. You should explain to a customer that systems today are interrelated and a fault in one system often causes performance issues in another system.

The following discussion focuses on the self-diagnostic routines on a dual climate control system used on Jeep to illustrate the diagnostic capabilities of these systems. The automatic zone control (AZC) system performs self-diagnostic routines as part of normal system operation similar to the diagnostic routines used in OBD II power train systems.

Classroom Manual
Chapter 10, page 361

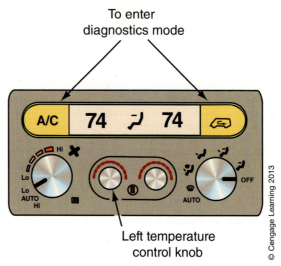

To enter diagnostics mode

Left temperature control knob

FIGURE 10-37 **The automatic zone control head buttons and knobs are used to enter the self-diagnostic mode.**

During normal operation the control module continuously monitors various system parameters. In the event a fault is detected, both a current and a history code will be stored. If the fault is intermittent, the current code will be erased after several successful drive cycles, but the history code will remain. The fault code may be retrieved by using a scan tool or entering the self-diagnostic mode through the front display panel of the control head. The control panel has three different self-diagnostic routines. They are the fault code test, the input circuit test, and the actuator test.

Like older automatic and semiautomatic climate control systems, self-diagnostics is entered by performing a sequence of steps within a specific period using the control head buttons or knobs. To enter self-diagnostics, turn the ignition key to the RUN position with engine OFF. Depress and HOLD both the A/C and the RECIRC buttons and rotate the left temperature control knob cw one detent (Figure 10-37). During this phase of the test all segments of the control panel should illuminate for as long as you hold the buttons. If the segments on the control panel fail to illuminate, replace the control head assembly.

Once control panel illumination has been verified, release the A/C and the RECIRC buttons. The control panel will clear and then display any fault codes stored. A zero on the display screen indicates that no codes are stored. If fault codes are available, they will be presented in ascending number order. Each code will be displayed for 1 second. When all codes have been displayed the code display cycle will be repeated until the left temperature control knob is rotated cw one detent or the ignition switch is turned to OFF.

In order to erase fault codes, hold down the A/C and the RECIRC buttons for 3 seconds while the system is in the fault code display mode. Fault code clearing is verified when two bars appear on the display. It should be noted that only history codes can be cleared. The fault must be found and repaired before current codes are removed.

After the fault code mode has been completed, the self-diagnostic select test mode phase may be entered. To select a test rotate the left temperature control knob until the desired test appears on the display screen. It will be necessary to refer to the manufacturer's service information to determine the correct test number for the fault being diagnosed. Once you have scrolled to the correct test number and it appears on the display, press the A/C button to activate the test. As an example, "21" will test the current mode position.

It should also be noted that on many automatic climate control systems today a relearn procedure is required if the control assembly, mode door actuators, or sensors are replaced. Also if the vehicle battery has been disconnected or completely discharged for an extended period, a relearn test may have to be conducted before the climate control system will function. With some makes and models a scan tool is required to activate this relearn process. Always consult specific vehicle and manufacturer service information as well as TSBs before attempting a repair to familiarize yourself with the system.

Separate Microprocessor-Controlled Systems

Most EATC systems that use a separate microprocessor for diagnosis have the microprocessor contained in the control assembly (Figure 10-38). Also, most of these early systems provide a means of self-diagnostics and have a method of retrieving trouble codes. Typical examples of such systems follow.

Nissan ATC Diagnostics. The ATC systems used on Nissan/Infinity are similar in operation and function to that of the domestic manufacturers. The following discussion is based on the self-diagnostic system of a 2008 Infinity M35/45 but is representative of many Nissan platforms and serves as an example of diagnostic systems used on Asian import vehicles. The self-diagnostic system diagnoses HVAC system sensors, actuator motors, blower motor, and any electrical component associated with the climate control system.

The self-diagnostic mode for this system is entered by first turning the key to the ON position and starting the engine. The OFF button on the A/C control panel must be pressed and held for at least 5 seconds. The OFF button must be pressed within 10 seconds of the engine starting. The self-diagnostic test will be cancelled if the AUTO button is pressed or the ignition is cycled to the OFF position. Moving from one test to the next is accomplished by pressing the temperature control buttons on the driver's side (Figure 10-39). By pressing the UP temperature button you will advance forward in the self-diagnostic steps, and by pressing the DOWN temperature button you will back up one step.

The following are the five basic steps of the self-diagnostic tests:

1. All of the LEDs and display segments are illuminated. If any of the segments or LEDs are not illuminated, the component is defective and the control head must be replaced.

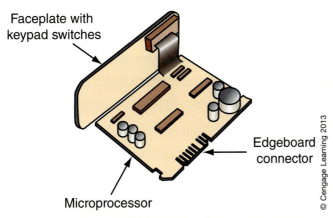

FIGURE 10-38 **Most EATC systems use a separate microprocessor located in the control assembly.**

© Cengage Learning 2013

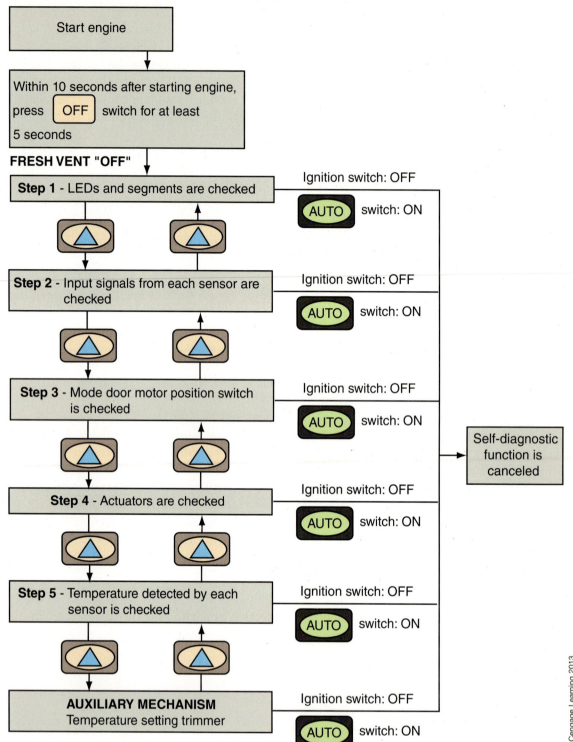

FIGURE 10-39 The five-step self-diagnostic process for Nissan/Infinity automatic temperature control systems.

2. Press the UP temperature control arrow to move to step 2. The input signals from each sensor are checked by the control module. If no faults are detected with the input sensors, the control head will display "20." If an input sensor fault is detected a code for that sensor will be displayed.

Code Number	Malfunctioning Sensor and Door Motor
21	Ambient Temperature Sensor
22	In-Car Temperature Sensor
24	Intake Sensor
25	Sun Load Sensor
26	Air Mix Door Motor PBR (Driver Side)
27	Air Mix Door Motor PBR (Passenger Side)

3. Press the UP temperature control arrow to move to step 3. The mode and intake door motor position switch(es) will be checked. If no faults are detected with the mode door or intake door motor position switch(es), the control head will display "30." If a mode door or intake door motor position switch fault is detected, a code for the defective switch will be displayed ranging from 31 to 39.

Code Number	Malfunctioning Mode Door Position Switch
31	VENT (driver's side)
32	DEFROST (driver's side)
33	VENT (passenger side)
34	DEFROST (passenger side)
35	VENT (open)
36	VENT (shut)
37	FRESH AIR
38	20% FRESH AIR
39	RECIRCULATION

4. Press the UP temperature control arrow to move to step 4. The actuators will be checked. The operation of each mode door motor is checked. The A/C control panel displays "41" and the control module positions the mode door in the vent position, then the intake door in the recirculation (REC) position, and the air mix door is moved to the full cold position. Press the DEF (defrost) button to move to the next mode in step 4 and "42" will be displayed on the control panel. There are six modes in step 4 and each mode is represented by a number on the A/C control panel. These numbers range from 41 to 46. The DEF button is used to select the next mode. In each mode, the A/C controller commands a specific door position, blower motor voltage, and compressor clutch operation. Door operation may be checked by the air discharge from the various ducts.

Code Number	41	42	43	44	45	46
Mode Door Position	Vent	Bi-Level 1	Bi-Level 2	Floor	Defrost/Floor	Defrost
Upper Ventilator Door Position	Open	Closed	Closed	Closed	Closed	Closed
Intake Door Position	Recirculated	Recirculated	20% Fresh	Fresh	Fresh	Fresh
Air Mix Door Position	Full Cold	Full Cold	Full Hot	Full Hot	Full Hot	Full Hot
Blower Motor Duty Ratio	37%	91%	65%	65%	65%	91%
Compressor	ON	ON	OFF	OFF	ON	ON
Electronic Control Valve Duty Ratio	100%	100%	0%	0%	50%	100%

5. Press the UP temperature control arrow to move to step 5. Temperature detection by each sensor is checked. After this mode is entered, "51" is displayed in the A/C panel display. If the DEF button is pressed, the temperature sensed by the ambient air sensor is displayed on the control panel display. Press the DEF button again and the temperature sensed by the in-car sensor is displayed on the control panel display. Press the DEF button a third time and the temperature sensed by the intake sensor is displayed on the control panel display (Figure 10-40). When the temperature displayed varies significantly from the actual temperature, the sensor and connections should be tested with an ohmmeter and a voltmeter.

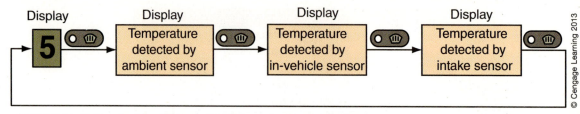

FIGURE 10-40 Step 5 of the five-step self-diagnostic process for Nissan/Infinity automatic temperature control systems displays the temperature detected by each sensor.

Press the Intake switch and CAN communication error between each unit is checked. If no faults with the CAN network are detected the control head will display "52."

After step 5 has completed, the blower speed button may be pressed to enter the auxiliary mode. After this mode is entered, "61" is displayed in the A/C panel display. Next, the temperature on the display may be adjusted so it is the same as the in-car temperature felt by the driver. After the auxiliary mode is entered, press the up or down temperature control buttons until the A/C control head displays the same temperature as that inside the vehicle. The temperature can be tailored ±6°F (±3°C) between the temperature displayed and that felt by the customer.

Chrysler EATC Troubleshooting. Before entering self-diagnostics, start the vehicle and allow it to reach normal operating temperature. Ensure that all exterior lights are off and press the PANEL button. If the display illuminates, the self-diagnostic mode can be entered. If, however, the display does not illuminate, check the fuses and circuits to the control assembly. If the fuses and circuits are good, replace the ATC computer.

If the display illuminates, the self-diagnostic mode may be entered by pressing the BI-LEVEL, FLOOR, and DEFROST buttons simultaneously (Figure 10-41). If no trouble codes are present, the self-test program will be completed within 90 seconds and display a "75."

During the process of running the self-diagnostic tests, make four observations that the computer is not able to make by itself:

1. When the test is first initiated, all of the display symbols and indicators should illuminate.
2. The blower motor should operate at its highest speed.
3. Air should flow through the panel outlets.
4. The air temperature should become hot, then cycle to cold.

The diagnostic flowchart may be used to determine the correct test to perform if any of these functions fail (Figure 10-42). The proper procedures for an observed failure are found in the table in Figure 10-43.

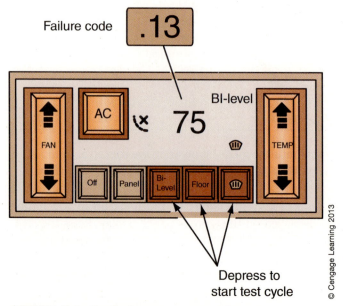

FIGURE 10-41 Use the buttons to enter diagnostics.

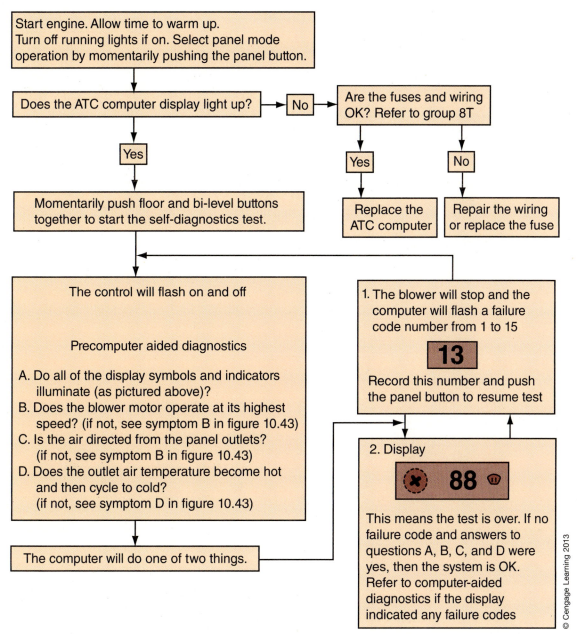

FIGURE 10-42 A typical Chrysler diagnostic flowchart for an ATC system in which a technician must answer four questions.

If a fault is detected in the system, a trouble code will be flashed on the display panel. To resume the test, record the trouble code then press the PANEL button. Refer to the service manual to diagnose the trouble codes received.

Ford EATC Troubleshooting. To correctly diagnose Ford's EATC system, the exact system description as well as the exact procedures for trouble code retrieval are required. This is because Ford uses different versions of EATC systems that have different diagnostic capabilities. The following is only a typical example of performing the self-test:

1. Turn the ignition switch to the RUN position.
2. Place the temperature selector to the "90" setting and select the OFF mode.
3. Wait 40 seconds while observing the display panel. If the VFD display begins to flash, there is a malfunction in the blend actuator circuit, the actuator, or the control assembly. If the LED light begins to flash, this indicates there is a malfunction in one of the other actuator circuits, the actuators, or the control assembly.
4. If no flashing of displays occurs, place the temperature selection to "60" and select the DEF mode.

DIAGNOSTIC CHART

NO	PROBABLE CAUSE	PROCEDURE
A	1. Control	
B	1. Wiring problem 2. Power vacuum module	a. Replace control module CAUTION: Take care when working around the the blower motor fan. The power/vacuum heat sink is hot (12 volts). DO NOT operate the module for a period longer than 10 minutes with the unit removed from the housing. b. Ensure that the connections are good at the blower motor and power/vacuum module. c. If diagnostic test results in a code 8 or 12, refer to the fault code page in the service manual. If no codes are present, check the blower motor fuse. d. Disconnect the blower motor and check for voltage. A reading of 3 to 12 volts for 1 to 8 bar segments on the display is correct. If correct, replace motor. e. If voltage is not correct, measure the voltage-to-vehicle ground. Voltage should read 12 volts with the ignition ON. If OK, replace the power/vacuum module.
C	1. Vacuum leakage 2. Power vacuum leakage	a. Service if any codes are found. b. Check all connections. c. Disconnect vacuum control and connect it to a manual control to test each port. To test the check valve, select Panel Mode, disconnect the engine vacuum, and see if mode changes quickly. d. Try a new power/vacuum module.
D	1. Refrigeration system 2. Heater system 3. Blend-air door	a. Complete diagnostic test. Refer to the Fault Code page in the Shop Manual if code appears. b. If a temperature difference of 40 °F (22.2 °C) or more is noted during the test, the blend-air door is engaged in the servomotor actuator. A lower temperature indicates a blend-air door operation problem c. Check heater system 85 °F setting is full heat; 65 °F is full cool. d. Check air-conditioning system.

FIGURE 10-43 If the technician's answer was NO to any of the self-diagnosis test questions of Figure 10-42, this diagnostic chart may be used to isolate the fault.

5. Wait 40 seconds while observing the VFD and LED displays. If there are no malfunctions in the actuator drive or feedback circuits, the displays will not flash.
6. Regardless of whether or not flashing displays were indicated, continue with self-diagnostics. Press the OFF and DEFROST buttons at the same time.
7. Within 2 seconds, press the AUTO button.

Once the self-diagnostics is entered, if an "88" is displayed, there are no trouble codes present. If there are any trouble codes retrieved, they will be displayed in sequence until the COOLER button is pressed. Always exit self-test mode by pressing the COOLER button before turning the ignition switch to the OFF position. Refer to the trouble code chart in Figure 10-44. When service repairs have been performed on the system, rerun the self-test to confirm that all faults have been corrected.

CODE	SYMPTOM	POSSIBLE CAUSE
1	Blend actuator is out of position. VFD flashes.	Open circuit in one or more actuator leads Actuator output arm jammed Actuator inoperative Control assembly inoperative
2	Mode actuator is out of position. LED flashes.	Same as 1
3	Pan/Def actuator is out of position. LED flashes.	Same as 1
4	Fresh air/recirculator actuator is out of position. LED flashes.	Same as 1
1, 5	Blend actuator output shorted. VFD flashes.	Output A or B shorted to ground, or to supply voltage, or together Actuator inoperative Control assembly inoperative
2, 6	Mode actuator is shorted. LED flashes.	Same as 1, 5
3, 7	Pan/Def actuator output shorted, LED flashes.	Actuator output is shorted to supply voltage Actuator inoperative Control assembly inoperative
4, 8	Fresh air/recirculator actuator output is shorted. LED flashes.	Same as 3, 7
9	No failures found; see supplemental diagnosis.	
10, 11	A/C clutch never ON.	Circuit 321 open BSC inoperative Control assembly inoperative
10, 11	A/C clutch always ON.	Circuit 321 shorted to ground BSC inoperative Control assembly inoperative
12	System stuck in full-heat. In-car temperature must be stable above 60 °F for this test to be valid.	Circuit 788, 470, 767, or 790 is open. Ambient or in-car sensor inoperative
13	System stays in full A/C.	Remove control assembly connectors. Measure resistance between pin 10 of connector #1 and pin 2 of connector #2. If the resistance is less than 3 KΩ, check wiring and in-car and ambient sensors. If resistance is greater than 3 KΩ, replace the control assembly.
14	Blower always at Max speed.	Turn OFF ignition. Remove connector #2 . Remove terminal #5. Replace connector, tape terminal and turn ON ignition. If blower still is at Max speed, check circuit 184 and the BSC. If the blower stops, control assembly inoperative.
15	Blower never runs.	Circuit 184 shorted to power supply BSC inoperative Control assembly inoperative

FIGURE 10-44 Trouble code chart for Ford EATC system.

GM ETCC Troubleshooting. General Motors (GM) uses several different versions of the microprocessor-controlled electronic touch climate control (ETCC) system. Depending on the GM division and system design, the door controls can be either by vacuum or by electric servomotor. Methods of entering diagnostics also vary between divisions and models. For this reason, the correct service manual for the system being serviced is needed to perform correct diagnostic procedures. Knowledge of one ETCC system type is no guarantee that you will be able to service other ETCC systems without the use of the proper service manual.

Many GM EATC systems can be checked for proper operation by using a functional chart (Figure 10-45). In addition, troubleshooting charts that correspond with fault code or symptoms (Figure 10-46) are a great help.

ELECTRONIC CLIMATE CONTROL (ECC) FUNCTIONAL TEST			
Air-conditioning system diagnostics should begin with a functional test. This test should be performed in the order listed in the chart below. If the answer to any test question is NO, proceed to the specirc trouble tree for further testing. Do not omit any steps in the test. Check the fuses and stop hazard operation to verify the stop/hazard fuse. Warm the engine up before performing the functional test, and check LED's above each button as the test is performed.			
TEST	SYSTEM CHECKS	CONTROL SETTING	TROUBLE TREE
1	Do the MPG and control head display?	All	1
2	Do COOLER and WARMER push buttons operate?	All	1
3	Set TEMP to 60 °F (42.2 °C)		
	a. Does blower operate?	LO-AUTO-HI	2A
	b. Is there low blower speed?	LO	2B
	c. Is there high blower speed?	HI	2C
	d. Is air flow from A/C outlets?	ECON-LO-AUTO-HI	3
	e. Does compressor engage?		
	1. engage?	LO-AUTO-HI	4
	2. disengage?	OFF-ECON	5
	f. Is A/C outlet air cold?	LO-AUTO-HI	6
	1. Is there only heat?		
	2. Is cooling adequate?		7
	g. Does recirc door fully open? (allow 1-2 minutes)	AUTO-HI	8
4	Set temperature to 90 °F (67.2 °C) a. Is heat adequate?		
5	Set temperature to 85 °F (67.2 °C) a. Is air warm or hot?	AUTO	9
6	Does front defroster operate?	LO-AUTO	10
7	Does rear defroster operate?	FRT DEF	11
8	Does rear defroster turn OFF?	RR DEF	12
		RR DEF OFF	13

FIGURE 10-45 General Motors' Electronic Automatic Temperature Control (EATC) function test.

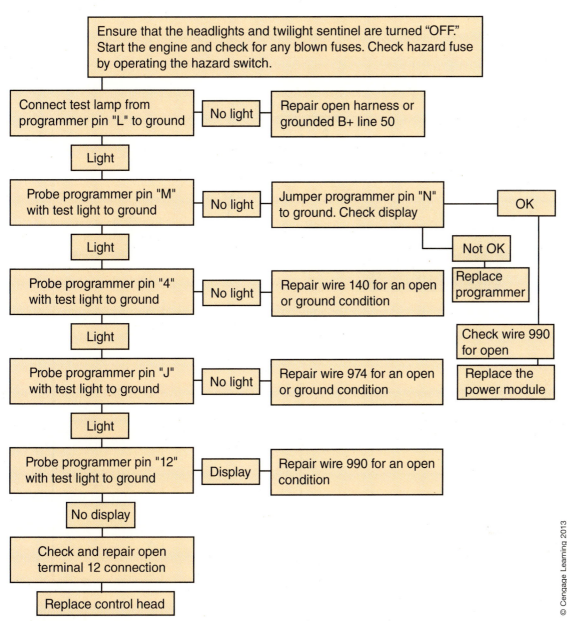

Ensure that the headlights and twilight sentinel are turned "OFF." Start the engine and check for any blown fuses. Check hazard fuse by operating the hazard switch.

Connect test lamp from programmer pin "L" to ground — No light → Repair open harness or grounded B+ line 50

Light

Probe programmer pin "M" with test light to ground — No light → Jumper programmer pin "N" to ground. Check display → OK

Not OK → Replace programmer

Light

Probe programmer pin "4" with test light to ground — No light → Repair wire 140 for an open or ground condition

Check wire 990 for open

Light

Probe programmer pin "J" with test light to ground — No light → Repair wire 974 for an open or ground condition

Replace the power module

Light

Probe programmer pin "12" with test light to ground — Display → Repair wire 990 for an open condition

No display

Check and repair open terminal 12 connection

Replace control head

FIGURE 10-46 A typical General Motors' Electronic Automatic Temperature Control (EATC) troubleshooting flowchart.

BCM-Controlled EATC Systems

Because the BCM-controlled EATC system incorporates many different microprocessors within its system, diagnostics can be very complex (Figure 10-47). Faults that seem to be unrelated to the EATC system may cause the system to malfunction. Since it was first introduced in 1986, BCM-controlled EATC systems have become increasingly popular on many GM vehicles. Each model year brings forth revisions and improvements in the system that also require different diagnostic procedures. In addition, system logic, as used by the different GM divisions, has changed through the years.

It is not possible to generally describe the diagnostic procedures required to service the many systems now in use. For this reason, one must have the correct service manual for the system being diagnosed. There are several different methods used to retrieve trouble codes, and it is important to follow the correct procedure. In all systems, PCM codes are

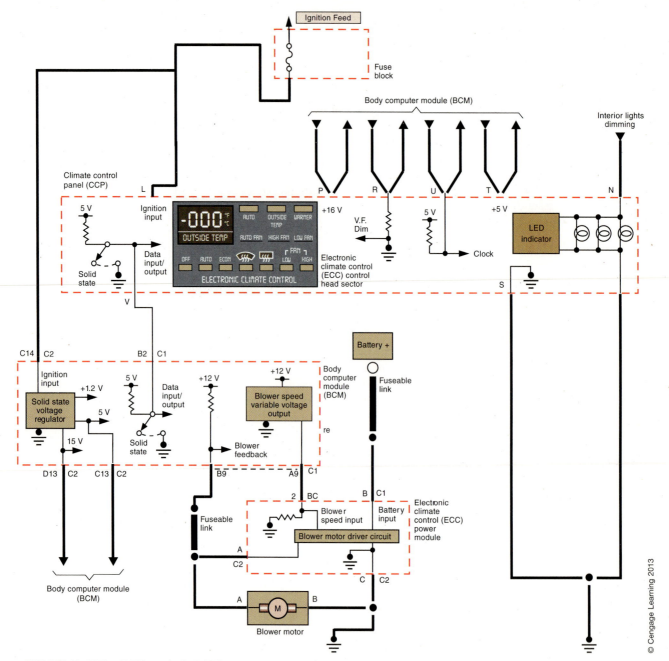

FIGURE 10-47 The BCM-controlled ATC system has several modules that use multiplexing to share information.

displayed first, followed by BCM codes. Codes associated with the EATC system can be in either set of codes, as well as network codes (e.g., PXXXX, BXXXX, UXXXX).

Once the codes have been retrieved, refer to the correct diagnostic chart. This test will pinpoint the fault in a logical manner. After all repairs to the system are complete, follow the service manual procedure for erasing codes and for resetting the system. Rerun the diagnostic test to confirm that the system is operating properly.

TROUBLE CODES

Most BCMs are capable of displaying the fault codes that were stored in memory. The procedure used to retrieve the codes varies greatly, and reference must be made to the appropriate service manual for the correct procedure.

BCM TROUBLE CODES

CODE	NOTES	PROBLEM
F10	1	Outside temperature sensor circuit
F11	1 - 2	A/C high-side temperature sensor circuit
F12	1 - 3	A/C low-side temperature sensor circuit
F13	1	In-car temperature sensor circuit
F30	1	CCP to BCM data circuit
F31	1	FDC to BCM data circuit
F32	1 - 4	Air mix door problem
F40	1	Heated windshield problem
F43	1	Low refrigerant problem
F46	2	Low refrigerant pressure
F47	2 - 5	High temperature clutch disengage
F48	2 - 5	BCM prom error
F49	1	
F51	1	

Notes: 1 Does not turn on any light
2 Turns on SERVICE A/C light
3 Disengages A/C clutch
4 Turns on cooling fans
5 Switches from AUTO to ECON

© Cengage Learning 2013

FIGURE 10-48 Body control module (BCM) diagnostic trouble codes lead the technician to the problem.

Classroom Manual

Chapter 10, page 359

Only some systems retain the code when the ignition is turned off and do not require test driving the vehicle to duplicate the fault. Once the fault is detected by the computer, the code must be retrieved before the ignition switch is turned off. The trouble code, however, does not necessarily indicate the faulty component. It only indicates that circuit of the system that is not operating properly (Figure 10-48). For example, the code displayed may be F11, indicating an air-conditioning system high-side temperature sensor problem. This does not mean, however, that the sensor is defective. It means that the fault is in that circuit, which includes the wiring, connections, and BCM as well as the sensor. To locate the problem, follow the diagnostic procedure in the service manual for the code received (Figure 10-49). There are two types of code that can be displayed: intermittent code and hard fault code.

Hard and Intermittent Codes

Some BCMs store trouble codes in their memory until they are erased by the technician or until a predetermined number of engine starts have occurred. Usually, the first set of fault codes to be displayed represent all of the fault codes that are stored in memory, including both hard and intermittent codes. The second set of fault codes to be displayed are only hard codes. The codes that are displayed in the first set but not displayed in the second set are intermittent codes.

Most diagnostic charts cannot be used to locate intermittent faults. This is because the testing at various points of the chart requires that the fault be present to locate the problem. Intermittent problems are often caused by poor electrical connections. Diagnosis, then, should start with a good visual inspection of the connectors, especially those involved with the trouble code.

Visual Inspection. One of the most important checks to be made before diagnosing a BCM-controlled system is a complete visual inspection. The inspection can

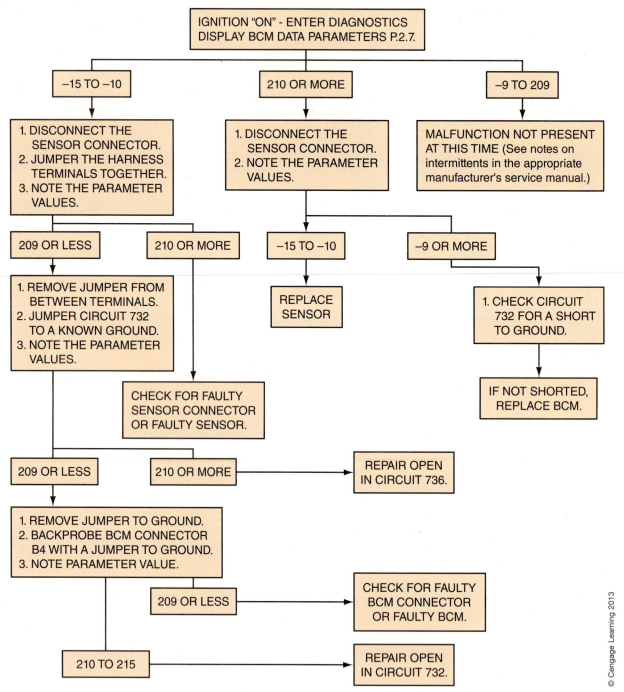

FIGURE 10-49 A typical diagnostic chart used to locate the cause of General Motors' trouble code using Tech 2 scan tool.

identify faults that could otherwise waste time in unnecessary diagnostics. Inspect the following:

1. Sensors and actuators for physical damage
2. Electrical connections to actuators, control modules, and sensors
3. All ground connections
4. Wiring for signs of broken or pinched wires or burned or chaffed spots indicating contact with sharp edges or hot exhaust manifolds
5. Vacuum hoses for breaks, cuts, disconnects, or pinches

NOTE: Check wires and hoses that are hidden under other components.

Entering BCM Diagnostics

There are, perhaps, as many methods of entering BCM diagnostics as there are vehicle makes and models. One thing that most have in common, however, is that a scan tool must be plugged into the diagnostic connector for the system to be tested. Always refer to the correct service manual for the vehicle being serviced, and use only the methods identified for retrieving trouble codes. Once the trouble codes are retrieved, consult the appropriate diagnostic chart for instructions on isolating the fault. It is also important to check the codes in the order required by the manufacturer.

Chrysler's DRB-III

The following procedure for using the DRB-III scanner is meant as a general guide only. It is intended to complement, not to replace, the service manual. Improper methods of trouble code retrieval may result in damage to the computer.

Early Chrysler systems use several modules that share information with the body controller through a multiplex system (Figure 10-50). Connecting the DRB-III into the diagnostic connector will access information concerning the operation of most vehicle

SERVICE TIP:
Before attempting to diagnose today's climate control systems, first retrieve any stored diagnostic codes. If an intermittent soft code is retrieved, road testing the vehicle may be required in order to duplicate the complaint and accessing codes before the ignition is cycled off.

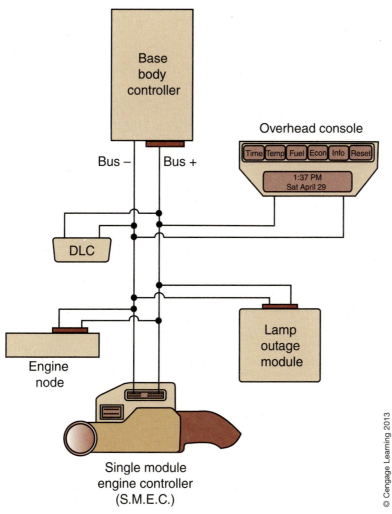

© Cengage Learning 2013

FIGURE 10-50 Multiplex system used to interface several different modules.

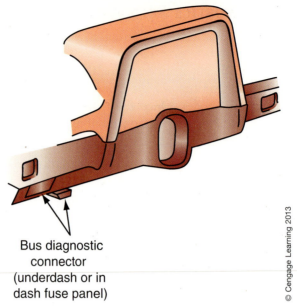

Bus diagnostic
connector
(underdash or in
dash fuse panel)

© Cengage Learning 2013

FIGURE 10-51 Diagnostic connector location.

systems. A typical procedure for entering body controller diagnostics using the DRB-III scanner is as follows:

1. Locate the diagnostic connection using the component locator (Figure 10-51).
2. Insert the correct program cartridge into the DRB-III scanner.
3. Connect the DRB-III to the vehicle by plugging its connector into the vehicle's diagnostic connector.
4. Turn the ignition switch to the RUN position. After the power-up sequence is completed, the copyright date and diagnostic program version should be displayed.
5. The display will change to a selection menu. The entire menu is not displayed; press the down arrow until the desired selection is found. In this example, press the down arrow twice.
6. Select 4 (SELECT SYSTEM) to enter the diagnostic test program. The display will change to a menu for selecting the system to be tested. Use of the down arrow reveals additional choices. Push the down arrow until the BODY option is shown.
7. Enter body system diagnostics by selecting 3 (BODY). The display will change to indicate that the BUS test is being performed (Figure 10-52). If the message is different from that shown in the figure, there is a problem in the CCD bus that must be corrected. No further testing is possible until this problem is corrected.
8. After a few seconds, the display will change and ask for input concerning the body style of the vehicle. Use the down arrow and scroll through the choices available.
9. Enter the number indicating the body style being diagnosed.
10. The display will then ask that a module be selected. Select BODY COMPUTER.

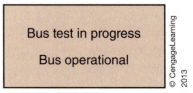

Bus test in progress

Bus operational

© CengageLearning 2013

FIGURE 10-52 This message must appear before proceeding with the diagnostics.

11. The display will indicate the name of the module selected, along with the version number of the module. Then, after a few seconds, the display will indicate BODY COMP MENU.

12. Use the down arrow key to scroll the menu selection, if needed. Press 2 (READ FAULTS). The DRB-III will either display that no faults were detected or provide the fault codes.

The first screen will indicate the number of fault codes found, the code for the first fault, and a description of the code. Scroll down the entire list of codes retrieved.

RETRIEVING CADILLAC BCM TROUBLE CODES

These procedures may vary among models, years, and the type of instrument cluster installed. Refer to the appropriate manufacturer's service manual for the vehicle being tested.

Pre-CAN bus network Cadillac systems allow access to trouble codes and other system operation information through the ECC panel. The BCM and the electronic control module (ECM) share information with each other so both system codes are retrieved through the ECC. The following procedure may typically be followed to enter diagnostics:

1. Place the ignition switch in the RUN position.

2. Depress the OFF and WARMER buttons on the ECC panel simultaneously (Figure 10-53). Hold the buttons until all display segments are illuminated.

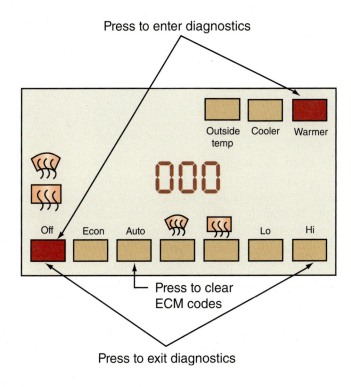

© Cengage Learning 2013

FIGURE 10-53 The buttons on the electronic climate control panel allow the technician to access information from the computer when it is in the diagnostic mode.

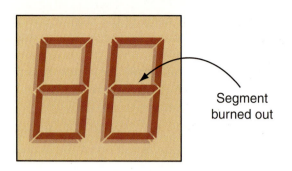

Segment
burned out

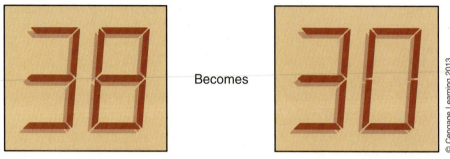

Becomes

© Cengage Learning 2013

FIGURE 10-54 Burned-out segments give a false code.

Cadillac uses the onboard ECC panel to display trouble codes, whereas other General Motors (GM) vehicles use a Tech II scan tool. Beginning in 1996, the Tech II scan tool was used to retrieve codes on certain models. That same year, Cadillac switched to the use of the Tech II scan tool to retrieve class 2 data. When diagnosing GM systems, make sure to follow the procedures specifically designated by GM for the vehicle being tested.

Diagnosis should not be attempted if all segments of the display do not illuminate. A problem may be misdiagnosed as the result of receiving an incorrect code. For example, if two segments of a display fail to illuminate, a code 24 could look like code 21 (Figure 10-54).

When the segment check is completed, the computer will display any trouble codes in its memory. An "8.8.8" will be displayed for about 1 second, then an "..E" will appear. This signals the beginning of engine controller trouble codes. The display will show all engine controller trouble codes beginning with the lowest number and progressing through the higher numbers. All codes associated with the engine controller will be prefixed with an "E." If there are no codes, however, "..E" will not be displayed.

Once all "E" codes are displayed, the computer will display BCM codes. The BCM codes are prefixed by an "F". An "F" will precede the first set of codes displayed. The first set will be all codes stored in memory for the last 100 engine starts. An ".F.F" will appear to signal the separation of the first pass and the second. The second set of trouble codes will be all hard codes.

When all codes are displayed, ".7.0" will be displayed, indicating that the system is ready for the next diagnostic feature to be selected. To erase the BCM trouble codes, press the OFF and LOW buttons simultaneously until "F.O.O" appears. Release the buttons and ".7.0" will reappear. Turn off the ignition switch and wait at least 10 seconds before reentering the diagnostic mode.

When in the diagnostic mode, exit the system without erasing the trouble codes by pressing AUTO on the ECC panel, and the temperature will reappear in the display.

CASE STUDY

A customer complains that an abnormal noise is coming from under the hood of her car. The service writer asks the customer the usual questions: "When did the noise start? When does it make the noise? How often is the noise noticeable?"

The customer answers the questions and notes that the noise seems to be growing louder. She first noticed the noise a few days ago.

After noting the mileage, the service writer checks the computer for the service record. According to the records, no major work has been performed on the car. Also, it seems to have been serviced regularly and is well maintained.

On starting the car for a test drive, the noise is immediately noted. The service writer raises the hood and, using a mechanic's stethoscope, is able to pinpoint the noise at the air-conditioner compressor.

"It couldn't be the compressor," the customer says. "I just had that repaired a week ago." The customer then explains that the repairs were made by an independent dealer in another city while the customer was out of town.

An inspection of the compressor clutch by the technician reveals that the wrong field coil snapring was installed. The snapring, which was too thin, allowed the field coil to barely touch the rotor, creating a noise.

Fortunately for the customer, no major damage was done to the clutch, and a proper snapring corrected the problem.

TERMS TO KNOW

Aspirator
Breakout box
Electronic automatic temperature control (EATC)
Fusible link
Ohmmeter
Open
Overload
Servomotor
Shorts
Solenoids
Spikes
Voltmeter

ASE-STYLE REVIEW QUESTIONS

1. *Technician A* says to measure current on the blower motor circuit by connecting the DVOM in parallel at any point between the switch and the motor ground.
   Technician B says the blower motor should be turned on while measuring current. Who is correct?
   A. A only
   B. B only
   C. Both A and B
   D. Neither A nor B

2. On an automatic temperature control system that will not hold the temperature set at the control panel, what should the technician do first?
   A. Replace the temperature mode door actuator.
   B. Replace the climate control panel.
   C. Check and replace the in-car temperature sensor if faulty.
   D. Refer to the appropriate manufacturers service information.

3. On a vehicle equipped with and automatic temperature control system the blower motor will not function when the vehicle is first started and heat is selected. Which of the following is the most likely cause?
   A. The blower motor has higher than specified resistance.
   B. The ambient air temperature is faulty.
   C. The system is functioning normally.
   D. The in-car temperature sensor is faulty.

4. *Technician A* says that a trouble code may be displayed on some master control heads.
   Technician B says that there are provisions for connecting an external scan tool on some systems. Who is correct?
   A. A only
   B. B only
   C. Both A and B
   D. Neither A nor B

5. *Technician A* says that a thermostat is used for temperature control in some systems.
   Technician B says that a low-pressure control is used for temperature control in some systems. Who is correct?
   A. A only
   B. B only
   C. Both A and B
   D. Neither A nor B

6. A rattling noise is heard when the engine is running and the air conditioning is off.
   Technician A says this noise may be caused by too narrow gap between the compressor's armature and rotor.
   Technician B says a quick accurate test is to see if the clutch is the source is to switch on the air conditioning to see if the noise stops. Who is correct?
   A. A only
   B. B only
   C. Both A and B
   D. Neither A nor B

7. Blower motor speed control is being discussed:

Technician A says that blower motor speed control is automatic in a temperature-controlled system.

Technician B says that blower motor speed may be manually selected in a temperature-controlled system. Who is correct?

A. A only
B. B only
C. Both A and B
D. Neither A nor B

8. All of the following air-conditioning system sensors are varying resistive sensors *except*:

A. Sun Load Sensor
B. Evaporator Temperature Sensor
C. Infrared Temperature Sensor
D. Ambient Temperature Sensor

9. Pressure switches are being discussed:

Technician A says that a compressor discharge pressure switch is a low-pressure switch.

Technician B says that the pressure switch used for temperature control is a low-pressure switch. Who is correct?

A. A only
B. B only
C. Both A and B
D. Neither A nor B

ASE CHALLENGE QUESTIONS

1. Low-pressure and high-pressure cutoff switches may be used for air-conditioning system:

A. Temperature control
B. Protection
C. Either A or B
D. Neither A nor B

2. The following statements regarding the relationships between the cooling system and air-conditioning system are all true, *except*:

A. An overheating engine will affect air-conditioning system performance.
B. Ambient air first passes through the radiator, then the condenser.
C. The same blower motor is used for the heater and air conditioner.
D. An air-conditioning system places an additional load on the cooling system.

3. Electrical testing is being discussed:

Technician A says that an analog ohmmeter should not be used to test an electronic circuit.

Technician B says that a digital ohmmeter may be used to test any electrical or electronic circuit. Who is correct?

A. A only
B. B only
C. Both A and B
D. Neither A nor B

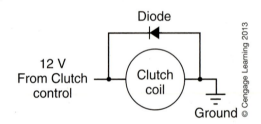

4. The diode shown in the illustration above is used to:

A. Increase voltage in the clutch coil circuit
B. Reduce voltage in the clutch coil circuit
C. Block the flow of current in the electrical system
D. Prevent voltage spikes in the electrical system

5. Which of the following is *not* a tool used by the air-conditioning system technician?

A. Scan tool
B. Breakout box
C. DRB-III scanner
D. OBD-5 scanner

Name _____ **Date** _____

TESTING AND REPLACING FUSES AND CIRCUIT BREAKERS

Upon completion of this job sheet, you should be able to test and replace fuses and circuit breakers.

NATEF Correlation

HEATING AND AIR CONDITIONING: Operating Systems and Related Controls Diagnosis and Repair; *Diagnose failures in the electrical controls of heating, ventilation, and A/C (HVAC) systems; determine necessary action.* **(P-2)**

Tools and Materials

Vehicle with accessible fuse/circuit breaker panel

Testlight (fused, non-powered)

Testlight (fused, powered)

Ohmmeter

Tools, as needed

Describe the vehicle being worked on.

Year _____ Make _____ Model _____

VIN _____ Engine type and size _____

Procedure

Using a non-powered testlight, perform the following tasks. Write a brief description of your procedure and of your findings.

1. Connect one probe of the testlight to ground.
   a. Touch the other probe to the positive side of the fuse panel bus bar.

   _____

   _____

   b. Touch the other probe to the other side of selected fuses and circuit breakers.
   Fuses: _____

   _____

   Circuit Breakers: _____

   _____

2. Remove a selected fuse and a circuit breaker. Make note of their location and amperage.

_____

_____

3. Touch the probes of the self-powered testlight to the two ends of the fuse.

_____

_____

4. Touch the probes of the self-powered test lamp to the two terminals of the circuit breaker.

_____

_____

NOTE: For steps 5 and 6, DO NOT connect the ohmmeter to the self-powered test lamp.

5. Hold the same test as step 3 using an ohmmeter.

_____

_____

6. Hold the same test as step 4 using an ohmmeter.

_____

_____

7. Perform any other tests as outlined by your instructor.

Test _____

Results _____

_____

Instructor's Response _____

_____

_____

Name _____ **Date** _____

DIAGNOSE TEMPERATURE CONTROL PROBLEMS IN THE HEATER/VENTILATION SYSTEM

Upon completion of this job sheet, you should be able to diagnose temperature control problems in the heater/ventilation system and determine the necessary action; diagnose blower system problems in the heater/ventilation system and determine the necessary action; and inspect, test, adjust, or replace climate control temperature and sun load sensors.

NATEF Correlation

HEATING AND AIR CONDITIONING: Heating, Ventilation, and Engine Cooling Systems Diagnosis and Repair; *Diagnose temperature control problems in the heater/ventilation system; determine necessary action.* **(P-2)**

Tools and Materials

Late-model vehicle

Service manual or information system

Safety glasses or goggles

Hand tools, as required

Tables 10-1 through 10-5 that follow in this job sheet

Describe the vehicle being worked on.

Year _____ Make _____ Model _____

VIN _____ Engine type and size _____

Procedure

Follow procedures outlined in the Shop Manual. Give a brief description of your procedure following each step. Ensure that the engine is cold, and wear eye protection. The following procedures are meant to be a guide for diagnosing an automatic temperature control system. For specific information on the vehicle you are working on, always refer to specific manufacturer's service information contained in the vehicle's service manual or other data system.

1. Referring to Table 10-1, check system for proper airflow and determine necessary action.

   Improper airflow
   - No airflow
   - No fresh air from vents
   - No cool air from vents
   - No warm air from vents

   _____

   _____

2. Referring to Table 10-2, check the system for unusual noise and determine necessary action.

   Unusual noise while HVAC system is in operation
   - Chattering sounds
   - Squealing sounds
   - Grinding sounds

   _____

   _____

3. Referring to Table 10-3, check the system for poor, intermittent, or no cooling and determine necessary action.

   Inadequate or no cooling
   - Little to no airflow from ducts
   - Air not cool
   - Air not warm

   Intermittent or no cooling
   - A/C system cycles rapidly
   - Cycles on high-pressure protector

   _____

   _____

4. Referring to Table 10-4, check the system for poor, intermittent, or no cooling and determine necessary action.

   Inadequate or no cooling
   - Little to no airflow from ducts
   - Air not cool with A/C selected
   - Air not warm

   Intermittent or no cooling
   - A/C system cycles rapidly
   - Cycles on high-pressure protector

   _____

   _____

5. Referring to Table 10-5, check the system for no cooling and determine necessary action.

   No cooling
   - Compressor will not engage
   - Compressor always engaged
   - High blower speed only
   - No blower
   - Improper air delivery
   - Insufficient heating
   - Insufficient cooling

   _____

   _____

Instructor's Response _____

_____

_____

TABLE 10-1

Diagnose the cause of temperature control problems in the heater/ventilation system; determine needed repairs.
Diagnose temperature control system problems; determine needed repairs.
Diagnose blower system problems; determine needed repairs.
Inspect, test, adjust, or replace climate control temperature and sun load sensors.

Problem Area	Symptoms	Possible Causes

IMPROPER AIRFLOW	No airflow	1. Defective master control
		2. No:
		   a. Vacuum, if pneumatic
		   b. Power, if electric
		3. Defective:
		   a. Motor, if electric
		   b. "Vacuum Pot," if pneumatic
		   c. Cable, if manual

	No fresh air from vents	1. Defective master control; electric, pneumatic, or manual
		2. Defective:
		   a. Wiring, if electric
		   b. Hose, if pneumatic
		   c. Cable, if manual
		3. Defective actuator:
		   a. Motor, if electric
		   b. "Vacuum Pot," if pneumatic
		   c. Retainer, if cable

	No cool air from vents	1. Defective master control; electric, pneumatic, or manual
		2. Defective:
		   a. Wiring, if electric
		   b. Hose, if pneumatic
		   c. Cable, if manual
		3. Defective actuator:
		   a. Motor, if electric
		   b. "Vacuum Pot," if pneumatic
		   c. Retainer, if cable
		4. Defective or inoperative air conditioner
		5. Defective (open) heater coolant flow control valve
		6. Duct disconnected or missing

	No warm air from vents	1. Defective master control; electric, pneumatic, or manual
		2. Defective:
		   a. Wiring, if electric
		   b. Hose, if pneumatic
		   c. Cable, if manual
		3. Defective actuator:
		   a. Motor, if electric
		   b. "Vacuum Pot," if pneumatic
		   c. Retainer, if cable
		4. Defective (closed) heater coolant flow control valve
		5. Heater (hoses) disconnected
		6. Duct disconnected or missing

TABLE 10-2

Diagnose the cause of unusual operating noises of the A/C system; determine needed repairs.

Problem Area	Symptoms	Possible Causes

NOISE	Chattering sound	1. Defective blower motor
		2. Blower loose on motor shaft
		3. Blower rubbing insulation on mode door gasket
		4. Debris in duct
		5. Low compressor:
		   a. Lubricant
		   b. Refrigerant
		6. Loose bracket or other part(s)

	Squealing sounds	1. Loose or glazed belt(s) 2. Worn belt(s) and or pulley(s) 3. Defective A/C clutch or idler pulley bearing(s) 4. Defective blower motor 5. Defective A/C compressor
	Grinding sounds	1. Defective A/C clutch 2. Defective A/C compressor 3. Defective bearing(s) in clutch or idler 4. Defective coolant (water) pump

TABLE 10-3

Diagnose the cause of failure in the electrical control system of heating, ventilating, and A/C systems, determine needed repairs. Inspect, test, repair, replace, and adjust load sensitive A/C compressor cutoff systems. Inspect, test, repair, and replace engine cooling/condenser fan motors, relays/modules, switches, sensors, wiring, and protection devices.

Problem Area	Symptoms	Possible Causes
INADEQUATE OR NO COOLING	Little to no airflow from ducts	1. Blown fuse or defective circuit breaker 2. Defective blower speed control or resistor 3. Defective master control 4. Defective relay 5. Defective wiring
	Air not cool	1. Defective clutch coil or ground connection 2. Defective low-or high-pressure control 3. Defective temperature control 4. Defective relay 5. Defective sensor
	Air not warm	1. Defective master control 2. Defective temperature control 3. Defective electric coolant flow control valve 4. Defective relay 5. Defective sensor
INTERMITTENT OR NO COOLING	A/C system cycles rapidly	1. Defective low-pressure control 2. Defective high-pressure control 3. Thermostat adjustment
	Cycles on high-pressure protector	1. Defective or inoperative cooling fan or motor 2. Defective high-pressure control 3. Engine overheating

TABLE 10-4

Inspect, test, repair, and replace A/C compressor clutch components or assembly. Inspect, test, repair, replace, and adjust A/C-related engine control systems. Inspect, test, adjust, repair, and replace electric actuator motors, relays/modules, sensors, wiring, and protection devices. Diagnose compressor clutch control system; determine needed repairs. Inspect, test, repair, and replace electric and vacuum motors, solenoids, and switches.

Problem Area	Symptoms	Possible Causes
INADEQUATE OR NO COOLING	Little to no airflow from ducts	1. Blown fuse or defective circuit breaker 2. Defective blower speed control or resistor 3. Defective master control 4. Defective relay 5. Defective wiring
	Air not cool with A/C selected	1. Defective clutch coil 2. Poor electrical connection 3. Defective low-pressure control 4. Defective high-pressure control 5. Defective temperature control 6. Defective master control 7. Defective relay 8. Defective sensor 9. Defective module 10. Shorted clutch diode
	Air not warm	1. Defective master control 2. Defective temperature control 3. Defective electric coolant flow control valve 4. Defective relay 5. Defective sensor 6. Defective module
INTERMITTENT OR NO COOLING	A/C system cycles rapidly	1. Defective low-pressure control 2. Defective high-pressure control 3. Thermostat adjustment 4. Defective module 5. Defective relay 6. Loose or defective wiring
	Cycles on high pressure protector	1. Defective or inoperative cooling fan motor 2. Defective high-pressure control 3. Engine overheating

TABLE 10-5

Inspect, test, and replace automatic temperature control (ATC) control panel. Inspect, test, adjust, or replace ATC microprocessor (climate control computer/programmer). Check and adjust calibration of ATC system.

Problem Area	Symptoms	Possible Causes
NO COOLING	Compressor will not engage	1. A/C clutch relay 2. Low-side temperature sensor 3. Defective low-pressure switch 4. Open clutch coil 5. Power steering pressure switch (if equipped) 6. Defective wiring 7. Defective BCM
	Compressor always engaged	1. Defective clutch 2. Defective clutch relay 3. Shorted control signal circuit 4. Mechanical binding
	High blower speed only	1. Open feedback circuit 2. Defective programmer 3. Open signal circuit 4. Defective BCM
	No blower	1. Blown fuse or circuit breaker 2. Defective blower motor 3. Loose or disconnected motor ground 4. Improper signal to programmer due to open or short 5. Power modules feed open 6. Defective programmer 7. Defective BCM
	Improper air delivery	1. Loss of vacuum source 2. Leak in vacuum circuit 3. Defective programmer 4. Defective BCM
	Insufficient heating	1. Air mix valve 2. Defective (closed) coolant flow control valve 3. Air mix valve linkage 4. Programmer arm adjustment 5. Programmer
	Insufficient cooling	1. Insufficient airflow 2. Refrigeration problems 3. Air mix valve linkage 4. Programmer arm adjustment 5. Programmer

Name _____ **Date** _____

USING A SCAN TOOL TO ACCESS HVAC SYSTEM

Upon completion of this job sheet, you should be able to connect and use a scan tool to access and record HVAC information and diagnostic trouble codes (DTCs).

NATEF Correlation

HEATING AND AIR CONDITIONING: A/C System Diagnosis and Repair; *Using scan tool, observe and record related HVAC data and trouble codes.* **(P-1)**

Tools and Materials

Late-model vehicle equipped with a BCM

Service manual or information system

Safety glasses or goggles

Scan tool

Describe the vehicle being worked on.

Year _____ Make _____ Model _____

VIN _____ Engine type and size _____

Procedure

1. Locate the data link connector. Where is the connector located? _____

   _____

2. Connect the scan tool to the DLC and turn the ignition switch to the RUN position.
3. Initialize the scan tool.
4. Select Displace DTC stored. Are there any current power train, body control, or network DTCs stored?
   a. List PXXXX code stored _____
   b. List BXXXX code stored _____
   c. List UXXXX code stored _____

5. Select History codes. Are there any history power train, body control, or network DTCs stored?
   a. List PXXXX code stored _____
   b. List BXXXX code stored _____
   c. List UXXXX code stored _____

6. Next, enter the body control module (BCM) functions. Select the HVAC system.

7. What information is displayed about the HVAC or BCM data stream? _____

_____

8. How is this information useful in monitoring the HVAC system? _____

_____

9. For the circuit with the fault code associated with it, look at all sensors and input/output values. Are any of the values out of specifications? If so, which sensors or actuators? _____

_____

_____

10. Follow the diagnostic chart for the fault code. What are your conclusions? _____

_____

Instructor's Response _____

_____

_____

Name _____ **Date** _____

TESTING AN AMBIENT AIR TEMPERATURE SENSOR

Upon completion of this job sheet, you should be able to check the operation of an ambient temperature sensor.

NATEF Correlation

HEATING AND AIR CONDITIONING: Operating Systems and Related Controls Diagnosis and Repair; *Diagnose malfunctions in the vacuum, mechanical, and electrical components and controls of the heating, ventilation, and A/C (HVAC) system; determine necessary action.* **(P-2)**

Tools and Materials

Late-model vehicle

Service manual or information system

Safety glasses or goggles

Hand tools, as required

Digital multimeter (DMM)

Thermometer

Describe the vehicle being worked on.

Year _____ Make _____ Model _____

VIN _____ Engine type and size _____

Procedure

Always follow the procedures outlined in the service manual for the specific make, model, and year vehicle being tested.

1. Describe the location of the ambient temperature sensor (ATS). _____

   _____

2. What color are the wires connected to the sensor? _____

   _____

3. Record the resistance specifications for a normal ATS for the vehicle being tested based on current ambient air temperature within 1 foot of the sensor. _____

   _____

4. Disconnect the harness connector from the sensor.
5. Measure the resistance value of the sensor and the ambient air temperature within 1 foot of sensor.
   a. Sensor resistance _____ Ω

   b. Ambient air temperature _____°F

6. Conclusion: Is the sensor within specification range? _____

   _____

Instructor's Response _____

_____

_____

JOB SHEET

Name _____ Date _____

AUTOMATIC TEMPERATURE CONTROL DIAGNOSTICS

Upon completion of this job sheet, you should be able to test the electronic automatic temperature control system using a scan tool or self-diagnostic mode of the climate control panel and determine needed repairs.

NATEF Correlation

HEATING AND AIR CONDITIONING: Operating Systems and Related Controls Diagnosis and Repair; *Diagnose malfunctions in the electrical controls of heating, ventilation, and A/C (HVAC) systems; determine necessary action.* **(P-2)**

HEATING AND AIR CONDITIONING: Operating Systems and Related Controls Diagnosis and Repair; *Check operation of automatic and semiautomatic HVAC control systems; determine necessary action.* **(P-3)**

Tools and Materials

Late-model vehicle

Service manual or information system

Safety glasses or goggles

Hand tools, as required

Scan tool

Describe the vehicle being worked on.

Year _____ Make _____ Model _____

VIN _____ Engine type and size _____

Procedure

Always follow the procedures outlined in the service manual for the specific make, model, and year vehicle being tested.

1. Describe the location of the ambient temperature sensor (ATS).
2. Activate the self-diagnostic mode or connect and activate a scan tool. What indications are provided to confirm that system self-diagnosis has been activated?

   _____

   _____

3. Perform a display panel segment test and record results. _____

   _____

4. Retrieve DTCs and record. _____

   _____

5. Are the fault codes current or history codes? _____

   _____

6. Perform all diagnostic tests the system is capable of running. What test is the system capable of performing? _____

   _____

   _____

7. Were any system faults detected? YES/NO
   If yes, use the service information to trace and diagnose the cause of the fault and record your findings. _____

   _____

   _____

8. Complete the repair and run the self-diagnosis test again. Was the repair completed successfully? YES/NO
   If no, consult your instructor for guidance. _____

   _____

9. Follow the procedure for erasing fault codes stored.

Instructor's Response _____

_____

_____

Name _____ Date _____

TESTING AN IN-CAR TEMPERATURE SENSOR

Upon completion of this job sheet you should be able to check the operation of an in-car temperature sensor.

NATEF Correlation

P-2: VII.D.4; HEATING AND AIR CONDITIONING: Operating Systems and Related Controls Diagnosis and Repair; *Diagnose malfunctions in the vacuum, mechanical, and electrical components and controls of the heating, ventilation, and A/C (HVAC) system; determine necessary action.*

Tools and Materials

Late-model vehicle

Service manual or information system

Safety glasses or goggles

Hand tools, as required

Digital Multimeter (DMM)

Thermometer

Describe the vehicle being worked on.

Year _____ Make _____ Model _____

VIN _____ Engine type and size _____

Procedure

Always follow the procedures outlined in the service manual for the specific make, model, and year vehicle being tested.

1. Describe the location of the in-car temperature sensor (ICTS).
2. What color are the wires connected to the sensor?_____

   _____

3. Record the resistance specifications for a normal ICTS for the vehicle being tested based on current ambient air temperature within one foot of sensor._____

   _____

4. Disconnect the harness connector from the sensor.
5. Measure the resistance value of the sensor and the ambient air temperature within one foot of sensor.

   a. Sensor resistance _____ Ω

   b. In-Car Temperature _____ °F

6. Conclusion: Is sensor within specification range? _____

   _____

Instructor's Response _____

_____

_____

Name _____ **Date** _____

TESTING AN EVAPORATOR TEMPERATURE SENSOR

Upon completion of this job sheet you should be able to check the operation of a two wire evaporator temperature sensor.

NATEF Correlation

P-2: VII.D.4; HEATING AND AIR CONDITIONING: Operating Systems and Related Controls Diagnosis and Repair; *Diagnose malfunctions in the vacuum, mechanical, and electrical components and controls of the heating, ventilation, and A/C (HVAC) system; determine necessary action.*

Tools and Materials

Late-model vehicle

Service manual or information system

Safety glasses or goggles

Hand tools, as required

Digital Multimeter (DMM)

Thermometer

Describe the vehicle being worked on.

Year _____ Make _____ Model _____

VIN _____ Engine type and size _____

Procedure

Always follow the procedures outlined in the service manual for the specific make, model, and year vehicle being tested.

1. Describe the location of the Evaporator Temperature Sensor (ETS).
2. What color are the two wires connected to the sensor?_____

   _____

3. Record the resistance specifications for a normal evaporator temperature sensor for the vehicle being tested based on the current temperature of the evaporator core?_____

4. Disconnect the harness connector from the sensor.

5. Measure the resistance value of the sensor and the air temperature within one foot of sensor.

   a. Sensor resistance _____ Ω

   b. Evaporator Temperature at or near sensor _____ °F

6. Conclusion: Is sensor within specification range? _____

Instructor's Response _____

_____

_____

Name _____ Date _____

TESTING A SUN LOAD SENSOR

Upon completion of this job sheet you should be able to check the operation of a sun load sensor by performing a quick test.

NATEF Correlation

P-2: VII.D.4; HEATING AND AIR CONDITIONING: Operating Systems and Related Controls Diagnosis and Repair; *Diagnose malfunctions in the vacuum, mechanical, and electrical components and controls of the heating, ventilation, and A/C (HVAC) system; determine necessary action.*

Tools and Materials

Late-model vehicle

Service manual or information system

Safety glasses or goggles

Hand tools, as required

Diagnostic scan tool

Describe the vehicle being worked on.

Year _____ Make _____ Model _____

VIN _____ Engine type and size _____

Procedure

Always follow the procedures outlined in the service manual for the specific make, model, and year vehicle being tested. During the sun load sensor quick test ventilation system air volume (blower speed) and temperature should both change.

1. Describe the location of the Sun Load Sensor (SLS).
2. What color are the two wires connected to the sensor? _____

   _____

3. Park the vehicle outside on a sunny day.
4. Turn the ignition to the ON position, but do not start the vehicle yet.
5. Connect the scan tool to the vehicle data link connector (DLC).
6. Select "Sensor Data Stream" from the menu.
7. Observe the sun load sensor input values.

   a. Record data _____

8. Start the engine and allow it to idle. Turn the air conditioning system on and select FULL AUTO mode.

9. Set the temperature to approximately 70°F (21°C) on the ATC system. Allow the vehicle to run and the system to stabilize for 10 minutes.

10. Cover the sensor with a dark piece of paper or cloth while observing data values on the scan tool.

   a. Record data before covering sensor _____

   b. Record data after covering sensor _____

11. Cover and uncover the sensor several times, waiting 30–60 seconds between each interval to give the system time to react. The sensor data values should respond quickly to changes in light levels. Did it? _____

12. If the sensor values do not change, further diagnosis will be required. Consult specific vehicle service information for detailed diagnostic steps.

13. Conclusion: Is sensor within specification range? _____

_____

Instructor's Response _____

_____

_____

Name _____ Date _____

TESTING AN INFRARED TEMPERATURE SENSOR

Upon completion of this job sheet you should be able to check the operation of a infrared temperature sensor by performing a quick test.

NATEF Correlation

P-2: VII.D.4; HEATING AND AIR CONDITIONING: Operating Systems and Related Controls Diagnosis and Repair; *Diagnose malfunctions in the vacuum, mechanical, and electrical components and controls of the heating, ventilation, and A/C (HVAC) system; determine necessary action.*

Tools and Materials

Late-model vehicle

Service manual or information system

Safety glasses or goggles

Hand tools, as required

2 hand towels

Small bucket of ice water

Small bucket of hot water

Diagnostic scan tool

Describe the vehicle being worked on.

Year _____ Make _____ Model _____

VIN _____ Engine type and size _____

Procedure

Always follow the procedures outlined in the service manual for the specific make, model, and year vehicle being tested. During the infrared temperature sensor quick test ventilation system air volume (blower speed) and temperature should both change.

1. Describe the location of the Infrared Temperature Sensor.
2. What color are the wires connected to the sensor? _____

_____

3. Turn the ignition to the ON position, but do not start the vehicle yet.
4. Connect the scan tool to the vehicle data link connector (DLC).
5. Select "Sensor Data Stream" from the menu.

6. Observe the infrared temperature sensor input values.
   a. Record data _____

7. Start the engine and allow it to idle. Turn the air conditioning system on and select FULL AUTO mode.

8. Set the temperature to approximately 70°F (21°C) on the ATC system. Allow the vehicle to run and the system to stabilize for 10 minutes.

9. Get two small buckets and two small towels. Fill one bucket with hot water and place a towel in it. Fill the other bucket with cold water and ice and place the other towel in this bucket.

10. Squeeze out the cold towel so as not to drip water on the interior of the car, and allow it to hang directly in front of the sensor while observing data values on the scan tool.
    a. Record data values for the infrared temperature sensor _____

Instructor's Response _____

_____

_____

Chapter 11

RETROFIT (R-12) [CFC-12] TO R-134A [HFC-134A]

BASIC TOOLS

Basic mechanic's tool set
Manifold and gauge set
Service hose set
Thermometer
Shop light(s)
Fender cover
Blanket
Vacuum pump
Can tap

UPON COMPLETION AND REVIEW OF THIS CHAPTER, YOU SHOULD BE ABLE TO:

- Recognize the difference between pure and impure refrigerant by interpreting gauge pressures relating to ambient temperature.

- Determine the purity of refrigerant in an air-conditioning system or container.

- Explain the necessity of using recovery-only equipment for contaminated refrigerant.

- Describe the method of affixing an access saddle valve onto an air-conditioning system.

- Determine when a system is void of refrigerant and air.

- Leak test the air-conditioning system.

- Recover R-12 refrigerant from a system.

- Diagnose and repair system components.

- Evacuate a system prior to charging.

- Flush the refrigerant system.

- Charge a system with R-134a refrigerant.

INTRODUCTION

General information is given in this chapter regarding the proper and safe practices and procedures for retrofitting an automotive air-conditioning system. It is most important, however, to follow the manufacturer's instructions when servicing any particular make and model vehicle. This chapter includes, under the appropriate heading, procedures for the following: purity test, access valve installation, recovery of contaminated refrigerant, and retrofit.

PURITY TEST

A refrigerant identifier (Figure 11-1) quickly and safely identifies the purity and type of refrigerant in a vehicle air-conditioning system or tank. The display will indicate the purity of the refrigerant being tested as a percentage of R-134a, R-12, and the percentage of air (noncondensable gas); in addition, it will indicate the presence of hydrocarbons (HC) in the sample being tested. Some analyzers will also detect the presence of R-22. A system is considered contaminated if it contains more than 2 percent of a "foreign" substance.

Use of a refrigerant identifier, often called a purity tester, should be the first step in servicing an automotive air-conditioning system. That way, one does not have to be concerned about customer dissatisfaction or damage to the vehicle that could occur if the wrong refrigerant is used. Further, testing refrigerant protects refrigerant supplies and recovery/recycling equipment. At today's prices, preventing just one tank of refrigerant from contamination can save several hundred dollars plus the high cost of disposing of the contaminated refrigerant.

Always follow the manufacturer's instructions for using any type of test equipment. The following procedure for using the Sentinel identifier is typical:

1. Turn on the MAIN POWER switch; the unit automatically clears the last refrigerant sample and is made ready for a new sample.

Classroom Manual

Chapter 11, page 375

A purity test should be held any time there is a concern about the quality of the refrigerant in the system.

SPECIAL TOOLS

Low-side (compound) gauge with gauge/hose adapter and service hose
Thermometer

FIGURE 11-1 Refrigerant identifier.

2. When READY appears on the display, connect a service hose from the tester to the vehicle air-conditioning system or tank of refrigerant being tested.
3. The tester automatically pulls in a sample and begins processing it; TESTING shows on the display.
4. Within about 1 minute, the display will show R-12, R-134a, or UNKNOWN. If UNKNOWN is displayed, the refrigerant is a mixture or is some other type of refrigerant. In either case, it should not be added to previously recovered refrigerant. Also, it should not be recycled or reused.
5. Turn off the MAIN POWER switch and disconnect the service hose.

If no other method of refrigerant identification is available and there is any doubt as to the condition of the refrigerant in an air-conditioning system, the following purity test may be used. It should be noted, however, that for a pressure-temperature test to be valid, there must be some liquid refrigerant in the system. If the refrigerant has leaked to the point that only vapor remains, the pressure will be below that specified at any given temperature. Proceed as follows:

1. Park the vehicle or place the tank inside the shop in an area that is free of drafts and where the ambient temperature is not expected to go below 70°F (21°C).
2. Raise the hood.
3. Determine the type of refrigerant that should be in the system or tank: R-12 or R-134a.
4. Attach a 0–150 psig (0–1,000 kPa) gauge of known accuracy, appropriate for the refrigerant type (Figure 11-2).
5. Place a thermometer of known accuracy (Figure 11-3) in the immediate area of the vehicle or tank to measure the ambient temperature.
6. First thing the following morning:
   a. Note and record the pressure reading shown on the gauge.
   b. Note and record the temperature reading shown on the thermometer.

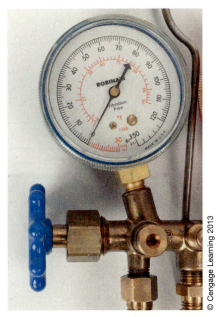

© Cengage Learning 2013

FIGURE 11-2 Attach an appropriate test gauge.

A   B

© Cengage Learning 2013

FIGURE 11-3 Typical (A) dial thermometer and (B) infrared temperature sensor.

7. Compare the gauge reading with the appropriate table:
   a. (Figure 11-4): (A) English; (B) metric for R-12.
   b. (Figure 11-5): (A) English; (B) metric for R-134a.

Temperature Fahrenheit	Pressure PSIG	kPa	Temperature Fahrenheit	Pressure PSIG	kPa	Temperature Celsius	Pressure PSIG	kPa	Temperature Celsius	Pressure PSIG	kPa
70	80	551	86	103	710	21.1	551	80	30.0	710	103
71	82	565	87	105	724	21.7	565	82	30.5	724	105
72	83	572	88	107	738	22.2	572	83	31.1	738	107
73	84	579	89	108	745	22.8	579	84	31.7	745	108
74	86	593	90	110	758	23.3	593	86	32.2	758	110
75	87	600	91	111	765	23.9	600	87	32.8	765	111
76	88	607	92	113	779	24.4	607	88	33.3	779	113
77	90	621	93	115	793	25.0	621	90	33.9	793	115
78	92	634	94	116	800	25.6	634	92	34.4	800	116
79	94	648	95	118	814	26.1	648	94	35.0	814	118
80	96	662	96	120	827	26.7	662	96	35.6	827	120
81	98	676	97	122	841	27.2	676	98	36.1	841	122
82	99	683	98	124	855	27.8	683	99	36.7	855	124
83	100	690	99	125	862	28.3	690	100	37.2	862	125
84	101	696	100	127	876	28.9	696	101	37.8	876	127
85	102	703	101	129	889	29.4	703	102	38.3	889	129

A   B

© Cengage Learning 2013

FIGURE 11-4 Temperature/pressure chart for R-12: (A) English and (B) metric.

Temperature Fahrenheit	Pressure PSIG	kPa	Temperature Fahrenheit	Pressure PSIG	kPa	Temperature Celsius	Pressure PSIG	kPa	Temperature Celsius	Pressure PSIG	kPa
70	76	524	86	102	703	21.1	524	76	30.0	703	102
71	77	531	87	103	710	21.7	531	77	30.5	710	103
72	79	545	88	105	724	22.2	545	79	31.1	724	105
73	80	551	89	107	738	22.8	551	80	31.7	738	107
74	82	565	90	109	752	23.3	565	82	32.2	752	109
75	83	572	91	111	765	23.9	572	83	32.8	765	111
76	85	586	92	113	779	24.4	586	85	33.3	779	113
77	86	593	93	115	793	25.0	593	86	33.9	793	115
78	88	607	94	117	807	25.6	607	88	34.4	807	117
79	90	621	95	118	814	26.1	621	90	35.0	814	118
80	91	627	96	120	827	26.7	627	91	35.6	827	120
81	93	641	97	122	841	27.2	641	93	36.1	841	122
82	95	655	98	125	862	27.8	655	95	36.7	862	125
83	96	662	99	127	876	28.3	662	96	37.2	876	127
84	98	676	100	129	889	28.9	676	98	37.8	889	129
85	100	690	101	131	903	29.4	690	100	38.3	903	131

A

B

© Cengage Learning 2013

FIGURE 11-5 Temperature/pressure chart for R-134a: (A) English and (B) metric.

Classroom Manual

Chapter 11, page 393

A **saddle valve** is a two-part accessory valve that may be clamped around the metal part of a system hose to provide access to the air-conditioning system for service or the installation of additional pressure switches.

ACCESS VALVES

A saddle clamp access valve may be installed if space does not permit converting the R-12 access valve to the R-134a valve configuration. Follow this procedure for the typical installation of the **saddle valve**.

1. Make certain that the system is free of refrigerant. Recover all of the refrigerant as outlined in this chapter if retrofitting the system.
2. Select the proper location for the valve.
   a. Will there be clearance for the hose access adapter?
   b. Will there be adequate clearance to close the hood and replace protective covers?
   c. Will access to other critical components be restricted or blocked?
   d. Is the tubing straight, clean, and sound?
3. Select the proper valve for the application.
   a. For low- or high-side use (the-low side valve is larger).
   b. The size of the tube the valve is to be installed on.
4. Position both halves of the saddle valve on the tube (Figure 11-6).

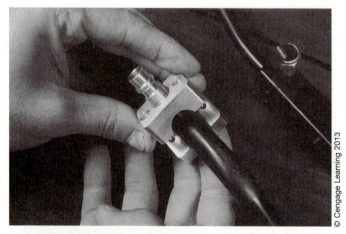

© Cengage Learning 2013

FIGURE 11-6 Position both halves of the valve on the tube.

5. Place the screws (usually socket head) and tighten them evenly. Do not overtighten them; 20–30 in.-lb. (2–3 N·m) is usually recommended.
6. Insert the piercing pin in the head of the access port fitting (Figure 11-7).
7. Tighten the pin until the head touches the top of the access port (Figure 11-8).
8. Remove the piercing pin and replace it with the valve core (Figure 11-9).

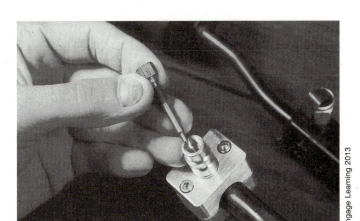

FIGURE 11-7 Insert the piercing pin.

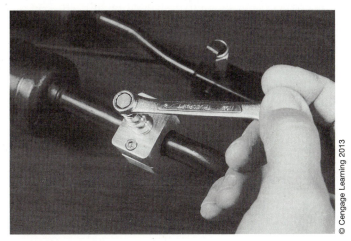

FIGURE 11-8 Tighten the pin.

SPECIAL TOOLS
In.-lb. torque wrench with socket (to match saddle valve screws)

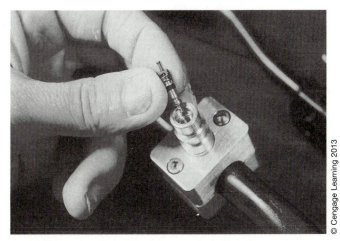

FIGURE 11-9 Replace the pin with the valve core.

FIGURE 11-10 Tighten the valve core.

9. Securely tighten the valve core (Figure 11-10).
10. Install the cap (or pressure switch) on the installed fitting.

RECOVER ONLY—AN ALTERNATE METHOD

This method of recovery is presented for information only. It should only be accomplished by, or under the direct supervision of, an experienced technician. The most important consideration is that the recovery cylinder will not have been filled to more than 80 percent of capacity (Figure 11-11) when the temperature is increased to ambient.

Refer to the illustration (Figure 11-12) and follow these instructions:

1. Place an identified recovery cylinder into a tub of ice on the floor beside the vehicle.
2. Add water and ice cream salt. This will lower the temperature to about 0°F (−17.7°C).
3. Connect a service hose from the high-side fitting of the system to the gas valve of the recovery cylinder.
4. Open all valves.
5. Cover the recovery cylinder and tub with a blanket to insulate them from the ambient air.
6. Place the shop light(s) or other heat source near the accumulator or receiver.
7. Allow 1 to 2 hours for recovery. The actual time that is required will depend on the ambient temperature and the amount of refrigerant to be recovered.

There are chemicals available that will lower the ice bath temperature to as low as −15°F (−26°C).

The Schrader valve in the service fitting is often referred to as a valve core.

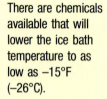

SERVICE TIP:
The recovery cylinder should be below the level of the air-conditioning system.

SPECIAL TOOLS
Recovery system
Tub (for ice bath)

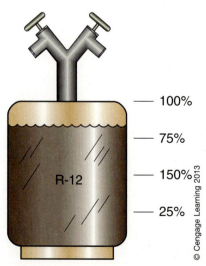

FIGURE 11-11 Recovery cylinders must not be filled more than 80 percent of capacity.

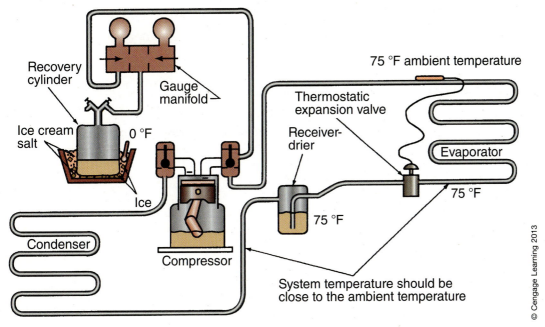

FIGURE 11-12 Setup for recovering refrigerant using a cold bath tank.

Labels in figure: Recovery cylinder; Gauge manifold; Ice cream salt; 0 °F; Ice; Condenser; Compressor; Receiver-drier; Thermostatic expansion valve; 75 °F ambient temperature; Evaporator; 75 °F; 75 °F; System temperature should be close to the ambient temperature

© Cengage Learning 2013

RETROFIT

Specific procedures to retrofit any particular make or model vehicle are provided by the respective vehicle manufacturers. Several aftermarket manufacturers also offer retrofit kits for more generic applications. For example, one such manufacturer claims that three kits are all that are required to retrofit all car lines. According to early information released by automotive manufacturers, however, the procedure, methods, and materials vary considerably from car line to car line.

For example, some require draining mineral oil, while others do not; some require flushing the system, others do not. Also, some require replacing components, such as the accumulator or receiver-drier or the condenser, and others do not.

In mid-June 1993, the Society of Automotive Engineers (SAE) issued its standard J1661 "Procedure for Retrofitting R-12 Mobile Air-Conditioning Systems to R-134a." The following service procedure, which is considered typical, is based on SAE's J1661.

Before attempting this procedure, be sure to review Chapter 11 of the Classroom Manual. This contains some very important information that must be understood to successfully retrofit a vehicle air-conditioning system.

PROCEDURE

The following step-by-step procedures are to be considered typical for retrofitting any vehicle from refrigerant R-12 to refrigerant R-134a. For specific procedures, however, follow the manufacturer's instructions.

Connect the Manifold and Gauge Set

Follow this procedure when connecting the R-12 manifold and gauge set into the system for service.

Prepare the System

1. Place fender covers on the car to avoid damage to the finish.
2. Remove the protective caps from the service valves. Some caps are made of light metal and can be removed by hand; others may require a wrench or pliers.

Classroom Manual

Chapter 11, page 393

CAUTION:
Do not attempt to use any other type refrigerants.

Use only R-134a to retrofit an automobile air-conditioning system.

CAUTION:
Before beginning the retrofit procedure, perform a purity test to determine the type and quality of the refrigerant in the air-conditioning system.

A **depressing pin** is a pin located in the end of a service hose to press (open) a Schrader-type valve.

Most hand valves are closed by turning in the clockwise (cw) direction.

 WARNING: Remove the caps slowly to ensure that refrigerant does not leak past the service valve.

Connect the Manifold Service Hoses

WARNING: The service hoses must be equipped with a Schrader valve depressing pin (Figure 11-13).

1. Make sure that the manifold hand shutoff valves (Figure 11-14) are closed.
2. Make sure that the hose shutoff valves (Figure 11-15) are closed.
3. Finger-tighten the low-side manifold hose to the suction side of the system.

© Cengage Learning 2013

FIGURE 11-13 R-12 service hoses equipped with Schrader valve depresser pin.

© Cengage Learning 2013

FIGURE 11-14 Make sure that manifold hand shutoff valves are closed.

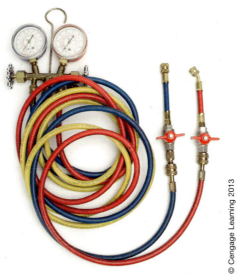

FIGURE 11-15 Make sure that the service hose shutoff valves are closed.

Flexible adapter

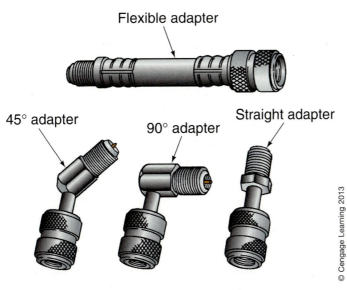

45° adapter   90° adapter   Straight adapter

FIGURE 11-16 Special high-side hose adapters.

4. Finger-tighten the high-side manifold hose to the discharge side of the system.
5. If retrofitting an older vehicle or heavy-duty, off-road equipment air-conditioning system having shutoff type service valves (see Figure 11-21), use a service valve wrench to rotate the stem two turns clockwise (cw).
6. Connect the service hose to the R-12 recovery system.

REFRIGERANT RECOVERY

Until the early 1990s, service technicians vented refrigerant into the atmosphere. Refrigerant was inexpensive, and the cost of recovery would probably have been greater than the cost of the refrigerant. The Clean Air Act (CAA) Amendments of 1990 changed that practice. The CAA enacted by the Environmental Protection Agency (EPA) required that, after July 1, 1992, no refrigerants may be intentionally vented.

SERVICE TIP:
The R-12 high-side fitting on most late-model car lines requires that a special adapter be connected to the hose (Figure 11-16) before being connected to the fitting.

SPECIAL TOOL
Recovery system

Unintentional venting in the performance of repairs is permitted under the CAA.

⚠️ **WARNING:** **Adequate ventilation must be maintained during this procedure. Do not discharge refrigerant near an open flame, as a hazardous toxic gas may be formed.**

Prepare the System

1. Start the engine and adjust its speed to 1,250–1,500 rpm.
2. Set all air-conditioning controls to the MAX cold position with the blower on HI speed.
3. Operate for 10–15 minutes to stabilize the system.

Recover Refrigerant

1. Return the engine speed to normal idle to prevent dieseling.
2. Turn off all air-conditioning controls.
3. Shut off the engine.
4. If not integrated in the recovery system, use a service hose and connect the recovery system to an approved recovery cylinder.
5. Open all hose shutoff valves.
6. Open both low- and high-side manifold hand valves.
7. Open the recovery cylinder shutoff valves, as applicable.
8. Connect the recovery system into an approved electrical outlet and turn on the main power switch.
9. Turn on the recovery system compressor switch.
10. Operate the vacuum pump until a vacuum pressure is indicated (Figure 11-17).
11. If the recovery system is not equipped with an automatic shutoff, turn off the compressor switch after achieving a vacuum (step 10).

FIGURE 11-17 Operate the pump until a vacuum is noted.

© Cengage Learning 2013

12. Be sure that the vacuum holds for a minimum of 5 minutes.
    a. If the vaccum does not hold, repeat the procedures starting with step 9 and continue until the system holds a stable vacuum for a minimum of 2 minutes.
    b. If the vacuum holds, proceed with step 13.
13. Close all valves: at the recovery cylinder, recovery system, service hoses, manifold, and compressor.
14. Disconnect all hoses previously connected.

 WARNING: **Some recovery systems have automatic shutoff valves. Be certain they are operating properly before disconnecting the hoses to avoid refrigerant loss that could result in personal injury.**

Repair or Replace Components

1. Determine what repairs, if any, are required.
2. If an oil change is required, proceed with step 3; if not, proceed with step 4.
3. Remove the necessary components to drain the oil from the component (Figure 11-18).
4. Flush the individual components while they are out of the vehicle. A typical setup for this procedure is shown in Figure 11-19.
5. Replace components such as the accumulator, receiver-drier, and/or condenser, if required. It may be necessary to replace the receiver-drier or accumulator-drier if the desiccant is not compatible with R-134a refrigerant.
6. Add or replace electrical fail-safe components, such as the refrigerant containment and high-pressure switch, if required.
7. Perform any other modifications and/or procedures required by the specific vehicle manufacturer.
8. Replace/reinstall all components serviced in steps 3 and 7.
9. If not accomplished by the requirements of steps 5, 6, or 7, repair any problems determined in step 1.

> Flushing is generally not recommended unless the component has first been removed from the vehicle.

> **CAUTION:**
> It is recommended that the receiver-drier or accumulator-drier be replaced any time the air-conditioning system is opened for major repairs.

FIGURE 11-18 Drain oil from an accumulator.

© Cengage Learning 2013

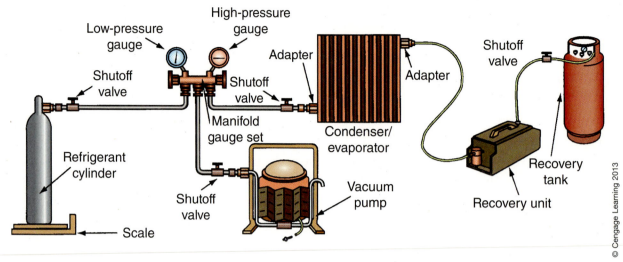

FIGURE 11-19 Typical setup for flushing a component.

FLUSH THE SYSTEM?

Flushing the air-conditioning system is not generally recommended. Because of the screens and strainers in the system, little if any debris will be removed. Also, most liquids (moisture and lubricant) in the low areas in the system—such as in the bottom of the evaporator, muffler, receiver, or accumulator—are not removed by flushing.

If flushing is to be performed, the flushing agent should be refrigerant—the same type used in the system. In the case of R-12, this is an expensive procedure. Also, system components should be removed for individual flushing after excess lubricant has been drained from them. Refrigerants used for flushing must be recovered.

There are a number of systems and techniques available for flushing an air-conditioning system. Some flush systems are attachments for the recovery/recycle machine and other systems are self-contained units. Some use refrigerant as a flushing agent and others use various fluids, even methylhydrate or naphtha (flammable fluids). Nitrogen is often used as a propellant for the cleaning fluid. Some suggest adding a filter to the liquid line after flushing to catch any remaining debris before the metering device.

It is important to follow the manufacturer's recommended procedures for the particular flushing system being used. It is also important not to neglect system lubrication after flushing, regardless of the method or system used. Any lubricant flushed out of the system must be replaced with clean, fresh lubricant of the proper type for the refrigerant being used.

Only PAG or POE lubricant should be used in an R-134a refrigerant system.

CAUTION: If the system was flushed, charge oil directly into the compressor to provide lubrication at startup.

Prepare the System for R-134a

1. Charge the system with the proper type and quantity of lubricant as recommended by the vehicle manufacturer for R-134a refrigerant.
2. Change service ports from R-12 to R-134a access type (Figure 11-20).
3. Check for leaks.
4. Affix decals to identify refrigerant type for future service (Figure 11-21).

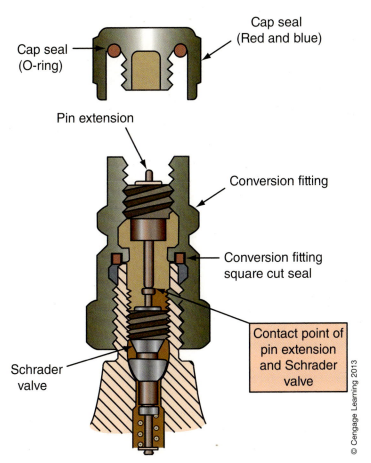

Cap seal
(O-ring)

Cap seal
(Red and blue)

Pin extension

Conversion fitting

Conversion fitting
square cut seal

Contact point of
pin extension
and Schrader
valve

Schrader
valve

© Cengage Learning 2013

FIGURE 11-20 Use adapter fittings to change R-12 service ports to R-134a service ports, and apply thread locking compound when installing.

NOTICE: RETROFITTED TO R-134a

RETROFIT PROCEDURE PERFORMED TO SAE J1661
USE ONLY R-134a REFRIGERANT AND SYNTHETIC
OIL TYPE: _____1_____ PN: _____2_____ OR
EQUIVALENT, OR A/C SYSTEM WILL BE DAMAGED

REFRIGERANT CHARGE/AMOUNT: _____3_____
LUBRICANT AMOUNT: ___4___ PAG ☐ ESTER ☐5

RETROFITTER NAME: _____6_____ DATE: ___7___
ADDRESS: _____8_____
CITY: _____9_____ STATE: ___10___ ZIP: ___11___

1 Type: Manufacturer of oil (Saturn, GM, Union Carbide, etc.).

2 PN: Part number assigned by manufacturer.

3 Refrigerant charge / amount: Quantity of charge installed.

4 Lubricant amount: Quantity of oil installed (indicate ounces, cc, ml).

5 Kind of oil installed (check either PAG or ESTER).

6 Retrofitter name: Name of facility that performed the retrofit.

7 Date: Date retrofit is performed.

8 Address: Address of facility that performed the retrofit.

9 City: City in which the facility is located.

10 State: State in which the facility is located.

11 Zip: Zip code of the facility.

© Cengage Learning 2013

FIGURE 11-21 A typical retrofit label.

SYSTEM VACUUM		TEMPERATURE	
in. Hg	kPa (ABS)	°F	°C
0.00	101.33	212	100
27.75	7.35	104	40
28.67	4.23	86	30
29.32	2.03	64	18
29.62	1.01	45	7
29.74	0.61	32	0
29.82	0.34	6	−14
29.91	0.03	−24	−31

© Cengage Learning 2013

FIGURE 11-22 Boiling point of water (H_2O) in a vacuum at sea level atmospheric pressure.

SERVICE TIP:
Retrofit conversion service fittings for R-134a refrigerant are available in several styles. Conversion fittings with extension pins offer improved performance by opening the existing Schrader valve more fully (see Figure 11-20). Be sure to use thread-lock adhesive on service fittings to hold them in place. Some retrofit conversion service fittings require the removal of the existing R-12 Schrader valve or damage to the new fitting could occur.

EVACUATING THE SYSTEM

Whenever it is serviced, an automotive air-conditioning system should be evacuated to the extent that the refrigerant has been removed. There are some who claim that moisture cannot be removed from an automotive air-conditioning system with a "standard" vacuum pump. SAE standard J1661, however, requires that a vacuum pump be capable of achieving a vacuum level of 29.2 in. Hg (2.7 kPa absolute) adjusted to altitude. The boiling point of water ($H2O$) at this level is 69°F (20.6°C) at sea level atmospheric pressure (Figure 11-22). That means that moisture cannot be removed from an air-conditioning system when the ambient temperature is below, say, 70°F (21.1°C).

If a vacuum pump is to be used to remove moisture from an automotive air-conditioning system, a quality two-stage, high-vacuum pump is recommended for adequate performance over a long period of time. Even the best vacuum pump, however, requires regular maintenance to ensure optimum performance.

Frequent oil changes are perhaps the single most important factor in a preventive maintenance (PM) program. A vacuum pump cannot handle moisture without some of it condensing in the lubricant. If this moisture is not removed by changing the oil, it can attack metal components within the pump. This will result in lockups or loss of pumping efficiency and capacity. For the average service shop, oil changes should be a normal part of the daily equipment maintenance program. It would be well, however, to change the oil after an extended pump down, especially after pumping down a system known to be wet. Specific instructions included with a vacuum pump should be followed for changing the oil.

Speed at Which a System Is Dehydrated

Several factors influence the "pumping speed" of a high-vacuum pump and thus the time required to remove all moisture from a refrigerant system. Some of the most important factors include:

- Size of the system, in cubic feet
- Amount of moisture to be removed
- Ambient temperature
- Internal restrictions within the system (Schrader valves and metering device)
- External restrictions (between the system and vacuum pump)
- Size of the pump
- Condition of the pump (clean, fresh oil)

REMOVING AND REPLACING A SCHRADER VALVE CORE IN A SERVICE VALVE

The following procedure may be considered typical for replacing a Schrader valve core in an R-134a air-conditioning system service valve. This procedure, with minor variations, may also be used to replace a Schrader valve core in a R-12 air-conditioning system service valve as well. Always follow specific procedures included by the equipment manufacturer when they differ from those given in this text.

P18-1 Install the high-side service valve access fitting.

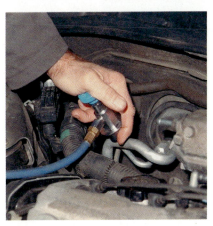

P18-2 Install the low-side service valve access fitting.

P18-3 With the recovery machine connected, open the appropriate gas or liquid valve (follow specific instructions for machine being used).

P18-4 Turn on the main power.

P18-5 Open the low- and high-side valve.

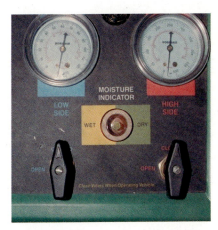

P18-6 Allow the recovery equipment to operate until both gauges indicate 0 psig (0 kPa) or less.

The elimination of restrictions in an air-conditioning system is not generally possible. The size of valves, manifold, and metering device cannot be altered during evacuation. The service lines, however, can be enlarged as well as shortened, and the Schrader valve cores can be removed during the evacuation process. Photo Sequence 18 illustrates the procedure for removing Schrader valve cores for evacuation and charging procedures.

P18-7 Remove the service valve access fitting from the leaking Schrader valve core.

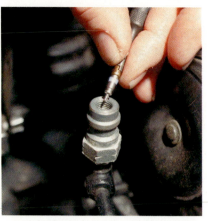

P18-8 Using a valve core tool, remove the leaking Schrader valve core.

P18-9 Install a new valve core using the valve core tool.

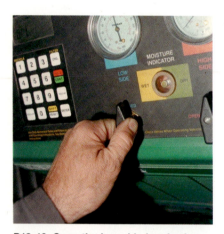

P18-10 Open the low-side hand valve.

P18-11 Start the vacuum pump.

P18-12 After 5 minutes, open the high side hand valve and proceed with procedures as outlined in Photo Sequence 7.

How Vacuum Is Measured

In the automotive air-conditioning industry, vacuum is generally measured with a standard Bourdon tube compound gauge. This type of gauge is suitable for standard vacuum reading, say 29 in. Hg. It cannot be used, however, to read millimeters or microns. For this reason, it is not suitable for use with high-vacuum pumps.

Electronic thermistor vacuum gauges (Figure 11-23) are available for use with high-vacuum pumps. They can accurately read a vacuum as low as 1 micron by using a sensing tube mounted at some point in the vacuum service line. The readout can be an analog meter scale, digital display, or a light-emitting diode (LED) sequential display. One advantage of a thermistor vacuum gauge is that it is sensitive to water vapor and other

A micron is a unit of linear measurement equal to $\frac{1}{25,400}$ of an inch.

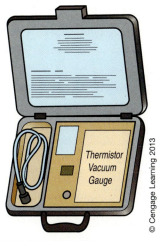

FIGURE 11-23 A thermistor vacuum gauge.

condensables and can give a good indication of the actual vacuum level within a system. A thermistor vacuum gauge, though not essential, is a worthwhile companion instrument for high-vacuum dehydration of an automotive air-conditioning system.

The location of the vacuum gauge will affect its reading in relation to the actual vacuum in the system. The closer the gauge is to the vacuum source, the lower the reading. When taking a final reading of the vacuum created in an air-conditioning system, one should isolate the vacuum pump with a vacuum valve and allow the pressure in the system to equalize.

If the pressure does not equalize, it is an indication of a leak. If it does equalize but only at a higher pressure, it is an indication that moisture remains in the system. If this is the case, more pumping time is required.

The following service procedure for evacuating the system may be used for the independent vacuum pump (Figure 11-24) or the dedicated charging station (Figure 11-25). The vacuum pump may be used for either R-12 or R-134a refrigerant systems. The charging station that is pictured contains a vacuum pump, manifold and gauge set, and calibrated charging cylinder and is for R-12 only. It is compatible with all R-12 recovery and recycling systems. Robinair (and others) also produces a similar dedicated charging station for R-134a refrigerant that is compatible with all R-134a recovery and recycle systems.

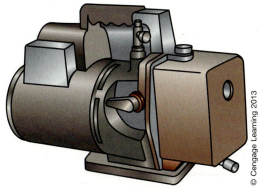

FIGURE 11-24 A typical vacuum pump.

FIGURE 11-25 A typical charging station.

Prepare the System

Follow the vacuum pump manufacturer's operating instructions if they differ from those given in this manual.

NOTE: Before performing any service procedure, ensure that both the low-side (compound) and high-side (pressure) gauges are zero calibrated.

1. Make sure that the high- and low-side manifold hand valves are in the closed position.
2. Make sure that the service hose shutoff valves are closed.
3. Remove the protective caps and covers from all service access fittings.
4. Connect the R-134a manifold and gauge set to the system in the same manner as was previously outlined for the R-12 manifold and gauge set.
5. Place the high- and low-side compressor service valves, if equipped, in the cracked position.
6. Remove the protective caps from the inlet and exhaust of the vacuum pump.
7. Connect the center manifold hose to the inlet of the vacuum pump.
8. Open all service hose shutoff valves.
9. Start the vacuum pump.
10. Open the low-side manifold hand valve.
11. Observe the low-side (compound) gauge needle. The needle should indicate a slight vacuum.
12. After 5 minutes, the compound gauge should indicate 20 in. Hg (33.8 kPa absolute) or less (Figure 11-26).
13. The high-side (pressure) gauge needle should be slightly below the zero index of the gauge.
14. If the high-side gauge does not drop below zero (Figure 11-27), unless restricted by a stop, a system blockage is indicated.
    a. If the system is blocked, discontinue the evacuation. Repair or remove the obstruction.
    b. If the system is clear, continue the evacuation with step 15.
15. Open the high-side manifold hand valve.
16. Operate the pump for 15 minutes and observe the gauges. The system should be at a vacuum of 24–26 in. Hg (20.3–13.5 kPa absolute) minimum if there is no leak.
17. If the system is not down to 24–26 in. Hg (20.3–13.5 kPa absolute), close the low-side hand valve and observe the compound gauge.
    a. If the compound gauge needle rises, indicating a loss of vacuum, there is a leak that must be repaired before the evacuation is continued.
    b. If no leak is evident, continue with the pump down.

If, after 5 minutes, there is not a reasonable vacuum noted, a leak is indicated.

indicated.

CAUTION:
Make sure the port cap is removed from the exhaust port to avoid damage to the vacuum pump.

FIGURE 11-26 The compound gauge (low) should indicate 20 in. Hg (33.8 kPa absolute) or below.

FIGURE 11-27 The high-side gauge should drop below zero.

18. Pump for a minimum of 30 minutes, as required by SAE J1661. A longer pump down is much better, if time permits. For maximum performance, a triple pump down is recommended by many.
19. After pump down, close the high- and low-side manifold hand valves.
20. Shut off the vacuum pump.
21. Close all valves (service hose, vacuum pump, and compressor, if equipped).
22. Disconnect the manifold hoses.
23. Replace any protective caps previously removed.

CHARGING AN R-134A AIR-CONDITIONING SYSTEM

It may be noted that R-134a is an **ozone-friendly** refrigerant and, as such, poses no known threat to the environment. Nonetheless, the EPA requires that this refrigerant also be recovered. This law became effective in the middle of November 1995.

Procedure

1. Place the vehicle in a draft-free work area. This is an aid in detecting small leaks.
2. Close all valves (service valves, if equipped, manifold gauge, service hose shutoff valves, and refrigerant cylinder or charging station shutoff valve).

A product is considered **ozone friendly** if it does not pose a hazard or danger to the ozone layer.

The system should now be under a deep vacuum.

A dual-probe electronic thermometer is ideal for measuring superheat.

CAUTION:
Do not open the manifold and gauge set hand valves until instructed to do so. Early opening could contaminate the system with moisture-laden air.

3. Connect the manifold and gauge set to the system following procedures previously outlined.
4. Connect the service hose to the refrigerant source. If a charging station is used, it is very important that the instructions provided by the manufacturer of the equipment are followed.
5. Open the service hose shutoff valves.
6. Open the system service valves, if equipped.
7. Observe the gauges.
   a. Confirm that the system is in a vacuum. If it is, proceed with step 8.
   b. If it is not, follow the procedure outlined for evacuating the system before proceeding.
8. Dispense one "pound" can (Figure 11-28) of R-134a refrigerant into the system.
   a. Invert the can for liquid dispensing (Figure 11-29).
   b. Open the high-side manifold hand valve.

© Cengage Learning 2013

FIGURE 11-28 Dispense one can of refrigerant into the system.

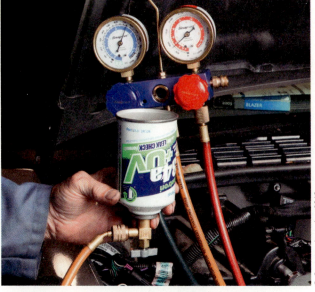

© Cengage Learning 2013

FIGURE 11-29 Invert the can for liquid dispensing.

FIGURE 11-30 Attach the electronic thermometer probes to the inlet and outlet of the evaporator.

    c. Empty the contents of the can into the system.

    d. Close the manifold high-side valve.

    e. Rotate the clutch armature several revolutions by hand to ensure that no liquid refrigerant is in the compressor.

9. Attach the electronic thermometer probes (Figure 11-30) to the inlet and outlet of the evaporator. Be sure that the end of the probe makes good contact with the metal tubes of the evaporator.

10. Open all windows.

11. Place a jumper wire across the terminals of the temperature or pressure control, usually found on the accumulator.

12. Start the engine.

13. Set all air-conditioner controls to HI.

14. Allow the engine to reach normal operating temperature.

15. Note and record the temperature of the two thermometers. Calculate the difference in temperature between the inlet and outlet tubes of the evaporator.

16. Wait a few minutes and record the temperatures again to confirm the readings.

17. Note and record the ambient temperature. Compare it with the chart in Figure 11-31, as applicable.

18. Follow the temperature differential chart (step 15) to determine how much refrigerant must be added to the system to ensure a proper charge.

19. Continue charging, as required. Tap a "pound" can of R-134a. With the can upright, open the manifold low-side valve. Dispense the contents of the can into the system. Close the low-side manifold valve. Repeat this step, as required.

20. Turn off the air conditioner.

21. Stop the engine.

22. Remove the jumper wire from the temperature/pressure switch (see step 11).

23. Close all valves (manifold, hose shutoff, and service, if equipped).

24. Recover refrigerant from the service hoses.

25. Disconnect all hoses from the system.

26. Replace all protective covers and caps.

The jumper prevents compressor short-cycling during charging procedures.

CAUTION: As a general rule, the capacity of a retrofit system for refrigerant R-134a capacity is about 90 percent of the original capacity for R-12 refrigerant.

Remember to replace the connector to the switch.

AMBIENT TEMPERATURE (°F)						AMOUNT OF R-134a TO ADD (OUNCES)
60	70	80	90	100	110	
Evaporator Inlet to Outlet Temperature Difference						
−8	−8	−8	−8	−8	−8	0
−7	−7	−7	−7	−7	−7	2
−6	−6	−6	−6	−6	−6	4
−5	−5	−5	−5	−5	−5	6
+13	+13	+13	+17	+20	+25	8
+21	+25	+29	+33	+37	+42	12
+40	+45	+50	+55	+60	+65	14

A

AMBIENT TEMPERATURE (°C)						AMOUNT OF R-134a TO ADD (mL)
16	21	27	32	38	43	
Evaporator Inlet to Outlet Temperature Difference						
−5	−5	−5	−5	−5	−5	0
−4	−4	−4	−4	−4	−4	59
−3	−3	−3	−3	−3	−3	118
−3	−3	−2	−1	0	0	177
+7	+7	+7	+9	+11	+14	237
+10	+14	+16	+18	+21	+23	335
+22	+25	+28	+31	+33	+36	414

B

© Cengage Learning 2013

FIGURE 11-31 Evaporator Delta T chart for inlet and outlet temperature.

CUSTOMER CARE: When performing under-hood service such as refrigerant retrofit, make a visual inspection of the engine cooling system. Advise the customer of any problems noticed that may lead to early failure of the cooling or heating system. These problems may include leaks, rotted or cracked radiator or heater hoses, or frayed or worn belt(s). In bringing these problems to the customer's attention, the customer is made aware of pending problems. Nothing is more frustrating than having a breakdown due to other failures just after having extensive (and expensive) repairs.

When customers are made aware of potential problems, they will generally approve repairs. While some may put off repairs, most will be thankful that your inspection may have prevented an expensive and inconvenient breakdown in the future.

In any event, the customer has been made aware of pending problems that are not covered by the current repair warranty. If the customer chooses not to have the repairs made, make a proper notation on the shop order form so it may be a matter of record.

CONCLUSION

Some final points:

- At the end of a successful retrofit, affix the proper label in a conspicuous place under the hood. The label (Figure 11-32) should at least contain the following information:
  1. Date of retrofit.
  2. Company or technician name and address.
  3. Type and amount of refrigerant (R-134a) in pounds (lb.), ounces (oz.), or milliliters (mL).
  4. Type and amount of lubricant (PAG or POE) in ounces (oz.) or milliliters (mL).

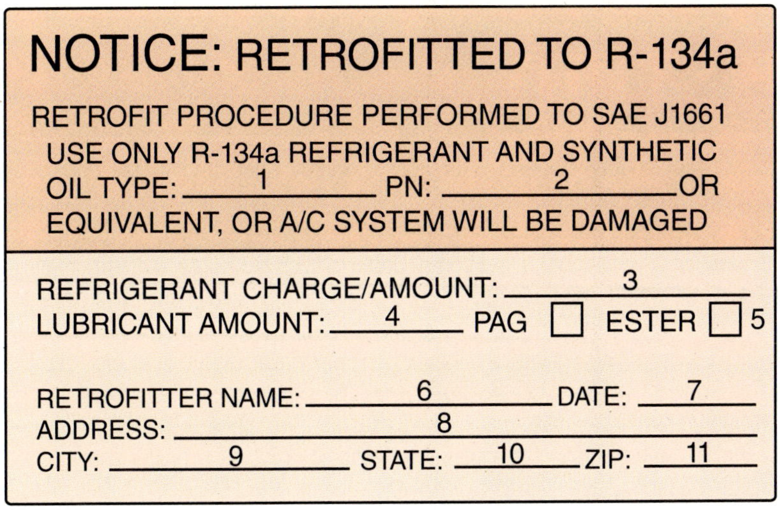

NOTICE: RETROFITTED TO R-134a

RETROFIT PROCEDURE PERFORMED TO SAE J1661
USE ONLY R-134a REFRIGERANT AND SYNTHETIC
OIL TYPE: ____1____ PN: ____2____ OR
EQUIVALENT, OR A/C SYSTEM WILL BE DAMAGED

REFRIGERANT CHARGE/AMOUNT: ____3____
LUBRICANT AMOUNT: ____4____ PAG ☐ ESTER ☐ 5

RETROFITTER NAME: ____6____ DATE: ____7____
ADDRESS: ____8____
CITY: ____9____ STATE: ____10____ ZIP: ____11____

1 Type: Manufacturer of oil (Saturn, GM, Union Carbide, etc.).

2 PN: Part number assigned by manufacturer.

3 Refrigerant charge / amount: Quantity of charge installed.

4 Lubricant amount: Quantity of oil installed (indicate ounces, cc, ml).

5 Kind of oil installed (check either PAG or ESTER).

6 Retrofitter name: Name of facility that performed the retrofit.

7 Date: Date retrofit is performed.

8 Address: Address of facility that performed the retrofit.

9 City: City in which the facility is located.

10 State: State in which the facility is located.

11 Zip: Zip code of the facility.

FIGURE 11-32 A retrofit label.

© Cengage Learning 2013

■ Do not remove the R-134a fitting adapters from the R-12 fittings. Once installed, they become a permanent part of the air-conditioning system.

■ Do not overcharge the air-conditioning system with refrigerant. The typical R-134a charge of refrigerant is about 90 percent of the original R-12 refrigerant charge. Refer to the chart in Figure 11-33 for the 90 percent rule.

R-12		R-134a	
OUNCES	MILLILITERS	OUNCES	MILLILITERS
48	1420	43.2	1278
44	1302	39.6	1171
40	1183	36.0	1065
36	1065	32.4	958
32	947	28.8	852
30	887	27.0	799
28	828	25.2	745
26	769	23.4	692
24	710	21.6	639
22	651	19.8	586
20	592	18.0	532
18	532	16.2	479
16	473	14.4	426
14	414	12.6	373

© Cengage Learning 2013

FIGURE 11-33 The 90 percent rule for R-134a versus R-12 refrigerant change.

CASE STUDY

A customer complained about an inoperative air-conditioning system. Questioning the customer revealed that the system had not worked since the end of last summer. "It was going to get cool in a few weeks and I would not need the air conditioner. I decided to put it off until spring."

A visual inspection of the system by the technician did not reveal any oil spots or ruptured hoses indicating a leak. When the manifold set was connected, the gauges revealed system pressure was equal on both gauges. Further, the temperature-pressure chart indicated that the system pressure was within acceptable limits for R-12 refrigerant for the ambient temperature.

The technician noticed that the lead wire to the clutch coil had been disconnected. Assuming that it had been intentionally disconnected, the technician reconnected it. Further questioning of the customer, however, revealed no knowledge of a disconnected wire.

Shortly after starting the engine and turning the air conditioner on, cool air was noted coming from the driver-side vent. The manifold gauges indicated proper pressures. A thermometer inserted in the passenger-side vent also indicated proper temperature.

Further discussion with the customer revealed that the problem apparently had begun while on vacation. The belts had been replaced and the mechanic must have pulled the wire loose during the repairs. The customer did not realize the problem for several days after the repairs since the climate was mild and the air conditioner was not turned on. The customer had not considered that the problem may have occurred during repairs. The customer suffered through the close of one summer and the start of another simply because of a mechanic's error and putting off repairs.

TERMS TO KNOW

Depressing pin

Ozone friendly

Saddle valve

ASE-STYLE REVIEW QUESTIONS

1. *Technician A* says that a shutoff-type service valve has a front-seated position.

   Technician B says that a Schrader-type service valve has a front-seated position.

   Who is correct?

   A. A only
   B. B only
   C. Both A and B
   D. Neither A nor B

2. After stabilizing the air-conditioning system, the engine speed is returned to normal.

   Technician A says this is to reduce airflow across the condenser.

   Technician B says this is to increase the cooling capacity of the evaporator.

   Who is correct?

   A. A only
   B. B only
   C. Both A and B
   D. Neither A nor B

3. Which of the following desiccants should not be used with R-134a refrigerant?

   A. XH5
   B. XH7
   C. XH9
   D. B and C

4. All of the following statements about system evacuation are true, *except*:

   A. All systems should be evacuated for a minimum of 30 minutes.

   B. If vacuum does not reach 24 to 26 in. Hg (20.3 to 23.5 kPa absolute), evacuate for a longer time.

   C. Vacuum should reach at least 24 to 26 in. Hg (20.3 to 23.5 kPa absolute) in the first 15 minutes.

   D. The longer the evacuation period, the better the moisture removal.

5. The recommended minimum efficiency of a vacuum pump at sea level atmospheric pressure is being discussed:

   Technician A says that atmospheric pressure has no effect on efficiency.

   Technician B says that the greater the vacuum achieved, the better the efficiency.

   Who is correct?

   A. A only
   B. B only
   C. Both A and B
   D. Neither A nor B

6. *Technician A* says that POE is compatible with R-134a.

   Technician B says that PAG oil is compatible with R-134a.

   Who is correct?

   A. A only
   B. B only
   C. Both A and B
   D. Neither A nor B

7. All of the following statements are true, *except*:
   A. Flushing the air-conditioning system is not recommended when retrofitting.
   B. R-134a retrofit capacity is about 90 percent of the original R-12 capacity.
   C. A refrigerant identifier will identify flammable refrigerants.
   D. A refrigerant identifier will identify R-134a at any purity.

8. The pressure required for leak testing is being discussed:

   Technician A says that a minimum of 60 psig (414 kPa) is required.

   Technician B says that a maximum of 40 psig (276 kPa) is required.

   Who is correct?

   A. A only
   B. B only
   C. Both A and B
   D. Neither A nor B

9. *Technician A* says that the new refrigerant installed must be identified on the retrofit label.

   Technician B says that the shop that performed the A/C system retrofit must be identified on the retrofit label.

   Who is correct?

   A. Technician A only
   B. Technician B only
   C. Both A and B
   D. Neither A nor B

10. If the compressor cycles while charging:

    Technician A says to place a jumper across the temperature switch.

    Technician B says to place a jumper across the low-pressure switch.

    Who is correct?

    A. A only
    B. B only
    C. Both A and B
    D. Neither A nor B

ASE CHALLENGE QUESTIONS

1. Under the Clean Air Act (CAA), any person who performs service to a motor vehicle air conditioner (MVAC) must:
   A. Be properly certified by an approved agency
   B. Use properly certified recovery equipment
   C. Both A and B
   D. Neither A nor B

2. Which of the following refrigerants may be vented?
   A. R-12
   B. R-134a
   C. Both A and B
   D. Neither A nor B

3. All of the following are removed from an air-conditioning system during evacuation, *except*:
   A. Air
   B. Moisture
   C. Refrigerant
   D. Lubricant

4. When retrofitting, the proper R-134a charge of refrigerant is about _____ of the original R-12 refrigerant charge.

   A. 95 percent
   B. 90 percent
   C. 85 percent
   D. 80 percent

5. The equipment shown in the illustration (right) may be used to:
   A. Evacuate and charge an air-conditioning system
   B. Recover refrigerant from an air-conditioning system
   C. Recycle refrigerant that has been recovered from an air-conditioning system
   D. All of the above

© Cengage Learning 2013

Name _____ **Date** _____

DETERMINING REFRIGERANT PURITY IN A MOBILE AIR-CONDITIONING SYSTEM BY VERBAL COMMUNICATION

Upon completion of this job sheet, you should be able to use good judgment regarding refrigerant purity.

NATEF Correlation

HEATING AND AIR CONDITIONING: Refrigerant Recovery, Recycling, and Handling; *Identify (by label application or use of a refrigerant identifier) and recover A/C system refrigerant.* **(P-1)**

Tools and Materials

Vehicle with air-conditioning system charged with refrigerant

Describe the vehicle being worked on.

Year _____ Make _____ Model _____

VIN _____ Engine type and size _____

Procedure

1. Question the customer:
   a. Has the vehicle air conditioner been serviced recently? _____

   b. When was it serviced? _____

   c. By whom? _____

   d. What type service was performed? _____

   e. What type refrigerant, if any, was used? _____

   f. What problems are you experiencing? _____

   Responses by the customer to the above questions help determine if the system should now be serviced. If there are any safety concerns, such as flammable refrigerant, DO NOT service the air-conditioning system. Give a brief summary of your interpretation of the customer's responses and tell why you:

2. a. Decided to service the air-conditioning system:

   _____

   _____

   b. Decided not to service the air-conditioning system:

   _____

   _____

Instructor's Response _____

_____

_____

Name _____ **Date** _____

DETERMINING REFRIGERANT PURITY IN A MOBILE AIR-CONDITIONING SYSTEM BY TESTING

Upon completion of this job sheet, you should be able to use a refrigerant tester to test for refrigerant purity.

NATEF Correlation

HEATING AND AIR CONDITIONING: Refrigerant Recovery, Recycling, and Handling; *Identify (by label application or use of a refrigerant identifier) and recover A/C system refrigerant.* **(P-1)**

Tools and Materials

Vehicle with air-conditioning system charged with refrigerant

Refrigerant purity tester

Instruction manual

Describe the vehicle being worked on.

Year _____ Make _____ Model _____

VIN _____ Engine type and size _____

Procedure

1. Determine, following Job Sheet 69, if you wish to proceed with refrigerant testing. Briefly explain:

   _____

   _____

2. Following procedures included with the purity tester, connect the tester to the air-conditioning system to draw a sample of refrigerant. Describe your procedure:

   _____

   _____

3. Was there an audible signal? _____ What would an audible signal indicate?

   _____

   _____

4. What is indicated on the readout? _____ What does this reading mean?

   _____

   _____

5. Based on the results of this test, what procedure will you use to recover the refrigerant?

_____

_____

_____

_____

Instructor's Response _____

_____

_____

Name _____ **Date** _____

IDENTIFYING RETROFIT COMPONENTS

Upon completion of this job sheet, you should be able to identify those components that must be replaced during retrofit procedures.

Tools and Materials

Vehicle with R-12 air-conditioning system to be retrofitted for R-134a

Service manual

Factory-approved retrofit kit

Describe the vehicle being worked on.

Year _____ Make _____ Model _____

VIN _____ Engine type and size _____

Procedure

Write a short report about component replacement during retrofit from R-12 to R-134a procedures. Explain why or why not the following components should be replaced:

1. Receiver or accumulator:

_____

_____

2. Hose or hoses:

_____

_____

3. Evaporator:

_____

_____

4. Condenser:

_____

_____

5. Pressure control switch:

_____

_____

6. Control thermostat:

_____

_____

7. Compressor:

_____

_____

8. Condenser fan and motor:

_____

_____

9. Evaporator blower and motor:

_____

_____

10. Other:

_____

_____

Instructor's Response _____

_____

_____

Name _____ **Date** _____

R-12 TO R-134A RETROFIT

Upon completion of this job sheet, you should be able to retrofit an R-12 air-conditioning system to an R-134a air-conditioning system.

Tools and Materials

Vehicle with R-12 air-conditioning system to be retrofitted for R-134a

Factory-approved retrofit kit

R-12 refrigerant recovery equipment

R-134a refrigerant charging equipment

Hand tools, as required

Describe the vehicle being worked on.

Year _____ Make _____ Model _____

VIN _____ Engine type and size _____

Procedure

After each step, write a brief summary of your procedure:

1. Remove the R-12 refrigerant by recovering it for future use.
   NOTE: DO NOT vent refrigerant to the atmosphere.

   _____

   _____

2. Remove and replace any defective air-conditioning system components.

   _____

   _____

3. Remove as much of the mineral oil as possible.

   _____

   _____

4. Add or replace components as required in the retrofit procedures.

   _____

   _____

5. Add or replace lubricant as required in the retrofit procedures.

   _____

   _____

6. Install R-134a service valve fittings and label.

_____

_____

7. Evacuate the air-conditioning system.

_____

_____

8. Leak test the air-conditioning system.

_____

_____

9. Charge the air-conditioning system with R-134a refrigerant.

_____

_____

10. Performance test the air-conditioning system.

_____

_____

11. What problems, if any, were encountered?

_____

_____

Instructor's Response _____

_____

_____

Final Exam Automotive Heating and Air Conditioning A7

1. What component part of the air-conditioning system causes the refrigerant to change from a liquid to a vapor?

   A. Evaporator

   B. Compressor

   C. Condenser

   D. Metering device

2. How does the air conditioning system remove excess humidity from the air entering the passenger compartment?

   A. Moisture collects on the duct walls.

   B. Moisture condenses on the condenser.

   C. Moisture condenses on the evaporator.

   D. Moisture is separated by the blower motor.

3. During a system performance test of the air-conditioning system operation both the high-side and low-side pressure readings are about the same and the compressor clutch is engaged. Which of the following is the most likely cause?

   A. A restriction in the low pressure line

   B. A faulty compressor valve plate

   C. Moisture contamination of the system

   D. A restricted expansion valve

4. Before discarding a disposable refrigerant tank, which of the following procedures should be performed?

   A. Make sure the tank valve is closed to prevent venting to the atmosphere.

   B. Flush the tank with refrigerant flushing agent.

   C. Recover any remaining refrigerant left in the tank.

   D. Open the valve to eliminate the pressure.

5. *Technician A* says that a retrofit label must identify the type and amount of refrigerant oil.

   Technician B says that a retrofit label must identify the amount of new refrigerant installed.

   Who is correct?

   A. A only

   B. B only

   C. Both A and B

   D. Neither A nor B

6. All of the following may cause a compressor clutch to slip, *except:*

   A. Overcharge of refrigerant

   B. Loose drive belt

   C. Improper air gap

   D. Low voltage

7. The voltmeter reading in the illustration below is 0. The *most* probable cause of this problem is that the:

   A. Windings are shorted

   B. Windings are open

   C. Relay is not energized

   D. Motor is seized

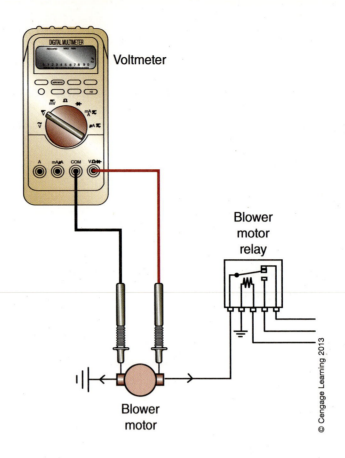

Voltmeter

Blower
motor
relay

Blower
motor

© Cengage Learning 2013

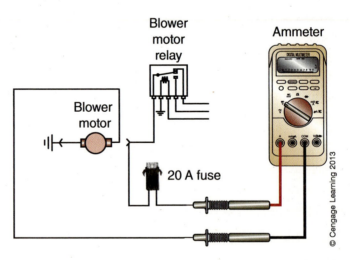

Blower
motor
relay

Ammeter

Blower
motor

20 A fuse

© Cengage Learning 2013

8. The motor does not operate when connected as shown in the illustration above. The fuse is good, and 15-A is displayed on the ammeter. The *most* likely cause of this problem is that the:

    A. Windings are shorted   **C.** Brushes are defective

    B. Windings are open   **D.** No power is going to the motor

9. High-voltage "spikes" are eliminated when the clutch is engaged and disengaged with the use of a:

    A. Thermistor   **C.** Transistor

    B. Resistor   **D.** Diode

10. A vehicle is brought in for service with the customer complaint that a foul odor is present when the air-conditioning system is on.

Technician A says that the cause may be ice forming on the evaporator core.

Technician B says that the cause may be mold and bacteria growing on the evaporator core.

Who is correct?

A. A only
B. B only
C. Both A and B
D. Neither A nor B

11. The refrigerant leaving the condenser is a _____.

A. Cold, low pressure vapor
B. Hot, high pressure vapor
C. Cold, low pressure liquid
D. Hot, high pressure liquid

12. A manifold and gauge set may be used for all of the following *except:*

A. Charging the system with refrigerant
B. Evacuating the refrigerant system
C. Checking system contamination
D. Checking system pressures

13. The loss of a vacuum signal at the control will *most* likely cause the system to "fail safe" to the _____ mode.

A. Heat
B. Defrost
C. Either A or B
D. Neither A nor B

14. A "blend" refrigerant means that:

A. It contains more than one component in its composition.
B. It may be mixed (blended with) another refrigerant.
C. Both A and B
D. Neither A nor B

15. The screen in the fixed orifice tube is found to be clogged. The recommended repair is to determine and correct the problem that caused the clogging and to:

A. Clean the screen.
B. Replace the screen.
C. Replace the fixed orifice tube.
D. Replace the fixed orifice tube and liquid line.

16. The pressure of the refrigerant in the condenser _____ as it gives up its heat to the ambient air.

A. Is increased
B. Remains about the same
C. Is reduced
D. Any of the above, depending on its temperature

17. The clutch air gap may be measured using a:

A. Dime
B. Wire-type feeler gauge
C. Stainless steel scale
D. Nonmagnetic feeler gauge

18. While charging a refrigerant system, if liquid refrigerant is added to the low side of the system of a running air conditioner, the following component could be damaged.
   A. The evaporator assembly
   B. The condenser assembly
   C. The compressor assembly
   D. Expansion valve assembly

19. All of the following statements about a system with a variable displacement compressor are true, *except:*
   A. There is no electromagnetic clutch.
   B. The cycling clutch is not used for temperature control.
   C. The swash plate angle determines the compressor displacement.
   D. When capacity demand is high, the swash plate is at its greatest angle.

20. Most vehicle manufacturers use which of the following refrigerant oil in their systems?
   A. Mineral oil
   B. PAG oil
   C. Ester oil
   D. PCV oil

21. All of the following are popular methods of leak detection, *except:*
   A. Halogen
   B. Halide
   C. Dye
   D. Vacuum

22. When the refrigerant container is inverted, as shown in the illustration above:
   A. The air-conditioning system must be charged through the high side with the compressor off.
   B. The air-conditioning system must be charged through the high side with the compressor running.
   C. The air-conditioning system must be charged through the low side with the compressor off.
   D. The air-conditioning system must be charged through the low side with the compressor running.

23. Low voltage at the clutch coil may cause:
   A. The clutch to slip
   B. A noisy clutch
   C. Both A and B
   D. Neither A nor B

24. All *except* which of the following statements about a vacuum pump are true? A vacuum pump may be used to remove:

   A. Moisture from an air-conditioning system

   B. Air from an air-conditioning system

   C. Trace refrigerant from an air-conditioning system

   D. Debris from an air-conditioning system

25. The low-side gauge on a manifold and gauge set indicates a vacuum below 29 in. Hg while the vacuum pump is running. Five minutes after the pump is turned off, the gauge indicates 25 in. Hg.

   Technician A says that the vacuum pump was not run long enough to remove residual refrigerant from the lubricant, and the rise in pressure is caused by refrigerant outgassing.

   Technician B says that the air-conditioning system may have a leak and the rise in pressure is caused by the introduction of ambient air.

   Who is correct?

   A. A only

   B. B only

   C. Both A and B

   D. Neither A nor B

26. On a refrigerant system equipped with an expansion valve the low-side pressure is 45 psi and the high-side pressure is 160 psi. The engine speed is 1800 rpm and the ambient air temperature is 88°F. What is the most likely cause of these pressure readings?

   A. System is operating correctly.

   B. Evaporator core is restricted.

   C. Receiver-dryer is restricted.

   D. Expansion valve is stuck in the open position.

27. The following statements are true about a check valve, *except:*

   A. Airflow is blocked in one direction only.

   B. Vacuum flow is blocked in one direction only.

   C. Air or vacuum is permitted to flow in one direction.

   D. A check valve is omnidirectional when used in a vacuum system.

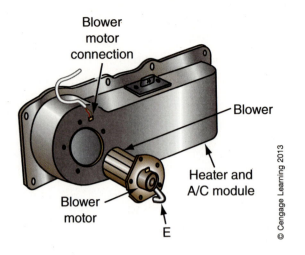

Blower motor connection

Blower

Heater and A/C module

Blower motor

E

© Cengage Learning 2013

28. What is callout E in the illustration above?

    A. Wiring harness   **C.** Ground wire

    B. Cooling tube   **D.** None of the above

29. All of the following may result in inadequate airflow, *except:*

    A. Duct or hose torn or disconnected

    B. Mode door binding, inoperative, or disconnected

    C. Defective or disconnected coolant flow control

    D. Outlet blocked or restricted

30. The most probable cause of windshield fogging is the:

    A. Heater coolant flow control valve is leaking.

    B. Heater coolant flow control valve is out of adjustment.

    C. Heater core is restricted.

    D. Heater core is leaking.

31. The high-side gauge needle is below 0 as shown in the illustration on the following page. This is an indication that the:

    A. Gauge is hooked up to an air-conditioning system that is under a vacuum.

    B. Gauge is out of calibration and should be adjusted to zero before being used.

    C. Both A and B

    D. Neither A nor B

© Cengage Learning 2013

32. During a system performance test of the air-conditioning system operation the technician notices that the low-side gauge is reading a vacuum. Which of the following is the most likely cause?

   A. An overcharged refrigerant system

   B. A flooded evaporator core

   C. Air contamination of the system

   D. A restricted expansion valve

33. To pass a purity test, the refrigerant being tested must be at least _____ pure.

   A. 99 percent

   B. 98 percent

   C. 97 percent

   D. 96 percent

34. The receiver-drier or accumulator should be replaced during all of the following services, *except* when replacing:

   A. A defective service valve core

   B. A compressor

   C. A condenser

   D. An evaporator

35. All of the following information is required on a retrofit label *except:*

   A. Date of retrofit

   B. Company or technician certificate number

   C. Type and amount of refrigerant

   D. Type and amount of lubricant

36. A thermistor's resistance is in proportion to:

   A. Its temperature

   B. The pressure applied to it

   C. The ambient light intensity

   D. The voltage applied to it

37. All of the following must be replaced when opened by an electrical overload, *except:*

   A. Fusible link

   B. Panel-mounted fuse

   C. In-line fuse

   D. Circuit breaker

38. What instrument can be used to test a fuse or circuit breaker?

  A. Ohmmeter   **C.** Both A and B

  B. Voltmeter   **D.** Neither A nor B

39. The master control is turned to maximum cooling. The blower motor does not run. A jumper wire is connected to the blower motor case and to body metal. There is a slight spark when connected, and the motor runs.

Technician A says that the problem is in the electrical control circuit.

Technician B says that the problem is in the electrical ground circuit.

Who is correct?

  A. A only   **C.** Both A and B

  B. B only   **D.** Neither A nor B

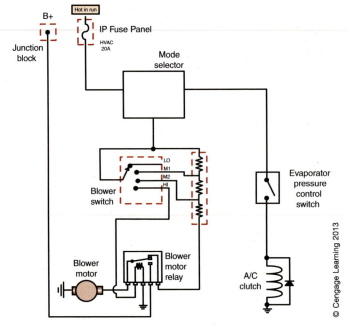

40. The blower motor in the schematic in the illustration above has ___ speeds.

  A. Two   **C.** Four

  B. Three   **D.** Variable

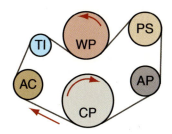

LEGEND:
AC - A/C Compressor Pulley
AP - Alternator Pulley
CP - Crankshaft Pulley

PS - Power Steering Pulley
TI - Tensioner/Idler Pulley
WP - Water Pump Pulley

41. If the crankshaft pulley turns clockwise (cw), all of the following statements about the illustration above are true, *except:*

   A. The compressor turns clockwise (cw).

   B. The power steering pulley turns clockwise (cw).

   C. The water pump turns clockwise (cw).

   D. The alternator pulley turns clockwise (cw).

42. A vehicle is equipped with dual front and rear evaporators. The rear outlet duct temperature is correct but the front duct temperature is too warm. Which of the following is the most likely cause?

   A. System is overcharged with refrigerant.

   B. System is undercharged with refrigerant.

   C. The rear expansion valve is stuck open.

   D. The front blend door is not in the correct position.

43. After driving for 20 miles the vehicle heater continues to blow cool air and the engine temperature is in the normal range. Which of the following is the most likely cause?

   A. A worn water pump impeller.

   B. Thermostat is stuck in the closed position.

   C. The heater core is restricted.

   D. The radiator is restricted.

44. The *least* likely cause of engine overheating is a defective:

   A. Temperature sending unit   **C.** Radiator cap

   B. Thermostat   **D.** Water pump

45. Extended-life antifreeze has a useful life of up to _____ miles/kilometers.

   A. 50,000/80,450   **C.** 150,000/241,350

   B. 100,000/160,900   **D.** 200,000/321,800

46. The environmental and health problems associated with the venting of refrigerants are being discussed.

Technician A says that some refrigerants vented near an open flame may produce a toxic vapor.

Technician B says that one may be subject to heavy penalties for unlawfully venting refrigerant.

Who is correct?

A. A only
B. B only
C. Both A and B
D. Neither A nor B

47. What is the minimum number of manifold and gauge sets required to comply with federal regulations and to ensure against refrigerant contamination?

A. Two
B. Three
C. Four
D. Five

48. During a system performance test of the air-conditioning system both the high-side and low-side pressure readings are in the normal operating range pressures but poor cooling is achieved. Which of the following is the most likely cause?

A. The compressor is faulty.
B. Airflow through the evaporator is restricted.
C. The receiver dryer is saturated.
D. Airflow through the condenser is restricted.

© Cengage Learning 2013

49. The information given on the label depicted in the above illustration is the vehicle's:

A. Serial and model number
B. Refrigerant and lubricant data
C. Emissions control data
D. Identification number

50. All of the following supply valuable service and technical information for the automotive air-conditioning service technician, *except:*

A. MACS
B. IATN
C. Mitchell/All Data
D. AAA

	to convert these	to these,	multiply by:
TEMPERATURE	Centigrade Degrees	Fahrenheit Degrees	1.8 then + 32
	Fahrenheit Degrees	Centigrade Degrees	0.556 after − 32
LENGTH	Millimeters	Inches	0.03937
	Inches	Millimeters	25.4
	Meters	Feet	3.28084
	Feet	Meters	0.3048
	Kilometers	Miles	0.62137
	Miles	Kilometers	1.60935
AREA	Square Centimeters	Square Inches	0.155
	Square Inches	Square Centimeters	6.45159
VOLUME	Cubic Centimeters	Cubic Inches	0.06103
	Cubic Inches	Cubic Centimeters	16.38703
	Cubic Centimeters	Liters	0.001
	Liters	Cubic Centimeters	1,000
	Liters	Cubic Inches	61.025
	Cubic Inches	Liters	0.01639
	Liters	Quarts	1.05672
	Quarts	Liters	0.94633
	Liters	Pints	2.11344
	Pints	Liters	0.47317
	Liters	Ounces	33.81497
	Ounces	Liters	0.02957
	Millileters	Ounces	0.3381497
	Ounces	Millileters	29.57
WEIGHT	Grams	Ounces	0.03527
	Ounces	Grams	28.34953
	Kilograms	Pounds	2.20462
	Pounds	Kilograms	0.45359
WORK	Centimeter Kilograms	Inch-Pounds	0.8676
	Inch-Pounds	Centimeter-Kilograms	1.15262
	Meter Kilograms	Foot-Pounds	7.23301
	Foot-Pounds	Newton-Meters	1.3558
PRESSURE	Kilograms/Square Centimeter	Pounds/Square Inch	14.22334
	Pounds/Square Inch	Kilograms/Square Centimeter	0.07031
	Bar	Pounds/Square Inch	14.504
	Pounds/Square Inch	Bar	0.0689
	Pounds/Square Inch	Kilopascals	6.895
	Kilopascals	Pounds/Square Inch	0.145

Bright Solutions, Inc.
Troy, MI

Carrier Corporation
Syracuse, NY

Clardy Manufacturing Corporation
Fort Worth, TX

Classic Tool Design, Inc.
New Windsor, NY

Component Assemblies, Inc.
Bryan, OH

Corrosion Consultants, Inc.
Roseville, MI

CPS Products, Inc.
Hialeah, FL

Envirotech Systems, Inc.
Niles, MI

FJC, Inc.
Davidson, NC

Floro Tech, Inc.
Pitman, NJ

Four Seasons
Division of Standard Motor Products, Inc.
Lewisville, TX 75057

Interdynamics, Inc.
Brooklyn, NY

K. D. Binnie Engineering Pty. Ltd.
Kirrawee, Australia NSW

KD Tools
Lancaster, PA

Kent Moore Division
SPX Corporation
Warren, MI

Lincor Distributors
N. Hollywood, CA

MAC Tools
Washington Courthouse, OH

Mastercool, Inc.
Rockaway, NJ

Neutronics, Inc.
Exton, PA

OTC Division
SPX Corporation
Owatonna, MN

Owens Research, Inc./Tubes 'N Hoses
Dallas, TX

P & F Technologies Ltd.
Mississauga, ONT, Canada

Ritchie Engineering Company Inc.
Garrett, IN

Robinair Division
SPX Corporation
Montpelier, OH

RTI Technologies, Inc.
York, PA

The S. A. Day Manufacturing Company, Inc.
Buffalo, NY

Snap-On Tools Corporation
Kenosha, WI

Superior Manufacturing Company
Morrow, GA

Technical Chemicals Company
Dallas, TX

Thermolab, Inc.
Farmersville, TX

TIF Instruments, Inc.
Miami, FL

Tracer Products Division
Spectronics Corporation
Westbury, NY

Uniweld Products
Ft. Lauderdale, FL

Uview Ultraviolet
Mississauga, ONT, Canada

Varian Vacuum Technologies
Lexington, MA

Viper/T-Tech Division
Century Manufacturing Company
Minneapolis, MN

Yokagawa Corporation of America
Newnan, GA

The following companies will accept contaminated refrigerant from your shop:

RemTec International
6150 Merger Dr.
Holland, OH 43528
(888) 873-6832

Full Circle
(With 11 Regional Offices)
121 S. Norwood Dr.
Hurst, TX 76053
(817) 282-0022 (x220)

Refrigerant Reclaim
805 Tile Dr.
Red Wing, MN 55066
(860) 651-6114 or
(800) 235-0705

EPA
List of Reclaimers
(800) 296-1996

Note: **Terms are highlighted in color**, followed by **Spanish translation in bold**.

Absolute Perfect in quality or nature, complete. Usually used in refrigeration context when referring to temperature or pressure.

Absoluto Perfecto en su calidad o naturaleza, completo. Suele usarse en contextos de refrigeración cuando se refiera a la temperatura o la presión.

Access valve See Service port and Service valve.

Valvula de acceso Ver Service port [Orificio de servicio] y Service value [Válvula de servicio].

Accumulator A tank located in the tailpipe to receive the refrigerant that leaves the evaporator. This device is constructed to ensure that no liquid refrigerant enters the compressor.

Acumulador Tanque ubicado en el tubo de escape para recibir el refrigerante que sale del evaporador. Dicho dispositivo está diseñado de modo que asegure que el refrigerante líquido no entre en el compresor.

Acme A type of fitting thread. The service hose connections to the R-134a manifold set have ½-16 acme threads.

Acme Un tipo de rosca de guarnición. Las conexiones del tubo de servio al conjunto de la manívela de R-134a tienen las roscas acme de 16.

Actuator A device that transfers a vacuum or electric signal to a mechanical motion. An actuator typically performs an on/off or open/close function.

Accionador Dispositivo que transfiere una serial de vacio o una señal eléctrica a un movimiento mecánico. Típicamente un accionador lleva a cabo la función de modulación de impulsos o la de abrir y cerrar.

Adapter A device or fitting that permits different size parts or components to be fastened or connected to each other.

Adaptador Dispositivo o ajuste que permite la sujeción o conexión entre sí de piezas de tamaños diferentes.

Aftermarket A term generally given to a device or accessory that is added to a vehicle by the dealer after original manufacture, such as an air-conditioning system.

Postmercado Término dado generalmente a un dispositivo o accesorio que el distribuidor de automóviles agrega al automóvil después de la fabricación original, como por ejemplo un sistema de acondi-cionamiento de aire.

Air gap The space between two components such as the rotor and armature of a clutch.

Espacio de aire El espacio entre dos componentes, como por ejemplo el rotor y la armadura de un embrague.

Ambient sensor A thermistor used in automatic temperature control units to sense ambient temperature. Also see Thermistor.

Sensor ambiente Termistor utilizado en unidades de regulación automática de temperatura para sentir la temperatura ambiente. Ver también Thermistor [Termistor].

Ambient temperature The temperature of the surrounding air.

Temperatura del ambiente La temperatura del aire alrededor.

Approved power source A power source that is consistent with the requirements of the equipment so far as voltage, frequency, and ampacity are concerned.

Fuente aprobada de potencia Fuente de potencia que cumple con los requisitos del equipo referente a la tensión, frecuencia, y ampacidad.

Armature The part of the clutch that mounts onto the crankshaft and engages with the rotor when energized.

Armadura La parte del embrague que se fija al cigüeñal y se engrans al exitarse el rotor.

Asbestos A silicate of calcium (Ca) and magnesium (Mg) mineral that does not burn or conduct heat. It has been determined that asbestos exposure is hazardous to health and must be avoided.

Asbesto Mineral de silicato de calcio (Ca) y magnesio (Mg) que no se quema ni conduce el calor. Se ha establecido que la exposición al asbesto es nociva y debe evitarse.

Aspirator A device that uses a negative (suction) pressure to move air.

Aspirador Un dispositivo que usa una presión negativa (la succión) para mover el aire.

Atmospheric pressure Air pressure at a given altitude. At sea level, atmospheric pressure is 14.696 psia (101.329 kPa absolute).

Presión atmosférica La presión del aire a una dada altitud. Al nivel del mar, la presión atmosférica es de 14,696 psia (101.329 kPa absoluto).

AUTO Abbreviation for automatic.

AUTO Abreviatura del automático.

Back seat (service valve) Turning the valve stem to the left (ccw) as far as possible back seats the valve. The valve outlet to the system is open and the service port is closed.

Asentar a la izquierda (válvula de servicio) El girar el vástago de la válvula al punto más a la izquierda posible asienta a la izquierda la válvula. La salida de la válvula al sistema está abierta y el orificio de servicio está cerrado.

Barb fitting A fitting that slips inside a hose and is held in place with a gear-type clamp. Ridges (barbs) on the fitting prevent the hose from slipping off.

Accesorio arponado Ajuste que se inserta dentro de una manguera y que se sujeta en su lugar con una abrazadera de tipo engranaje. Proyecciones (puas) en el ajuste impiden que se deslice la manguera.

Barrier hose A hose having an impervious lining to prevent refrigerant leakage through its wall. Air-conditioning systems in vehicles have had barrier hoses since 1988.

Manguera de barrera Una manguera que tiene un forro impervio que previene el goteo del refrigerante a través de su muro. Los sistemas de aire acondicionado en los vehículos han incluido las mangueras de barrera desde el 1988.

BCM An abbreviation for blower control module.

BCM Abreviatura de módulo regulador del soplador.

Belt See V-belt and Serpentine belt.

Correa Ver V-belt [Correa en V], y Serpentine belt [Correa serpentina].

Belt tension Tightness of a belt or belts, usually measured in foot-pounds (ft.-lb.) or Newton-meters (N·m).

Tensión de la correa Tensión de una correa o correas, medida nor-malmente en libras-pies (ft.-lb.) o metros-Newton (N·m).

Blower See Squirrel-cage blower.

Soplador Ver Squirrel-cage blower [Soplador con jaula de ardilla].

Blower motor See Motor.

Motor de soplador Ver Motor.

Blower relay An electrical device used to control the function or speed of a blower motor.

Rele del soplador Dispositivo eléctrico utilizado para regular la fun-ción o velocidad de un motor de soplador.

Boiling point The temperature at which a liquid changes to a vapor.

Punto de ebullición Temperatura a la que un líquido se convierte en vapor.

Break a vacuum The next step after evacuating a system. The vac-uum should be broken with refrigerant or other suitable dry gas, not ambient air or oxygen.

Romper un vacío El paso que inmediatamente sigue la evacuación de un sistema. El vacío debe de romperse con refrigerante u otro gas seco apropiado, y no con aire ambiente u oxígeno.

Breakout box A tool in which the probes of a digital volt-ohmmeter (DVOM) may be inserted to access various sensors and actuators through pin connectors to the computer.

Accesorio detector Una herramienta en la cual las sondas de un ohmímetro digital (DVOM) pueden insertarse para ganar la entrada a varios sensores e actuadores de la computadora por medio de las conexiones a las espigas de contacto.

Bypass An alternate passage that may be used instead of the main passage.

Desviación Pasaje alternativo que puede utilizarse en vez del pasaje principal.

Bypass hose A hose that is generally small and is used as an alter-nate passage to bypass a component or device.

Manguera desviadora Manguera que generalmente es pequeña y se utiliza como pasaje alternativo para desviar un componente o dispositivo.

CAA Clear Air Act.

CAA Ley para Aire Limpio.

Calibration To check, adjust, or determine the accuracy of an instrument used for measuring, for example, temperature or pressure.

Calibración Revisar, ajustar o determinar la precisión de un instru-mento que se usa para medir, por ejemplo, la temperatura o la presión.

Can tap A device used to pierce, dispense, and seal small cans of refrigerant.

Macho de roscar para latas Dispositivo utilizado para perforar, dis-tribuir y sellar pequeñas latas de refrigerante.

Can tap valve A valve found on a can tap that is used to control the flow of refrigerant.

Válvula de macho de roscar para latas Válvula que se encuentra en un macho de roscar para latas utilizada para regular el flujo de refrigerante.

Cap A protective cover. Also used as an abbreviation for capillary (tube) or capacitor.

Tapadera Cubierta protectiva. Utilizada tambien como abreviatura del tubo capilar o capacitador.

Cap tube A tube with a calibrated inside diameter and length used to control the flow of refrigerant. In automotive air-conditioning sys-tems, the tube connecting the remote bulb to the expansion valve or to the thermostat is called the capillary tube.

Tubo capilar Tubo de diámetro interior y longitud calibrados; se utiliza para regular el flujo de refrigerante. En sistemas automotrices para el acondicionamiento de aire el tubo que conecta la bombilla a distancia con la válvula de expansión o con el termostato se llama el tubo capilar.

Carbon monoxide (CO) A major air pollutant that is potentially lethal if inhaled, even in small amounts. An odorless gas composed of carbon (C) and hydrogen (H) formed by the incomplete combus-tion of any fuel containing carbon.

Monóxido de carbono (CO) Un contaminante de aire principal que puede ser letal si se inhala, aún en pequeñas cantidades. Un gas sin olor compuesto de carbono (C) e hidrógeno (H) formado por la combus-tión incompleta de cualquier combustible que contiene el carbono.

Carbon seal face A seal face made of a carbon composition rather than from another material such as steel or ceramic.

Frente de carbono de la junta hermética Frente de la junta her-mética fabricada de un compuesto de carbono en vez de otro mate-rial, como por ejemplo el acero o material cerámico.

Caution A notice to warn of potential personal injury situations and conditions.

Precaución Aviso para advertir situaciones y condiciones que podrían causar heridas personales.

CCW Counterclockwise.

CCW Sentido inverso al de las agujas del reloj.

Celsius A metric temperature scale using zero as the freezing point of water. The boiling point of water is 100°C (212°F).

Celsio Escala de temperatura métrica en la que el cero se utiliza como el punto de congelación de agua. El punto de ebullición de agua es 100°C (212°F).

Ceramic seal face A seal face made of a ceramic material instead of steel or carbon.

Frente cerámica de la junta hermética Frente de la junta hermé-tica fabricada de un material cerámico en vez del acero o carbono.

Certified Having a certificate. A certificate is awarded or issued to those who have demonstrated appropriate competence through test-ing or practical experience.

Certificado El poseer un certificado. Se les otorga o emite un certifi-cado a los que han demostrado una cierta capacidad por medio de exámenes y/o experiencia practica.

CFC-12 See Refrigerant-12.

CFC-12 Ver [Refrigerante-12].

Charge A specific amount of refrigerant or oil by volume or weight.

Carga Cantidad específica de refrigerante o de aceite por volúmen o peso.

Check valve A device located in the liquid line or inlet to the drier. The valve prevents liquid refrigerant from flowing the opposite way when the unit is shut off.

Válvula de retención Dispositivo ubicado en la línea de líquido o en la entrada al secador. Al cerrarse la unidad, la válvula impide que el refrigerante líquido fluya en el sentido contrario.

Clean Air Act (CAA) A Title IV amendment signed into law in 1990 that established national policy relative to the reduction and elimination of ozone-depleting substances.

Ley para Aire Limpio (CAA) Enmienda Titulo IV firmado y aprobado en 1990 que estableció la política nacional relacionada con la reducción y eliminación de sustancias que agotan el ozono.

Clockwise A term referring to a clockwise (cw), or left-to-right rotation or motion.

Sentido de las agujas del reloj Término que se refiere a un movimiento en el sentido correcto de las agujas del reloj (cw por sus siglas en ingles), es decir, rotación o movimiento desde la izquierda hacia la derecha.

Clutch An electromechanical device mounted on the air-conditioning compressor used to start and stop compressor action, thereby controlling refrigerant circulating through the system.

Embrague Dispositivo electromecánico montado en el compresor del acondicionador de aire y utilizado para arrancar y detener la acción del compresor, regulando así la circulación del refrigerante a través del sistema.

Clutch coil The electrical part of a clutch assembly. When electrical power is applied to the clutch coil, the clutch is engaged to start and stop compressor action.

Bobina del embrague La parte eléctrica del conjunto del embrague. Cuando se aplica una potencia eléctrica a la bobina del embrague, éste se engrana para arrancar y detener la acción del compresor.

Clutch pulley A term often used for "clutch rotor"; that portion of the clutch in which the belt rides.

Polea del embrague Un término que se suele usa para el "rotor del embrague." Esa porción del embrague en la cual viaja la correa.

Compound gauge A gauge that registers both pressure and vacuum (above and below atmospheric pressure); used on the low side of the systems.

Manómetro compuesto Calibrador que registra tanto la presión como el vacio (a un nivel superior e inferior a la presión atmosférica); utilizado en el lado de baja presión de los sistemas.

Compression fitting A type of fitting used to connect two or more tubes of the same or different diameter together to form a leakproof joint.

Ajuste de compresión Tipo de ajuste utilizado para sujetar dos o más tubos del mismo tamaño o de un tamaño diferente para formar una junta hermética contra fugas.

Compression nut A nut-like device used to seat the compression ring into the compression fitting to ensure a leakproof joint.

Tuerca de compresión Dispositivo parecido a una tuerca utilizado para asentar el anillo de compresión dentro del ajuste de compresión para asegurar una junta hermética contra fugas.

Compression ring A ring-like part of a compression fitting used for a seal between the tube and fitting.

Aro de compresión Pieza parecida a un anillo del ajuste de compresión utilizada como una junta hermética entre el tubo y el ajuste.

Compressor shaft seal An assembly consisting of springs, snap-rings, O-rings, shaft seal, seal sets, and gasket. The shaft to be turned without a loss of refrigerant or oil.

Junta hermética del arbol del compresor Conjunto que consiste de muelles, anillos de muelles, juntas tóricas, una junta hermética del arbol, conjuntos de juntas herméticas, y una guarnición. La junta hermética del arbol está montada en el cigüeñal del compresor y permite que el arhol se gire sin una pérdida de refrigerante o aceite.

Constant tension hose clamp A hose clamp, often referred to as a "spring clamp," so designed that it is under constant tension.

Grapa de manguera de tensión constante Una grapa de manguera, que suele referirse como "grapa de resorte," diseñada en tal manera para estar bajo una tensión constante.

Contaminated A term generally used when referring to a refrigerant cylinder or a system that is known to contain foreign substances such as other incompatible or hazardous refrigerants.

Contaminado Témino generalmente utilizado al referirse a un cilindro para refrigerante o a un sistema que es reconocido contener sustan-cias extrañas, como por ejemplo otros refrigerantes incompatibles o peligrosos.

Contaminated refrigerant Any refrigerant that is not at least 98 percent pure. Refrigerant may be considered to be contaminated if it contains excess air or another type of refrigerant.

Refrigerante contaminado Cualquier refrigerante que no es al menos el 98 por ciento puro. El refrigerante puede considerarse contaminado si contiene un exceso del aire o cualquier otro tipo de refrigerante.

Counterclockwise (ccw) A direction, right to left, opposite to that of a clock.

Sentido contrario al de las agujas del reloj Dirección de la derecha hacia la izquierda contraria a la correcta de las agujas del reloj.

Cracked position A mid-seated or open position.

Posición parcialmente asentada Posición abierta o media asentada.

Customer The vehicle owner or a person who orders or pays for goods or services.

Cliente El dueño de un vehículo o una persona que pide y/o paga para las mercancías y servicios.

CW Abbreviation for clockwise. Also cw.

CW Abreviatura del sentido de las agujas del reloj. Tambien cw.

Cycle clutch time (total) Time from the moment the clutch engages until it disengages, then reengages. Total time is equal to on time plus off time for one cycle.

Duración del ciclo del embrague (total) Espacio de tiempo medido desde el momento en que se engraña el embrague hasta que se desengrañe y se engrañe de nuevo. El tiempo total es equivalente al trabajo efectivo más el trabajo no efectivo por un ciclo.

Cycling-clutch pressure switch A pressure-actuated electrical switch used to cycle the compressor at a predetermined pressure.

Automata manométrico del embrague con funcionamiento cíclico Interruptor eléctrico accionado a presión utilizado para ciclar el compresor a una presión predeterminada.

Cycling-clutch system An air-conditioning system in which the air temperature is controlled by starting and stopping the compressor with a thermostat or pressure control.

Sistema de embrague con funcionamiento cíclico Sistema de acondicionamiento de aire en el cual la temperatura del aire se regula al arrancarse y detenerse el compresor con un termostato o regulador de presión.

Cycling time A term often used for "cycling-clutch time." The total time from when the clutch engages until it disengages and again engages; equal to one on time plus one off time for one cycle.

Funcionamiento cíclico Un término que se usa para "tiempo de cliclaje del embrague." El total del tiempo desde que se engancha el embrague hasta que se desengancha y engancha de nuevo; iguala a un tiempo prendido más un tiempo apagado por un ciclo.

Debris Foreign matter such as the remains of something broken or deteriorated.

Escombro Materia extranjera tal como los restos de algo roto o deteriorado.

Decal A label that is designed to stick fast when transferred. A decal affixed under the hood of a vehicle is used to identify the type of refrigerant used in a system.

Calcomania Etiqueta diseñada para pegarse fuertemente al ser trans-ferido. Una calcomania pegada debajo de la capota se utiliza para identificar el tipo de refrigerante utilizado en un sistema.

Department of Transportation The U.S. Department of Transportation is a federal agency charged with regulation and control of the shipment of all hazardous materials.

Departamento de Transportes El Departamento de Transportes de los Estados Unidos de America es una agencia federal que tiene a su cargo la regulación y control del transporte de todos los materiales peligrosos.

Dependability Reliability; trustworthiness.

Caracter responsible Digno de confianza; integridad.

Depressing pin A pin located in the end of a service hose to press (open) a Schrader-type valve.

Pasador depresor Pasador ubicado en el extremo de una manguera de servicio para forzar que se abra una válvula de tipo Schrader.

Diagnosis The procedure followed to locate the cause of a malfunction.

Diagnosis Procedimiento que se sigue para localizar la causa de una disfunción.

Disarm To turn off; to disable a device or circuit.

Desarmar Apagar, incapacitar un dispositivo o un circuito.

Disinfectant A cleansing agent that destroys bacteria and other microorganisms.

Desinfectante Un agente de limpieza que destruye la bacteria u otros microorganismos.

Dissipate To reduce, weaken, or use up; to become thin or weak.

Disipar Reducir, debilitar o agotar; aclarar o ser débil.

Drive pulley The pulley that transmits the input force to other pulleys or devices.

Polea motriz La polea que transmite la fuerza de entra a las otras poleas o dispositivos.

Dry nitrogen The element nitrogen (N) that has been processed to ensure that it is free of moisture.

Nitrógeno seco El elemento nitrógeno (N) que ha sido procesado para asegurar que esté libre de humedad.

Dual Two

Doble Dos.

Dual system Two systems; usually refers to two evaporators in an air-conditioning system, one in the front and one in the rear of the vehicle, driven off a single compressor and condenser system.

Sistema doble Dos sistemas; se refiere normalmente a dos evaporadores en un sistema de acondicionamiento de aire; uno en la parte delantera y el otro en la parte trasera del vehículo; los dos son accionados por un solo sistema compresor condensador.

Duct A tube or passage used to provide a means to transfer air or liquid from one point or place to another.

Conducto Tubo o pasaje utilizado para proveer un medio para trans-ferir aire o líquido desde un punto o lugar a otro.

EATC Electronic automatic temperature control.

EATC Regulador automático y electrónico de temperatura.

ECC Electronic climate control.

ECC Regulador electrónico de clima.

Electronic charging meter A term often used for "electronic scale," a device used to accurately dispense or monitor the amount of refrigerant being charged into an air-conditioning system.

Medidor electrónico de carga Un término que se suele usar para una "escala electrónica," un dispositivo que se usa para repartir y/o regular la cantidad del refrigerante que se usa para cargar un sistema de acondicionado de aire.

English fastener Any type fastener with English size designations, numbers, decimals, or fractions of an inch.

Asegurador inglés Cualquier tipo de asegurador provisto de indicaciones, números, decimales, o fracciones de una pulgada del sistema inglés.

Environmental Protection Agency (EPA) An agency of the U.S. government that is charged with the responsibility of protecting the environment and enforcing the Clean Air Act (CAA) of 1990.

Agencia para la Protección del Medio Ambiente (EPA) Agencia del gobierno estadounidense que tiene a su cargo la responsabilidad de proteger el medio ambiente y ejecutar la Ley para Aire Limpio (CAA por sus siglas en inglés) de 1990.

EPA Environmental Protection Agency.

EPA Agencia para la Protección del Medio Ambiente.

Etch An intentional or unintentional erosion of a metal surface generally caused by an acid.

Atacar con ácido Desgaste previsto o imprevisto de una superficie metálico, ocasionado generalmente por un ácido.

Etching See Etch.

Ataque con acido Ver Etch [Atacar con ácido].

Evacuate To create a vacuum within a system to remove all traces of air and moisture.

Evacuar El dejar un vacío dentro de un sistema para remover comple-tamente todo aire y humedad.

Evacuation See Evacuate.

Evacuacion Ver Evacuate [Evacuar].

Evaporator core The tube and fin assembly located inside the evaporator housing. The refrigerant fluid picks up heat in the evaporator core when it changes into a vapor.

Núcleo del evaporador El conjunto de tubo y aletas ubicado dentro del alojamiento del evaporador. El refrigerante acumula calor en el núcleo del evaporador cuando se convierte en vapor.

Expansion tank An auxiliary tank that is usually connected to the inlet tank or a radiator and that provides additional storage space for heated coolant. Often called a coolant recovery tank.

Tanque de expansión Tanque auxiliar que normalmente se conecta al tanque de entrada o a un radiador y que provee almacenaje adicional del enfriante calentado. Llamado con frecuencia tanque para la recuperación del enfriante.

External On the outside.

Externo A exterior.

External snapring A snapring found on the outside of a part such as a shaft.

Anillo de muelle exterior Anillo de muelle que se encuentra en el exterior de una pieza, como por ejemplo un árbol.

Facilities Something created and equipped to serve a particular function, such as a specialty garage used to service motor vehicles.

Instalación Ago creado y equipado para servir en una función particular, tal como un taller especializado que mantiene los vehículos motorizados.

Fan relay A relay for the cooling or auxiliary fan motors.

Relé del ventilador Relé para los motores de enfriamiento y/o los auxiliares.

Federal Clean Air Act See Clean Air Act.

Ley Federal para Aire Limpio Ver Clean Air Act [Ley para Aire Limpio].

Fill neck The part of the radiator on which the pressure cap is attached. Most radiators, however, are filled via the recovery tank.

Cuello de relleno La parte del radiador a la que se fija la tapadera de presión. Sin embargo, la mayoría de radiadores se llena por medio del tanque de recuperación.

Filter A device used with the drier or as a separate unit to remove foreign material from the refrigerant.

Filtro Dispositivo utilizado con el secador o como unidad separada para extraer material extraño del refrigerante.

Filter drier A device that has a filter to remove foreign material from the refrigerant and a desiccant to remove moisture from the refrigerant.

Secador del filtro Dispositivo provisto de un filtro para remover el material extraño del refrigerante y un desecante para remover la humedad del refrigerante.

Flammable refrigerant Any refrigerant that contains a flammable material and is not approved for use. Any refrigerant, however, may be considered flammable under certain abnormal operating conditions.

Refrigerante inflamable Cualquier refrigerante que contiene una materia inflamable y que no es aprobado su uso. Cualquier refrigerante, no obstante, puede considerarse inflamable bajo ciertas condiciones de operación anormales.

Flange A projecting rim, collar, or edge on an object used to keep the object in place or to secure it to another object.

Brida Cerco, collar, o extremo proyectante ubicado sobre un objeto uti-lizado para mantener un objeto en su lugar o para fijarlo a otro objeto.

Flare A flange or cone-shaped end applied to a piece of tubing to provide a means of fastening to a fitting.

Abocinado Brida o extremo en forma cónica aplicado a una pieza de tubería para proveer un medio de asegurarse a un ajuste.

Fluorescent tracer dye A dye solution introduced into the air-conditioning system for leak-testing procedures. An ultraviolet (UV) lamp is used to detect the site of the leak.

Colorante fluorescente trazadora Una solución de tinta que se introduce en un sistema de aire acondicionado con el fin de comprobar los procedimientos contra fugas. Una lámpara ultravioleta (UV) se usa para detectar el sitio de la fuga.

Forced air Air that is moved mechanically such as by a fan or blower.

Aire forzado Aire que se mueve mecánicamente, como por ejemplo por un ventilador o soplador.

Fringe benefits The extra benefits aside from a salary that an employee may expect, such as vacation, sick leave, insurance, or employee discounts.

Beneficios extras Los beneficios extras además del sueldo que un empleado puede esperar recibir, incluyendo las vacaciones, tiempo libre para enfermedad, la aseguranza o los discuentos de empleados.

Front seat Closing of the line, leaving the compressor open to the service port fitting. This allows service to the compressor without purging the entire system. Never operate the system with the valves front seated.

Asentar a la derecha El cerrar la línea, dejando abierto el compresor al ajuste del orificio de servicio, lo cual permite prestar servicio al compresor sin purgar todo el sistema. Nunca haga funcionar el sistema con las válvulas asentadas a la derecha.

Functional test See Performance Test.

Prueba funcional Ver Performance Test [Prueba de rendimiento].

Fungi Plural of "fungus," an organism, such as mold, that grows in the damp atmosphere inside an evaporator plenum, often producing an undesirable odor.

Hongos Plural de "hongo" un organismo tal como el hongo, que crece en la atmósfera húmeda de un pleno evaporador, produciendo muchas veces un olor no agradable.

Fusible link A type of fuse made of a special wire that melts to open a circuit when current draw is excessive.

Cartucho de fusible Tipo de fusible fabricado de un alambre especial que se funde para abrir un circuito cuando ocurre una sobrecarga del circuito.

Gasket A thin layer of material or composition that is placed between two machined surfaces to provide a leakproof seal between them.

Guarnición Capa delgada de material o compuesto que se coloca entre dos superficies maquinadas para proveer una junta hermética para evitar fugas entre ellas.

Gauge A tool of a known calibration used to measure components. For example, a feeler gauge is used to measure the air gap between a clutch rotor and armature.

Calibrador Herramienta de una calibración conocida utilizada para la medición de componentes. Por ejemplo, un calibrador de espesores se utiliza para medir el espacio de aire entre el rotor del embrague y la armadura.

Graduated container A measure such as a beaker or measuring cup that has a graduated scale for the measurement of a liquid.

Recipiente graduado Una medida, como por ejemplo un cubilete o una taza de medir, provista de una escala graduada para la medición de un líquido.

Gross weight The weight of a substance or matter that includes the weight of its container.

Peso bruto Peso de una sustancia o materia que incluye el peso de su recipiente.

Ground A general term given to the negative (−) side of an electrical system.

Tierra Término general para indicar el lado negativo (−) de un sistema eléctrico.

Grounded An intentional or unintentional connection of a wire, positive (+) or negative (−), to the ground. A short circuit is said to be grounded.

Puesto a tierra Una conexión prevista o imprevista de un alambre, positiva (+) o negativa (−), a la tierra. Se dice que un cortocircuito es puesto a tierra.

Hazard A possible source of danger that may cause damage to a structure or equipment or that may cause personal injury.

Peligro Un causante posible de un peligro que puede dañar una estructura o un equipo o puede causar daño personal.

HCFC Hydrochlorofluorocarbon refrigerant.

HCFC Refrigerante de hidroclorofluorocarbono.

HFC-134a A hydrofluorocarbon refrigerant gas used as a refrigerant. The refrigerant of choice to replace R-12 in automotive air-conditioning systems. Often referred to as R-134a, this refrigerant is not harmful to the ozone.

HFC-134a Un refrigerante de gas hidrofluorocarbono que se usa como refrigerante. Es el refrigerante preferido para remplazar el CFC-12 en los sistemas de aire acondicionador automotrices. Suele referirse como el R134-a, este refrigerante no es dañino al medio ambiente.

Header tanks The top and bottom tanks (downflow) or side tanks (crossflow) of a radiator. The tanks in which coolant is accumulated or received.

Tanques para alimentación por gravedad Los tanques superiores e inferiores (flujo descendente) o los tanques laterales (flujo transversal) de un radiador. Tanques en los cuales el enfriador se acumula o se recibe.

Heater That part of the climate control comfort system consisting of the heater core, hoses, coolant flow control valve, and related controls used to provide air to the vehicle interior.

Calentador Esa parte del sistema de confort de aclimatizaje que con-siste del núcleo de calefacción, las mangueras, la válvula de flujo del fluido refrigerante, y los controles parecidos que sirven para suminis-trar aire al interior de un vehículo.

Heater core A radiator-like heat exchanger located in the case/duct system through which coolant flows to provide heat to the vehicle interior.

Núcleo del calentador Intercambiador de calor parecido a un radiador y ubicado en el sistema de caja/conducto a través del cual fluye el enfriador para proveer calor al interior del vehículo.

Heat exchanger An apparatus in which heat is transferred from one medium to another on the principle that heat moves to an object with less heat.

Intercambiador de calor Aparato en el que se transfiere el calor de un medio a otro, lo cual se basa en el principio que el calor se atrae a un objeto que tiene menos calor.

HI The designation for high as in blower speed or system mode.

HI Indicación para indicar marcha rápida, como por ejemplo la veloci-dad de un soplador o el modo de un sistema.

High-side gauge The right-side gauge on the manifold used to read refrigerant pressure in the high side of the system.

Calibrador del lado de alta presión El calibrador del lado derecho del múltiple utilizado para medir la presión del refrigerante en el lado de alta presión del sistema.

High-side hand valve The high-side valve on the manifold set used to control flow between the high side and service ports.

Válvula de mano del lado de alta presión Válvula del lado de alta presión que se encuentra en el conjunto del múltiple, utilizada para regular el flujo entre el lado de alta presión y los orificios de servicio.

High-side service valve A device located on the discharge side of the compressor; this valve permits the service technician to check the high-side pressures and perform other necessary operations.

Válvula de servicio del lado de alta presión Dispositivo ubicado en el lado de descarga del compresor; dicha válvula permite que el mecánico verifique las presiones en lado de alta presión y lleve a cabo otras funciones necesarias.

High-side switch See Pressure switch.

Autómata manométrico del lado de alta presión Ver Pressure switch Autómata manométrico.

High-torque clutch A heavy-duty clutch assembly used on some vehicles known to operate with higher-than-average head pressure.

Embrague de alto par de torsión Conjunto de embrague para servicio pesado utilizado en algunos vehículos que funcionan con una altura piezométrica más alta que la normal.

Hot A term given the positive (+) side of an energized electrical system. Also refers to an object that is heated.

Cargado/caliente Término utilizado para referirse al lado positivo (+) de un sistema eléctrico excitado. Se refiere tambien a un objeto que es calentado.

Hot knife A knife-like tool that has a heated blade. Used for separating objects, for example, evaporator cases.

Cuchillo en caliente Herramienta parecida a un cuchillo provista de una hoja calentada. Utilizada para separar objetos; p.e. las cajas de evaporadores.

Housekeeping A system of keeping the shop floors clean, lighting adequate, tools in good repair and operating order, and storing materials properly.

Aseo Un sistema de mantener limpios los pisos del taller, la alumbración adecuada, las herramientas reparadas y funcionando, y el almacenaje adecuada de los materiales.

Hub The central part of a wheel-like device such as a clutch armature.

Cubo Parte central de un dispositivo parecido a una rueda, como por ejemplo la armadura del embrague.

Hygiene A system of rules and principles intended to promote and preserve health.

Higiene Sistema de normas y principios cuyo propósito es promover y preservar la salud.

Hygroscopic Readily absorbing and retaining moisture.

Higroscópico Lo que absorbe y retiene facilmente la humedad.

Idler A pulley device that keeps the belt whip out of the drive belt of an automotive air conditioner. The idler is used as a means of tightening the belt.

Polea loca Polea que mantiene la vibración de la correa fuera de la cor-rea de transmisión de un acondicionador de aire automotriz. Se utiliza la polea loca para proveerle tensión a la correa.

Idler pulley A pulley used to tension or torque the belt(s).

Polea tensora Polea utilizada para proveer tensión o par de torsión a la(s) correa(s).

Idle speed The speed (rpm) at which the engine runs while at rest (idle).

Marcha mínima Velocidad (rpm) a la que no hay ninguna carga en el motor (marcha mínima).

In-car temperature sensor A thermistor used in automatic temperature control units for sensing the in-car temperature. Also see Thermistor.

Sensor de temperatura del interior del vehículo Termistor utilizado en unidades de regulación automática de temperatura para sentir la temperatura del interior del vehículo. Ver también Thermistor [Termistor].

Indexing tab A mark or protrusion on mating components to ensure that they will be assembled in their proper position.

Fijación indicadora Una marca o una parte sobresaliente de los com-ponentes parejadas para asegurar que se asamblean en su posición correcta.

Insert fitting A fitting that is designed to fit inside, such as a barb fitting that fits inside a hose.

Ajuste inserto Ajuste diseñado para insertarse dentro de un objeto, como por ejemplo un ajuste arponado que se inserta dentro de una manguera.

Internal Inside; within.

Interno Al interior, dentro de una cosa.

Internal snapring A snapring used to hold a component or part inside a cavity or case.

Anillo de muelle interno Anillo de muelle utilizado para sujetar un componente o una pieza dentro de una cavidad o caja.

Jumper A wire used to temporarily bypass a device or component for the purpose of testing.

Barreta Alambre utilizado para desviar un dispositivo o componente de manera temporal para llevar a cabo una prueba.

Kilogram A unit of measure in the metric system. One kilogram is equal to 2.2010-2.615 pounds in the English system.

Kilogramo Unidad de medida en el sistema métrico. Un kilogramo equivale a 2205 libras en el sistema inglés.

KiloPascal A unit of measure in the metric system. One kiloPascal (kPa) is equal to 0.145 pound per square inch (psi) in the English system.

Kilopascal Unidad de medida en el sistema métrico. Un kilopascal (kPa) equivale a 0,145 libra por pulgada cuadrada en el sistema inglés.

kPa KiloPascal.

kPa Kilopascal.

Liquid A state of matter; a column of fluid without solids or gas pockets.

Líquido Estado de materia; columna de fluido sin sólidos ni bolsillos de gas.

Low-refrigerant switch A switch that senses low pressure due to a loss of refrigerant and stops compressor action. Some alert the operator and set a trouble code.

Interruptor para advertir un nivel bajo de refrigerante Interruptor que siente una presión baja debido a una perdida de refrigerante y que detiene la acción del compresor. Algunos interruptores advierten al operador y/o fijan un código indicador de fallas.

Low-side gauge The left-side gauge on the manifold used to read refrigerant pressure in the low side of the system.

Calibrador de lado de baja presión El calibrador en el lado izquierdo del múltiple utilizado para medir la presión del refrigerante del lado de baja presión del sistema.

Low-side hand valve The manifold valve used to control flow between the low side and service ports of the manifold.

Válvula de mano del lado de baja presión Válvula de distribución utilizada para regular el flujo entre el lado de baja presión y los orificios de servicio del colector.

Low-side service valve A device located on the suction side of the compressor that allows the service technician to check low-side pressures and perform other necessary service operations.

Válvula de servicio del lado de baja presión Dispositivo ubicado en el lado de succión del compresor; dicha válvula permite que el mecánico verifique las presiones del lado de baja presión y lleve a cabo otras funciones necesarias de servicio.

Manifold A device equipped with a hand shutoff valve. Gauges are connected to the manifold for use in system testing and servicing.

Múltiple Dispositivo provisto de una válvula de cierre accionada a mano. Calibradores se conectan al múltiple para ser utilizados para llevar a cabo pruebas del sistema y para servicio.

Manifold and gauge set A manifold complete with gauges and charging hoses.

Conjunto del múltiple y calibrador Múltiple provisto de calibradores y mangueras de carga.

Manifold hand valve Valves used to open and close passages through the manifold set.

Válvula de distribución accionada a mano Válvulas utilizadas para abrir y cerrar conductos a través del conjunto del múltiple.

Manufacturer A person or company whose business is to produce a product or components for a product.

Fabricante Persona o empresa cuyo propósito es fabricar un producto o componentes para un producto.

Manufacturer's procedures Specific step-by-step instructions provided by the manufacturer for the assembly, disassembly, installation, replacement, or repair of a particular product manufactured by them.

Procedimientos del fabricante Instrucciones específicas a seguir paso por paso; dichas instrucciones son suministradas por el fabricante para montar, desmontar, instalar, remplazar, y/o reparar un producto específico fabricado por él.

MAX A mode, maximum, for heating or cooling. Selecting MAX generally overrides all other conditions that may have been programmed.

MAX (Máximo) Modo máximo para calentamiento o enfriamiento. El seleccionar MAX generalmente anula todas las otras condiciones que pueden haber sido programadas.

Metric fastener Any type fastener with metric size designations, numbers, or millimeters.

Asegurador métrico Cualquier asegurador provisto de indicaciones, números, o milímetros.

Mid-positioned The position of a stem-type service valve where all fluid passages are interconnected. Also referred to as "cracked."

Ubicación-central Posición de una válvula de servicio de tipo vástago donde todos los pasajes que conducen fluidos se interconectan. Lla-mado también "parcialmente asentada".

Mildew A form of fungus formed under damp conditions.

Hongo Un tipo de hongo que se desarrolla en las condiciones húmedas.

Mold A fungus that causes disintegration of organic matter.

Hongo Un hongo que causa la disintegración de la materia orgánica.

Motor An electrical device that produces a continuous turning motion. A motor is used to propel a fan blade or a blower wheel.

Motor Dispositivo eléctrico que produce un movimiento giratorio continuo. Se utiliza un motor para impeler las aletas del ventilador o la rueda del soplador.

Mounting See Flange.

Brida de montaje Ver Flange [Brida].

Mounting boss See Flange.

Protuberancia de montaje Ver Flange [Brida].

MSDS Material safety data sheet.

MSDS Hojas de información sobre la seguridad de un material.

Mushroomed A condition caused by pounding of a punch or a chisel, producing a mushroom-shaped end that should be ground off to ensure maximum safety.

Hinchado Condición ocasionada por el golpeo de un punzón o cincel, lo cual hace que el extremo vuelva en forma de un hongo y que debe ser afilado para asegurar máxima seguridad.

Net weight The weight of a product only; container and packaging not included.

Peso neto Peso de solo el producto mismo; no incluye el recipiente y encajonamiento.

Neutral On neither side; the position of gears when force is not being transmitted.

Neutro Que no está en ningún lado; posición de los engranajes cuando no se transmite la potencia.

Noncycling clutch An electromechanical compressor clutch that does not cycle on and off as a means of temperature control; it is used to turn the system on when cooling is desired and off when cooling is not desired.

Embrague sin funcionamiento cíclico Embrague electromecánico del compresor que no se enciende y se apaga como medio de regular la temperatura; se utiliza para arrancar el sistema cuando se desea enfriamiento y para detener el sistema cuando no se desea enfriamiento.

Observe To see and note; to perceive; to notice.

Observar Ver y anotar; percibir; fijarse en algo.

OEM Original equipment manufacturer.

OEM Fabricante original del equipo.

Off-road A term often used for an "off-the-road" vehicle.

Fuera de carretera Un término que se usa para referirse a un vehículo "fuera de la carretera."

Off-the-road Generally refers to vehicles that are not licensed for road use, such as harvesters, bulldozers, and so on.

Fuera de carretera Generalmente se refiere a vehículos que no son permitidos operar en la carretera, como por ejemplo cosechadoras, rasadoras, etcétera.

Ohmmeter An electrical instrument used to measure the resistance in ohms of a circuit or component.

Ohmiometro Instrumento eléctrico utilizado para medir la resistencia en ohmios de un circuito o componente.

Open Not closed. An open switch, for example, breaks an electrical circuit.

Abierto No cerrado. Un interruptor abierto corta un circuito eléctrico, por ejemplo.

Orifice A small hole. A calibrated opening in a tube or pipe to regulate the flow of a fluid or liquid.

Orificio Agujero pequeño. Apertura calibrada en un tubo o cañería para regular el flujo de un fluido o de un líquido.

O-ring A synthetic rubber or plastic gasket with a round- or square-shaped cross-section.

Junta tórica Guarnición sintética de caucho o de plástico provista de una sección transversal en forma redonda o cuadrada.

OSHA Occupational Safety and Health Administration.

OSHA Direccion para la Seguridad y Salud Industrial.

Outside temperature sensor See Ambient sensor.

Sensor de la temperatura ambiente Ver Ambient sensor [Sensor ambiente].

Overcharge Indicates that too much refrigerant or refrigeration oil is added to the system.

Sobrecarga Indica que una cantidad excesiva de refrigerante o aceite de refrigeración ha sido agregada al sistema.

Overload Anything in excess of the design criteria. An overload will generally cause the protective device such as a fuse or pressure relief to open.

Sobrecarga Cualquier cosa en exceso del criterio de diseño. General-mente una sobrecarga causará que se abra el dispositivo de protección, como por ejemplo un fusible o alivio de presión.

Ozone friendly Any product that does not pose a hazard or danger to the ozone.

Sustancia no dañina al ozono Cualquier producto que no es peligroso o amenaza al ozono.

Park Generally refers to a component or mechanism that is at rest.

Reposo Generalmente se refiere a un componente o mecanismo que no está funcionando.

PCM Power control module.

PCM Módulo regulador del transmisor de potencia.

Performance test Readings of the temperature and pressure under controlled conditions to determine if an air-conditioning system is operating at full efficiency.

Prueba de rendimiento Lecturas de la temperatura y presión bajo condiciones controladas para determinar si un sistema de acondicionamiento de aire funciona a un rendimiento completo.

Piercing pin The part of a saddle valve that is used to pierce a hole in the tubing.

Pasador perforador Parte de la válvula de silleta utilizada para perforar un agujero en la tubería.

Pin type A single or multiple electrical connector that is round or pin shaped and fits inside a matching connector.

Conectador de tipo pasador Conectador eléctrico único o múltiple en forma redonda o en forma de pasador que se inserta dentro de un conectador emparajado.

Poly belt See Serpentine belt.

Correa poli Ver Serpentine belt [Correa serpentina].

Polyol ester (ESTER) A synthetic oil-like lubricant that is occasionally recommended for use in an R-134a system. This lubricant is compatible with both R-134a and R-12.

Poliolester Lubrificante sintético parecido a aceite que se recomienda de vez en cuando para usar en un sistema HFC-134a. Dicho lubrificante es compatible tanto con HFC-134a como CFC-12.

Positive pressure Any pressure above atmospheric.

Presión positiva Cualquier presión sobre la de la atmosférica.

Pound A weight measure, 16 oz. A term often used when referring to a small can of refrigerant, although the can does not necessarily contain 15 oz.

Libra Medida de peso, 16 onzas. Término utilizado con frecuencia al referirse a una lata pequeña de refrigerante, aunque es posible que la lata contenga menos de 16 onzas.

Pound cans This is a generic term used when referring to small disposable cans of refrigerant that contain less than 16 oz. of refrigerant.

Latas de una libra Éste es un término general que se usa para referirse a las latas pequeñas desechables que contienen menos de 16 onzas del refrigerante.

"Pound" of refrigerant A term used by some technicians when referring to a small can of refrigerant that actually contains less than 16 oz.

Libra de refrigerante Término utilizado por algunos mecánicos al referirse a una lata pequeña de refrigerante que en realidad contiene menos de 16 onzas.

Power module Controls the operation of the blower motor in an automatic temperature control system.

Transmisor de potencia Regula el funcionamiento del motor del soplador en un sistema de control automático de temperatura.

Predetermined A set of fixed values or parameters that have been programmed or otherwise fixed into an operating system.

Predeterminado Valores fijos o parámetros que han sido programados o de otra manera fijados en un sistema de funcionamiento.

Pressure The application of a continuous force by one body onto another body. Force per unit area or force divided by area usually expressed in pounds per square inch (psi) or kilopascal (kPa).

Presión La aplicación de una fuerza contínua de un cuerpo contra otro cuerpo. La fuerza por unidad de un área o la fuerza dividida por el área suele expresarse en libras por pulgada cuadrada (psi) o kilopascal (kPa).

Pressure gauge A calibrated instrument for measuring pressure.

Manometro Instrumento calibrado para medir la presión.

Pressure switch An electrical switch that is activated by a predetermined low or high pressure. A high-pressure switch is generally used for system protection; a low-pressure switch may be used for temperature control or system protection.

Autómata manométrico Interruptor eléctrico accionado por una baja o alta presión predeterminada. Generalmente se utiliza un autómata manométrico de alta presión para la protección del sistema; puede utilizarse uno de baja presión para la regulación de temperatura o protección del sistema.

Propane A flammable gas used as a propellant for the halide leak detector.

Propano Gas inflamable utilizado como propulsor para el detector de fugas de halogenuro.

Psig Pounds per square inch gauge.

Psig Calibrador de libras por pulgada cuadrada.

Purge To remove moisture and air from a system or a component by flushing with a dry gas such as nitrogen (N) to remove all refrigerant from the system.

Purgar Remover humedad y/o aire de un sistema o un componente al descargarlo con un gas seco, como por ejemplo el nitrogeno (N), para remover todo el refrigerante del sistema.

Purity test A static test that may be performed to compare the suspect refrigerant pressure with an appropriate temperature chart to determine its purity.

Prueba de pureza Prueba estática que puede llevarse a cabo para com-parar la presión del refrigerante con un gráfico de temperatura apro-priado para determinar la pureza del mismo.

Radiation The transfer of heat without heating the medium through which it is transmitted.

Radiación La transferencia de calor sin calentar el medio por el cual se transmite.

Ram air Air that is forced through the radiator and condenser coils by the movement of the vehicle or the action of the fan.

Aire Admitido en sentido de la marcha. Aire forzado a través de las bobinas del radiador y del condensador por medio del movimiento del vehículo o la acción del ventilador.

Rebuilt To build after having been disassembled, inspected, and worn and after damaged parts and components are replaced.

Reconstruído Fabricar después de haber sido desmontado y revisado, y luego remplazar las piezas desgastadas y averiadas.

Receiver-drier A tank-like vessel having a desiccant and used for the storage of refrigerant.

Receptor-secador Recipiente parecido a un tanque provisto de un desecante y utilizado para el almacenaje de refrigerante.

RECIR An abbreviation for the recirculate mode, as with air.

RECIR Abreviatura del modo recirculatorio, como por ejemplo con aire.

Recovery system A term often used to refer to the circuit inside the recovery unit used to recycle or transfer refrigerant from the air-conditioning system to the recovery cylinder.

Sistema de recuperación Término utilizado con frecuencia para referirse al circuito dentro de la unidad de recuperación interior utilizado para reciclar y/o transferir el refrigerante del sistema de acondi-cionamiento de aire al cilindro de recuperación.

Recovery tank An auxiliary tank, usually connected to the inlet tank of a radiator, which provides additional storage space for heated coolant.

Tanque de recuperación Tanque auxiliar que normalmente se conecta al tanque de entrada de un radiador, lo cual provee alma-cenaje adi-cional para el enfriante calentado.

Refrigerant A chemical compound, such as R-134a, used in an air-conditioning system to achieve the desired refrigerating effect.

Refrigerante Un compuesto químico, tal como el R-134a, que se usa en un sistema de aire acondicionado para realizar un efecto refrigerante deseado.

Refrigerant-12 The refrigerant used in automotive air condi-tioners, as well as other air-conditioning and refrigeration systems. The chemical name of Refrigerant-12 is dichlorodifluoromethane. The chemical symbol is $CC_{12}F_2$.

Refrigerante 12 Refrigerante utilizado tanto en acondicionadores de aire automotrices como en otros sistemas de acondicionamiento de aire y refrigeración. Ele nombre químico del Refrigerante-12 es diclorodifluorometano, y el símbolo químico es $CC_{12}F_2$.

Relay An electrical switch device that is activated by a low-current source and controls a high-current device.

Relé Interruptor eléctrico que es accionado por una fuente de cor-riente baja y regula un dispositivo de corriente alta.

Reserve tank A storage vessel for excess fluid. See Recovery tank; Receiver-drier, and Accumulator.

Tanque de reserva Recipiente de almacenaje para un exceso de fluido. Ver Recovery tank [Tanque de recuperación], Receiver-drier [Receptor-secador], y Accumulator [Acumulador].

Resistor A voltage-dropping device that is usually wire wound and that provides a means of controlling fan speeds.

Resistor Dispositivo de caída de tensión que normalmente es deva-nado con alambre y provee un medio de regular la velocidad del ventilador.

Respirator A mask or face shield worn in a hazardous environment to provide clean fresh air and oxygen.

Mascarilla Máscara o protector de cara que se lleva puesto en un ambi-ente peligroso para proveer aire limpio y puro y/o oxígeno.

Responsibility Being reliable and trustworthy.

Responsabilidad Ser confiable y fidedigno.

Restricted Having limitations. Keeping within limits, confines, or boundaries.

Restringido Que tiene limitaciones. Mantenerse dentro de límites, confines, o fronteras.

Restrictor An insert fitting or device used to control the flow of refrigerant or refrigeration oil.

Limitador Pieza inserta o dispositivo utilizado para regular el flujo de refrigerante o aceite de refrigeración.

Rotor The rotating or freewheeling portion of a clutch; the belt slides on the rotor.

Rotor Parte giratoria o con marcha a rueda libre de un embrague; la correa se desliza sobre el rotor.

RPM Revolutions per minute; also rpm or r/min.

RPM Revoluciones por minuto; tambien rpm o r/min.

Running design change A design change made during a current model/year production.

Cambio al diseño corriente Un cambio al diseño hecho durante la fabricación del modelo/año actual.

RV Recreational vehicle.

RV Vehículo para el recreo.

Saddle valve A two-part accessory valve that may be clamped around the metal part of a system hose to provide access to the air-conditioning system for service.

Válvula de silleta Válvula accesoria de dos partes que puede fijarse con una abrazadera a la parte metálica de una manguera del sistema para proveer acceso al sistema de acondicionamiento de aire para lle-var a cabo servicio.

SAE Society of Automotive Engineers.

SAE Sociedad de Ingenieros Automotrices.

Safety Freedom from danger or injury; the state of being safe.

Seguridad Libre de peligro o daño, calidad o estado de seguro.

Scan tool A portable computer that may be connected to the vehi-cle's diagnostic connector to read data from the vehicle's onboard computer.

Dispositivo explorador Una computadora portátil que se puede conectar al conector diagnóstico del vehículo para leer los datos de la computadora abordo del vehículo.

Schrader valve A spring-loaded valve similar to a tire valve. The Schrader valve is located inside the service valve fitting and is used on some control devices to hold refrigerant in the system. Special adapters must be used with the gauge hose to allow access to the system.

Válvula Schrader Válvula con cierre automático parecida al vástago del neumático. La válvula Schrader está ubicada dentro del ajuste de la válvula de servicio y se utiliza en algunos dispositivos de regula-ción para guardar refrigerante dentro del sistema. Deben utilizarse adaptadores especiales con una manguera calibrador para permitir acceso al sistema.

Seal Generally refers to a compressor shaft oil seal; matching shaft-mounted seal face and front head-mounted seal seat to prevent refrigerant and oil from escaping. May also refer to any gasket or O-ring used between two mating surfaces for the same purpose.

Junta hermética Generalmente se refiere a la junta hermética del árbol del compresor; la frente de junta hermética montada en el árbol y el asiento de junta hermética montado en el cabezal delan-tero empare-jados para evitar la fuga de refrigerante y/o de aceite. Puede referirse también a cualquier guarnición o junta tórica utili-zada entre dos superficies emparejadas para el mismo propósito.

Seal seat The part of a compressor shaft seal assembly that is station-ary and matches the rotating part, known as the seal face or shaft seal.

Asiento de la junta hermética Parte del conjunto de la junta her-mética del árbol del compresor que es inmóvil y que se empareja a la parte rotativa; conocido como la frente de junta hermética o la junta hermética del árbol.

Serpentine belt A flat or V-groove belt that winds through all of the engine accessories to drive them off the crankshaft pulley.

Correa serpentina Correa plana o con ranuras en V que atraviesa todos los accesorios del motor para forzarlos fuera de la polea del cigüeñal.

Service port A fitting found on the service valves and some control devices; the manifold set hoses are connected to this fitting.

Orificio de servicio Ajuste ubicado en las válvulas de servicio y en algunos dispositivos de regulación; las mangueras del conjunto del colector se conectan a este ajuste.

Service procedure A suggested routine for the step-by-step act of troubleshooting, diagnosing, and repairs.

Procedimiento de servicio Rutina sugerida para la acción a seguir paso a paso para detectar fallas, diagnosticar, y/o reparar.

Service valve See High-side (Low-side) service valve.

Válvula de servicio Ver High-side (Low-side) service valve [Válvula de servicio del lado de alta presión (baja presión)].

Servomotor An electrical motor that is used to control a mechanical device, such as a heater coolant flow control valve.

Motor servo Un motor eléctrico que se usa para controlar un disposi-tivo mecánico, tal como una válvula de control de flujo del fluido refrigerante del calentador.

Shaft A long cylindrical-shaped rod that rotates to transmit power, such as a compressor crankshaft.

Eje Una biela cilíndrica larga que gira para transferir la potencia, tal como un cigüeñal del compresor.

Shaft key A soft metal key that secures a member on a shaft to prevent it from slipping.

Chaveta del árbol Chaveta de metal blando que fija una pieza a un árbol para evitar su deslizamiento.

Shaft seal See Compressor shaft seal.

Junta hermética del árbol Ver Compressor shaft seal [Junta hermética del árbol del compresor].

Short Of brief duration, for example, short cycling. Also refers to an intentional or unintentional grounding of an electrical circuit.

Breve/corto De una duración breve; p.e., funcionamiento cíclico breve. Se refiere también a un puesto a tierra previsto o imprevisto de un circuito eléctrico.

Shutoff valve A valve that provides positive shutoff of a fluid or vapor passage.

Válvula de cierre Válvula que provee el cierre positivo del pasaje de un fluido o un vapor.

Sliding resistor A resistor having the provision of varying its resistance depending on the position of a sliding member. Aso may be referred to as a "rheostat" or "pot."

Resistor deslizante Un resistor que tiene la provisión de variar su resistencia según la posición de una parte deslizante. Tambien puede referirse como un reóstato o un potenciómetro/reductor.

SNAP An acronym for "Significant New Alternatives Policy."

SNAP Una sigla del término "Póliza de Alternativas Nuevas Significantes."

Snapring A metal ring used to secure and retain a component to another component.

Anillo de muelle Anillo metálico utilizado para fijar y sujetar un com-ponente a otro.

Snapshot A feature of OBD II that shows, on various scanners, the conditions that the vehicle was operating under when a particular trouble code was set. For example, the vehicle was at 125°F, ambient temperature was 55°F, throttle position was part throttle at 1.45 volts, rpm was 1450, brake was off, transmission was in third gear with torque converter unlocked, air-conditioning system was off, and so on.

Instantáneo Una característica del OBD II que muestra, en varios detectores, las condiciones bajo las cuales operaba el vehículo cuando se registró un código de fallo. Por ejemplo, el vehículo regis-traba 125°F, la temperatura del ambiente era el 55°F, la posición del regulador estaba en una posición parcial de 1.45 voltios, el rpm era 1450, el freno estaba desenganchada, la transmisión estaba en la ter-cera velocidad con el convertidor del par desenclavado, el sistema de acondicionador de aire estaba apagado, y etcétera.

Society of Automotive Engineers A professional organization of the automotive industry. Founded in 1905 as the Society of Automobile Engineers, the SAE is dedicated to providing technical information and standards to the automotive industry.

Sociedad de Ingenieros automotrices Organización profesional de la industria automotriz. Establecido en 1905 como la Sociedad de Ingenieros de Automoviles (SAE por sus siglas en ingles), dicha sociedad se dedica a proveerle información técnica y normas a la indústria automotriz.

Socket The concave part of a joint that receives a concave member. A term generally used for "socket wrench," referring to a female 6-, 8-, or 12-point wrench so designed to fit over a nut or bolt head.

Casquillo La parte cóncava de una junta que recibe una parte cóncava. El término "socket" generalmente se usa para indicar una "llave de tubo" y se refiere a la parte hembra de una llave de tamaño de 6, 8, o 12 puntos diseñada para quedar en una tuerca o una cabeza de un perno.

Solenoid See Solenoid valve.

Solenoide Ver Solenoid valve [Válvula de solenoide].

Solenoid valve An electromagnetic valve controlled remotely by electrically energizing and de-energizing a coil.

Válvula de solenoide Válvula electromagnética regulada a distancia por una bobina electronicamente.

Solid state Referring to electronics consisting of semiconductor devices and other related nonmechanical components.

Estado sólido Se refiere a componentes electrónicos que consisten en dispositivos semiconductores y otros componentes relacionados no mecánicos.

Spade-type connector A single or multiple electrical connector that has flat spade-like mating provisions.

Conectador de tipo azadón Conectador único o múltiple provisto de dispositivos planos de tipo azadón para emparejarse.

Specifications Design characteristics of a component or assembly noted by the manufacturer. Specifications for a vehicle include fluid capacities, weights, and other pertinent maintenance information.

Especificaciones Características de diseño de un componente o con-junto indicadas por el fabricante. Las especificaciones para un vehículo incluyen capacidades del fluido, peso y otra información pertinente para el mantenimiento del vehículo.

Spike In our application, an electrical spike. An unwanted momentary high-energy electrical surge.

Impulso afilado En nuestro campo, un impulso afilado eléctrico. Una elevación repentina eléctrica de alta energia no deseada.

Spring lock Part of a spring lock fitting. A special fitting used to form a leakproof joint.

Obturador de resorte Parte de una fijación de obturador de resorte. Una fijación especial que sirve para formar una junta a prueba de fugas.

Spring lock fitting A special fitting using a spring to lock the mating parts together, forming a leakproof joint.

Ajuste de cierre automático Ajuste especial utilizando un resorte para cerrar piezas emparejadas para formar así una junta hermética contra fugas.

Squirrel-cage blower A blower wheel designed to provide a large volume of air with a minimum of noise. The blower is more compact than the fan and air can be directed more efficiently.

Soplador con jaula de ardilla Rueda de soplador diseñada para proveer un gran caudal de aire con un mínimo de ruido. El soplador es más compacto que el ventilador y el aire puede dirigirse con un mayor rendimiento.

Stabilize To make steady.

Estabilizar Quedarse detenida una cosa.

Standing vacuum test A leak test performed on an air-conditioning system by pulling a vacuum and then determining, by observation, if the vacuum holds for a predetermined period of time to ensure that there are no leaks.

Prueba de vacío fijo Una prueba de fugas que se efectúa en el sistema de aire acondicionado por medio de establecer un vacío y luego determinar, por observación, si se mantiene el vacío por un periodo prescrito de tiempo para asegurar que no hay fugas.

Stratify Arrange or form into layers. To fully blend.

Estratificar Arreglar o formar en capas. Mezclar completamente.

Subsystem A system within a system.

Subsistema Sistema dentro de un sistema.

Sun load Heat intensity or light intensity produced by the sun.

Carga del sol Intensidad calorífica y/o de la luz generada por el sol.

Sun-load sensor A device that senses heat or light intensity that is placed on the dashboard to determine the amount of sun entering the vehicle.

Sensor de carga de sol Un dispositivo que detecta la intensidad del calor o de luz que hay en la tabla de instrumentos para determinar la cantidad de luz del sol entrando al vehículo.

Superheat switch An electrical switch activated by an abnormal temperature-pressure condition (a superheated vapor); used for system protection.

Interruptor de vapor sobrecalentado Interruptor eléctrico accionado por una condición anormal de presión y temperatura (vapor sobrecalentado); utilizado para la protección del sistema.

System All of the components and lines that make up an air-conditioning system.

Sistema Todos los componentes y líneas que componen un sistema de acondicionamiento de aire.

Tank See Header tanks and Expansion tank.

Tanque Ver Header tanks [Tanques para alimentación por gravedad] y Expansion tank [Tanque de expansión].

Tare weight The weight of the packaging material. See Net weight and Gross weight.

Taraje Peso del material de encajonamiento. Ver Net weight [Peso neto] y Gross weight [Peso bruto].

Technical service bulletin (TSB) Periodic information provided by the vehicle manufacturer regarding any problems and offering solutions to problems encountered in their vehicles.

Boletín de servicio técnico (TSB) La información periódica proveído por el fabricante del vehículo concernante cualquier problema que se encuentra en sus vehículos y proporcionando las soluciones.

Technician One concerned and involved in the design, service, or repair in a specific area, such as an automotive service technician or, more specifically, automotive air-conditioning service technician.

Técnico Uno que se concierna y se involucra en el diseño, el servicio, o la reparación en un área específico, o más específicamente, un técnico de servicio de aire acondicionado automotivo.

Temperature door A door within the case/duct stem to direct air through the heater and evaporator core.

Puerta de temperatura Puerta ubicada dentro del vástago de caja/conducto para conducir el aire a través del nucleo del calentador y/o del evaporador.

Temperature switch A switch actuated by a change in temperature at a predetermined point.

Interruptor de temperatura Interruptor accionado por un cambio de temperatura a un punto predeterminado.

Tensioner A device used to impart tension, such as an automatic belt tensioner.

Tensionador Un dispositivo que se usa para mantener la tensión, tal como un tensionador de correa automático.

Tension gauge A tool for measuring the tension of a belt.

Manómetro para tensión Herramienta para medir la tensión de una correa.

Thermistor A temperature-sensing resistor that has the ability to change values with changing temperature.

Termistor Resistor sensible a temperatura que tiene la capacidad de cambiar valores al ocurrir un cambio de temperatura.

Torque A turning force, for example, the force required to seal a connection; measured in (English) foot-pounds (ft.-lb.) or inch-pounds (in.-lb.); (metric) Newton-meters (N·m).

Par de torsión Fuerza de torcimiento; por ejemplo, la fuerza requerida para sellar una conexión; medido en libras-pies (ft.-lb.) (inglesas) o en libras pulgadas (in.-lb.); metros-Newton (N·m [metricos]).

Triple evacuation A process of evacuation that involves three pump downs and two system purges with an inert gas such as dry nitrogen (N).

Evacuación triple Proceso de evacuación que involucra tres envíos con bomba y dos purgas del sistema con un gas inerte, como por ejemplo el nitrógeno seco (N).

Troubleshoot The act of diagnosing the cause of various system malfunctions.

Detección de fallas Procedimiento o arte de diagnosticar la causa de varias fallas del sistema.

TXV Thermostatic expansion valve.

TXV Válvula de expansión termostática.

Ultraviolet (UV) The part of the electromagnetic spectrum emitted by the sun that lies between visible violet light and X-rays.

Ultravioleta Parte del espectro electromagnético generado por el sol que se encuentra entre la luz violeta visible y los rayos X.

Ultraviolet dye A fluid that may be injected into the air-conditioning system for leak-testing purposes. An ultraviolet (UV) lamp is used to locate the leak.

Tinta ultravioleta Un fluido que se puede inyectar en un sistema de aire acondicionado para comprobar contra las fugas. Una lámpara ultravioleta se usa para localizar la fuga.

Vacuum gauge A gauge used to measure below atmospheric pressure.

Vacuómetro Calibrador utilizado para medir a una presión inferior a la de la atmósfera.

Vacuum motor A device designed to provide mechanical control by the use of a vacuum.

Motor de vacío Dispositivo diseñado para proveer regulación mecánica mediante un vacío.

Vacuum pump A mechanical device used to evacuate the refrigeration system to rid it of excess moisture and air.

Bomba de vacío Dispositivo mecánico utilizado para evacuar el sistema de refrigeración para purgarlo de un exceso de humedad y aire.

Vacuum signal The presence of a vacuum.

Señal de vacío Presencia de un vacío.

V-belt A rubber-like continuous loop placed between the engine crankshaft pulley and accessories to transfer rotary motion of the crankshaft to the accessories.

Correa en V Bucle continuo parecido a caucho ubicado entre la polea del cigüeñal del motor y los accesorios para transferir el movimiento giratorio de aquel a estos.

Ventilation The act of supplying fresh air to an enclosed space such as the inside of an automobile.

Ventilación Proceso de suministrar el aire fresco a un espacio cerrado, como por ejemplo al interior de un automóvil.

V-groove belt See V-belt.

Correa con ranuras en V Ver V-belt [Correa en V].

VIN An acronym for "vehicle identification number."

VIN Una sigla para "numero de identificación del vehículo."

Voltmeter A device used to measure volt(s).

Voltímetro Dispositivo utilizado para la medición de voltios.

Wiring harness A group of wires wrapped in a shroud for the distribution of power from one point to another point.

Cableado preformado Grupo de alambres envuelto por una gualdera para distribuir potencia de un punto a otro.